Modelling Operational Risk Using Bayesian Inference

Pavel V. Shevchenko

Modelling Operational Risk Using Bayesian Inference

Springer

Dr. Pavel V. Shevchenko
CSIRO
Mathematics, Informatics and Statistics
Locked Bag 17, North Ryde
NSW, 1670
Australia
pavel.shevchenko@csiro.au

ISBN 978-3-642-15922-0 e-ISBN 978-3-642-15923-7
DOI 10.1007/978-3-642-15923-7
Springer Heidelberg Dordrecht London New York

Library of Congress Control Number: 2010938383

Cover design: WMXDesign GmbH, Heidelberg

Printed on acid-free paper

Springer is part of Springer Science+Business Media (www.springer.com)

Dedicated to my wife, daughter and parents

Preface

You don't write because you want to say something,
you write because you have something to say.
F. Scott Fitzgerald

The management of operational risk in the banking industry has undergone significant changes over the last decade due to substantial changes in the operational environment. Globalization, deregulation, the use of complex financial products and changes in information technology have resulted in exposure to new risks very different from market and credit risks. In response, the Basel Committee on Banking Supervision (BCBS) has developed a new regulatory framework for capital measurement and standards for the banking sector, referred to as Basel II, aimed at sound risk sensitive capital requirements. Basel II formally defined operational risk and introduced corresponding capital requirements. BCBS began discussions on operational risk management in 1998, leading to the inclusion of operational risk capital requirements into the latest Basel II developed during 2001–2006.

Currently, major banks are undertaking quantitative modelling of operational risk to satisfy these requirements under the so-called Basel II Advanced Measurement Approaches (AMA). A popular method under the AMA is the Loss Distribution Approach (LDA) based on statistical quantification of the frequency and severity distributions for operational risk losses. The LDA is the main focus of this book. Over the last 3 years, major banks in most parts of the world have received accreditation under the Basel II AMA by adopting the LDA, despite there being a number of unresolved methodological challenges in its implementation. Overall, the area of quantitative operational risk is very new and different methods are under hot debate.

Since 2000, I have been involved in consulting projects for several major banks, assisting with the development of their operational risk models and software systems to comply with the new Basel II requirements. The development of a consistent mathematical framework for operational risk treatment, addressing all aspects required in practical implementation, is a challenging task. Due to the absence of a coherent framework different ad-hoc solutions are often used in practice.

As a result of consulting projects for banks, discussions with regulators and academic research, I feel that there is a need for a textbook on quantitative issues in modelling operational risk that should be resolved and addressed in practice. This book, in particular, will focus on the LDA and will advocate the use of a Bayesian inference method (some alternative methods will be described and referenced too).

Though it is very new in this area, I believe that the Bayesian approach is well suited for modelling operational risk as it allows for a consistent and convenient statistical framework to quantify the uncertainties involved. It also allows for the combination of expert opinions with historical internal and external data in estimation procedures. These are critical, especially for operational risks that have small datasets. During the last 5 years many aspects and problems in the quantitative modelling of operational risk have been addressed in monographs, research papers and reports from loss data collection exercises. These will be referred to within this book. The Bayesian approach advocated here is very new for operational risk and is certainly not fully covered in the available spectrum of books and papers within the area.

Unfortunately, it was not possible to include examples of the real operational risk data into this book due to confidentiality issues. As a result, only illustrative examples with realistic parameter values are used and the book might look too 'academic'. However, I hope that discussed results and methodologies will make a positive contribution to a reliable estimation of capital charge for operational risk.

This book is aimed at practitioners in risk management, academic researchers in financial mathematics, banking industry regulators and advanced graduate students in the area. One aim is to have a book that can be used as a reference text for practitioners interested in a clear and concise treatment of concepts and methods needed in practice. Another aim is to have chapters that can be used for teaching university courses on quantitative risk management. The book also provides a comprehensive list of references to guide more advanced readers through the vast literature and will takes the reader to the frontier of practically relevant research. I hope that the book will facilitate communication between regulators, end-users and academics.

This project would not be possible without a great community of researchers in the area of operational risk. I would like to particularly mention publications by K. Böcker, A. Chernobai, M. Cruz, P. Embrechts, A. Frachot, C. Klüppelberg, G. Mignola, O. Moudoulaud, S. Rachev, T. Roncalli and R. Ugoccioni which have greatly impacted on and influenced the composition of this work. This book would also not be possible without help from many colleagues and coworkers.

Overall, I am very grateful to my employer (CSIRO Mathematics, Informatics and Statistics of Australia), where, over the past 10 years, I have gained knowledge and practical experience in modelling financial risk. Special thanks goes to my CSIRO colleagues: M. Cameron and F. de Hoog for the support and encouragement to write the book; G. Peters for his expert advice on Markov chain Monte Carlo techniques; X. Luo and J. Donnelly for stimulating discussions; J. Donnelly, F. de Hoog, M. Westcott, A. Tobin and G. Peters for reviewing the manuscript.

A great *thank you* goes to the ETH Zurich researchers who have had a significant influence on my research: H. Bühlmann, P. Embrechts and M. Wüthrich. Also, I am very grateful to my other colleagues: M. Delasey, D. Farmer, J. McManus, U. Schmock and P. Thomson who have influenced my knowledge of the subject; and R. McGregor for making recommendations on some formal aspects of English expression and presentation. And finally, I am wholly indebted to my wife and daughter for the understanding and support they have granted me throughout this time consuming project.

Sydney Pavel V. Shevchenko
April 2010

Contents

List of Abbreviations and Symbols

ABC	approximate Bayesian computation
AMA	Advanced Measurement Approaches
BBA	British Bankers Association
BCBS	Basel Committee on Banking Supervision
BIS	Bank for International Settlements
Cov	covariance
DFT	discrete Fourier transform
EVT	extreme value theory
GEV	generalised extreme value distribution
GCD	generalised Champernowne distribution
GPD	generalised Pareto distribution
FFT	Fast Fourier Transform
FRS	Federal Reserve System
IAA	International Actuarial Association
iid	independent and identically distributed
LDA	Loss Distribution Approach
LDCE	Loss Data Collection Exercise
MC	Monte Carlo
MCMC	Markov chain Monte Carlo
MLE	maximum likelihood estimator
MSE	mean squared error
PCB	Planning and Coordination Bureau
QIS	Quantitative Impact Study
stdev	standard deviation
Var	variance
VaR	Value-at-Risk
Vco	variational coefficient
RW-MH	random walk Metropolis-Hastings algorithm

Chapter 1
Operational Risk and Basel II

Far better an approximate answer to the right question, which is often vague, than an exact answer to the wrong question, which can always be made precise.

J.W. Tukey

Abstract The management of operational risk is not a new concept in the banking industry. Operational risks such as external fraud, internal fraud, and processing errors have had to be managed since the beginning of banking. Traditionally, these risks were managed using insurance protection and audit. Globalisation, complex financial products and changes in information technology, combined with a growing number of high-profile operational loss events worldwide have increased the importance of operational risk management for the banking industry. This has prompted regulators to decide that banks have to set aside risk capital to cover operational risk losses. The Basel Committee on Banking Supervision (BCBS) began the discussions on operational risk management in 1998 leading to the inclusion of operational risk capital requirements into the new international regulatory framework, Basel II, developed during 2001–2006. Currently, major banks are undertaking quantitative modelling of operational risk to satisfy these requirements. This chapter gives an overview of the Basel II requirements for operational risk management, several important loss data collection exercises conducted by regulators in different countries and proposed modelling approaches.

1.1 Introduction to Operational Risk

Banks are required by the regulators to allocate capital against potential losses. It can be viewed as some sort of self-insurance. The main risk categories attracting capital charge in financial institutions are credit risk, market risk and operational risk (Fig. 1.1). The last, operational risk, did not require explicit capital allocation until recently; previously, it was implicitly covered by the capital charge for credit risk. The concept of operational risk is generic for organizations of all types. In general, it is related to the losses caused by the way a firm operates rather than those caused by market movements or credit downgrades. Operational risk is significant in many financial institutions. It accounts for approximately 15–25% of the total capital in many large banks that requires allocation of the order of USD

P. Shevchenko, *Modelling Operational Risk Using Bayesian Inference*,
DOI 10.1007/978-3-642-15923-7_1, © Springer-Verlag Berlin Heidelberg 2011

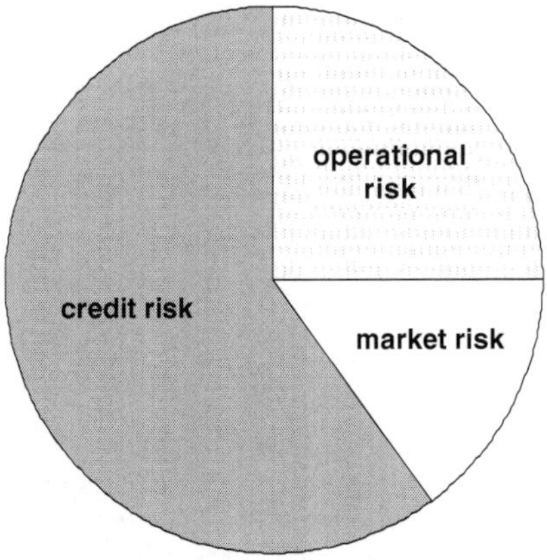

Fig. 1.1 Illustration of the capital allocation for credit, operational and market risks in a major bank

2–10 billion. Typically, for banks, operational risk is the largest risk after credit risk. The management of operational risk is not a new concept in the banking industry, but only within the last decade has it been identified as a category that should be actively measured and managed. It has always been important for banks to try to prevent some operational risks such as external fraud, internal fraud and processing errors. Traditionally, banks relied almost exclusively upon insurance protection and internal control mechanisms within business lines supplemented by audit to manage operational risks. To illustrate the concept of operational risk processes consider the following few examples.[1]

Example 1.1 (An Automobile Journey) Imagine that you have to take a trip from City A to City B by car. You have done such a trip before and based on previous experience you plan that the trip will take two days including an overnight stop when you have travelled about half way. The total travel distance is approximately 1,000 km. Estimating the cost of petrol, meals, and hotel you plan to spend AUD 400. After the trip you compare the actual cost with the original plan and observe that:

- The trip took two days more than planned due to a breakdown, some traffic delays, taking a wrong route when trying to drive over a closed road, and bad weather.
- The total trip cost was AUD 4,000 due to the required repairs, extra hotel nights, and traffic fines for speeding. Also, due to the delay, you were late for an important business meeting so you missed a business opportunity, which is an indirect loss.

You have undertaken such a trip before, several times, and have never experienced such losses and delays. The losses experienced this time were unexpected but

[1] Similar and other illustrative examples, and detailed consideration of the largest historical operational risk losses can be found in e.g. Cruz [65], King [134].

could be reduced if you undertook a proper risk management. For example, if you took a car with more advanced controls (cruise control, GPS navigation, proximity and maintenance alarms) and considered the weather forecast, you could reduce the losses substantially.

Of course it is not an example from a financial institution but the essence is similar. That is, you can reduce your unexpected losses (money and time) by improving the monitoring and control systems for the process. In this case the process is an automobile journey; the control systems are, for example, GPS and ABS braking; the risk factors are weather, traffic accidents and compliance with traffic rules. Of course, a risk management planning of this trip should consider many other questions, such as consideration of alternative travel arrangements, comparison of hotel and petrol prices, and the measurement of losses (that is, how do you measure the indirect loss due to the missed business opportunity). While some of the risks are outside your control, you can reduce the potential losses using mitigation strategies.

Example 1.2 (Foreign Exchange Deal) Consider a foreign exchange deal where a trader:

- buys USD 70 million for AUD 100 million (i.e. AUD 1=USD 0.7); and
- sells USD 70 million for AUD 100,071,480 (i.e. AUD 1=USD 0.6996)

with the total profit AUD 71,480. However, due to mistakes in the back office, there was a settlement delay of several days, and the bank had to pay AUD 101,000 in penalties to the counterparties. Overall, due to operational error there was a loss of AUD 29,520.

Example 1.3 (Large Historical Losses)

- *The Barings Bank.* One of the most famous operational loss events was the bankruptcy of the Barings Bank (loss GBP 1.3 billion in 1995). This is alleged to have occurred because the trader, Nick Leeson, took an enormous position in futures and options, significantly exceeding his trading limits without approval. This case has been widely discussed in many papers, books and by Nick Leeson himself. Being in charge of the trade and the back office enabled Leeson to hide his position and create an illusion of large profits. He was motivated by large bonuses and the desire for status within the bank. It could be argued that this loss occurred due to a lack of controls (i.e. inadequate separation of the front and back office duties; and the absence of an accounting system enabling the settlements department in London to reconcile trades with clients' orders made worldwide).
- Other examples of extremely large operational risk losses include Sumitomo Corporation (USD 2.6 billion in 1996), Enron (USD 2.2 billion in 2001), National Australia Bank (AUD 360 million in 2004) and Société Générale (Euro 4.9 billion in 2008).
- *The economic crisis 2008–2009.* Many events of the recent global economic crisis had their root causes in operational failures within financial firms: mortgage fraud, inadequate assessment of model risk, failure to implement and maintain adequate systems and controls, "bonus culture" motivating high short-term sales regardless of the long-term consequences for the company and its clients.

1.2 Defining Operational Risk

Globalisation, complex financial products and changes in information technology, combined with a growing number of high-profile operational loss events worldwide, have increased the importance of operational risk for the banking industry. In response to these changes, new international regulatory requirements (Basel II) have been developed for the banking industry. Currently, major financial institutions are undertaking quantitative modelling of risk to satisfy the requirements. There was no widely accepted definition of operational risk when the Basel Committee on Banking Supervision (BCBS) began discussions on operational risk management at the end of the 1990s; see BCBS [13]. Often, operational risk was defined as any risk not categorised as market or credit risk. Some banks defined it as the risk of loss arising from various types of human or technical error.

Some earlier definitions can be found in a 1997 survey conducted by the British Bankers Association (BBA); see BBA [39]. In January 2001, the BCBS issued a proposal for a New Basel Capital Accord (referred to as Basel II) where operational risk was formally defined as a new category of risk, in addition to market and credit risks, attracting a capital charge. In the working paper BCBS [19] on the regulatory treatment of operational risk and in the revised Basel II framework BCBS [16], the following definition of operational risk was adopted.[2]

> Operational risk is defined as the risk of loss resulting from inadequate or failed internal processes, people and systems or from external events. This definition includes legal risk but excludes strategic and reputational risk.

This definition did not change in the latest version of Basel II framework, BCBS ([17], p. 144). The International Actuarial Association, IAA [126], has adopted the same definition of operational risk in the capital requirements for insurance companies.

In this book we focus on modelling potential operational risk losses using *statistical techniques* for calculation of the economic and regulatory capital. It is important to mention that operational risk management includes many activities [140] such as:

- developing policies and internal standards;
- developing key risk indicators;
- planning management of major business disruptions; and
- maintaining a database of operational risk incidents.

1.3 Basel II Approaches to Quantify Operational Risk

The Basel II framework is based on a three-pillar concept.

---

[2] The original text is available free of charge on the BIS website www.BIS.org/bcbs/publ.htm

- *Pillar I: Minimum capital requirements.* This pillar requires an explicit minimum capital allocated for operational risk that can be calculated using different approaches.
- *Pillar II: Supervisory review process.* This pillar focuses on the supervision of banks' systems and capital adequacy by regulatory authorities.
- *Pillar III: Market discipline.* The objective of this pillar is to establish market discipline through public disclosure of risk measures and other relevant information on risk management.

This book focuses on Pillar I and considers probabilistic models for operational risk losses. Under the Basel II framework, three approaches can be used to quantify the operational risk annual capital charge $\mathbb{C}$:

- *The Basic Indicator Approach:*

$$\mathbb{C} = \alpha \frac{1}{n} \sum_{j=1}^{3} \max(\mathrm{GI}(j), 0), \quad n = \sum_{j=1}^{3} 1_{\{\mathrm{GI}(j)>0\}}, \tag{1.1}$$

  where $\mathrm{GI}(j)$, $j = 1, 2, 3$ are the annual gross incomes over the previous three years, n is the number of years with positive gross income, and $\alpha = 0.15$.
- *The Standardised Approach:*

$$\mathbb{C} = \frac{1}{3} \sum_{j=1}^{3} \max \left[\sum_{i=1}^{8} \beta_i \mathrm{GI}_i(j), 0 \right], \tag{1.2}$$

  where β_i, $i = 1, \ldots, 8$ are the factors for eight business lines (BL) listed in Table 1.1 and $\mathrm{GI}_i(j)$, $j = 1, 2, 3$ are the annual gross incomes of the i-th BL in the previous 3 years.
- *The Advanced Measurement Approaches (AMA):* a bank can calculate the capital charge using an internally developed model subject to regulatory approval.

Hereafter we consider AMA only. A bank intending to use the AMA should demonstrate accuracy of the internal models within the matrix of Basel II risk cells (eight business lines by seven event types, see Tables 1.1, 1.2, and 1.3) relevant to the bank and satisfy criteria, including:

- The use of internal data, relevant external data, scenario analysis and factors reflecting the business environment and internal control systems;
- The risk measure used for capital charge should correspond to the 99.9% confidence level for a one-year holding period;
- Diversification benefits are allowed if dependence modelling is approved by the regulator;
- Capital reduction due to insurance is capped at 20%.

Expected and unexpected losses. The initial Basel II proposal suggested that the capital charge should cover unexpected losses (UL), while expected losses (EL) should

Table 1.1 Basel II business lines (BL) Level 1. $\beta_1, \ldots, \beta_8$ are the business line factors used in the Basel II standardised approach; see BCBS ([17], pp. 147, 302). The original texts and data are available free of charge on the BIS website www.BIS.org/bcbs/publ.htm

i	Business line, BL(i)	β_i
1	Corporate finance	0.18
2	Trading and sales	0.18
3	Retail banking	0.12
4	Commercial banking	0.15
5	Payment and settlement	0.18
6	Agency services	0.15
7	Asset management	0.12
8	Retail brokerage	0.12

Table 1.2 Basel II event types (ET) Level 1; see BCBS ([17], pp. 305–307). The original text is available free of charge on the BIS website www.BIS.org/bcbs/publ.htm

j	Event type, ET(j)
1	Internal fraud
2	External fraud
3	Employment practices and workplace safety
4	Clients, products and business practices
5	Damage to physical assets
6	Business disruption and system failures
7	Execution, delivery and process management

Table 1.3 Basel risk matrix of business lines (BL) and event types (ET)

	ET(1)	ET(2)	⋯	ET(j)	⋯	ET(7)
BL(1)						
BL(2)						
⋮						
BL(i)			annual losses to be			
⋮			predicted over a one-year time horizon			
BL(8)						

be covered by the bank through internal provisions. The reasoning was that many bank activities have regular losses (e.g. credit card fraud). However, the accounting rules for provisions may not reflect the true EL. As a result, the final Basel II version proposed that regulatory capital is calculated as the sum of EL and UL, unless the bank can demonstrate an adequate capture of EL through its internal business practices; see BCBS ([17], p. 151). Hereafter, for simplicity, we consider the capital to be a sum of the EL and UL which is the 99.9% Value-at-Risk (VaR). The latter is the 0.999 quantile of the annual loss distribution that will be formally defined and discussed in the next chapter. The loss exceeding the 99.9% VaR does not require a capital charge. This is a so-called *catastrophic loss* or *stress loss*, also often called a *one in 1,000 year event*. Figure 1.2 gives an illustration of the EL, UL and VaR quantities.

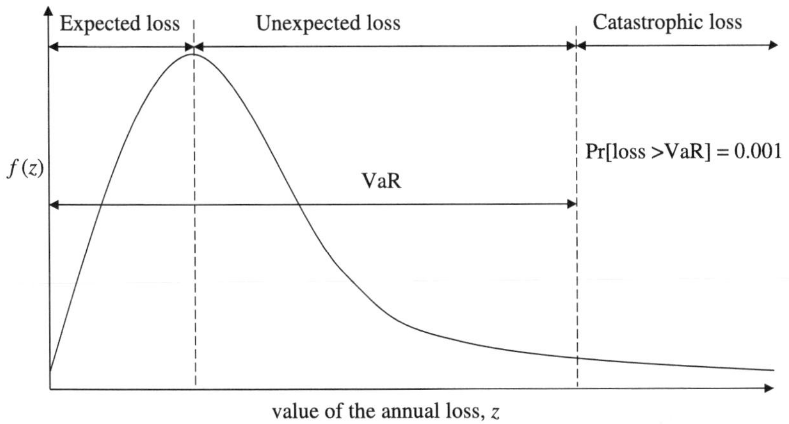

Fig. 1.2 Illustration of the expected and unexpected losses in the capital requirements at the 99.9% confidence level for a 1-year holding period. $f(z)$ is the probability density function of the annual loss

Regulatory and economic capital. The main purpose of the capital charge required by banking industry regulators is to protect a bank against potential losses; it can be viewed as a form of self-insurance. In simple terms, *regulatory capital* is the minimum amount imposed by the regulators, while *economic capital* is the amount that market forces imply for the risk. While regulatory capital for operational risk is based on the 99.9% confidence level over a 1-year holding period, economic capital is often higher. For example, some banks use the 99.95–99.97% confidence levels for economic capital.

Mapping of the activities into the Basel II matrix. The seven risk event types and eight business lines (referred to as Level 1) in Tables 1.1 and 1.2 are split by Basel II further (see Tables 1.4 and 1.5) providing a mapping of activities (where the losses may occur) into Level 1 risk cells. One can think of this as a hierarchical tree structure of business lines where each business line node has a branch of event types attached to it. For simplicity, often we consider Level 1 risk cells only, although in practice, it is not unusual for banks to quantify risks at the lower levels. Note, the number of risk cells at the lower levels is of the order of a hundred. Due to lack of data, banks quantify operational risk at the higher level.

1.4 Loss Data Collections

Several Quantitative Impact Studies (QIS) have been conducted to gain a better understanding of the potential effects of the Basel II capital requirements. QIS 2, QIS 2.5 and QIS 3 were conducted by the Basel Committee in 2001 and 2002. These impact studies gathered data on an international basis across many countries. Several participating countries decided to conduct further national impact studies (QIS 4). In 2005, to review the Basel II framework, BCBS undertook QIS 5. Detailed information on these studies can be found on the BCBS web site www.bis.org/bcbs/qis.

Table 1.4 Basel II mapping into level 1 business lines; see BCBS ([17], p. 302). The original text is available free of charge on the BIS website www.BIS.org/bcbs/publ.htm

Level 1	Level 2	Activity groups (level 3)
Corporate finance	Corporate finance	Mergers and acquisitions, underwriting, privatisations, securitisation, research, debt (government, high yield), equity, syndications, IPO, secondary private placements
	Municipal/Government finance	
	Merchant banking	
	Advisory services	
Trading and sales	Sales	Fixed income, equity, foreign exchanges, commodities, credit, funding, own position securities, lending and repos, brokerage, debt, prime brokerage
	Market making	
	Proprietary positions	
	Treasury	
Retail banking	Retail banking	Retail lending and deposits, banking services, trust and estates
	Private banking	Private lending and deposits, banking services, trust and estates, investment advice
	Card services	Merchant/commercial/corporate cards, private labels and retail
Commercial banking	Commercial banking	Project finance, real estate, export finance, trade finance, factoring, leasing, lending, guarantees, bills of exchange
Payment and settlement	External clients	Payments and collections, funds transfer, clearing and settlement
Agency services	Custody	Escrow, depository receipts, securities lending (customers) corporate actions
	Corporate agency	Issuer and paying agents
	Corporate trust	
Asset management	Discretionary fund management	Pooled, segregated, retail, institutional, closed, open, private equity
	Non-discretionary fund management	Pooled, segregated, retail, institutional, closed, open
Retail brokerage	Retail brokerage	Execution and full service

Some quantitative impact studies have been accompanied by operational loss data collection exercises (LDCE). The first two exercises conducted by the Risk Management Group of the BCBS on an international basis are referred to as the 2001 LDCE and 2002 LDCE. These were followed by the national 2004 LDCE in USA and the 2007 LDCE in Japan. Below we provide a summary for these LDCEs.[3]

[3] Recently, the BCBS conducted the 2008 LDCE and a public report summarising the results of the exercise appeared in July 2009; see BCBS [18].

Table 1.5 Basel II loss event type classification, BCBS ([17], pp. 305–307). The original text is available free of charge on the BIS website www.BIS.org/bcbs/publ.htm

Level 1	Level 2	Activity (level 3)
Internal fraud	Unauthorised activity	Transactions not reported (intentional); transaction type unauthorised (w/monetary loss); mismarking of position (intentional)
	Theft and fraud	Fraud/credit fraud/worthless deposits; theft/extortion/embezzlement/robbery; misappropriation of assets, malicious destruction of assets; forgery; check kiting; smuggling; account take-over/impersonation/etc.; tax non-compliance/evasion (wilful); bribes/kickbacks; insider trading (not on firm's account)
External fraud	Theft and fraud	Theft/robbery; forgery; check kiting
	Systems security	Hacking damage; theft of information (w/monetary loss)
Employment practices and workplace safety	Employee relations	Compensation, benefit, termination issues; organised labour activity
	Safe environment	General liability (slip and fall, etc.); employee health and safety rules events; workers compensation
	Diversity and discrimination	All discrimination types
Clients, products and business practices	Suitability, disclosure and fiduciary	Fiduciary breaches/guideline violations; suitability/disclosure issues (KYC, etc.); retail customer disclosure violations; breach of privacy; aggressive sales; account churning; misuse of confidential information; lender liability
	Improper business or market practices	Antitrust; improper trade/market practices; market manipulation; insider trading (on firm's account); unlicensed activity; money laundering
	Product flaws	Product defects (unauthorised, etc.); model errors
	Selection, sponsorship and exposure	Failure to investigate client per guidelines; exceeding client exposure limits
	Advisory activities	Disputes over performance of advisory activities
Damage to physical assets	Disasters and other events	Natural disaster losses; human losses from external sources (terrorism, vandalism)
Business disruption and system failures	Systems	Hardware; software; telecommunications; utility outage/disruptions
Execution, delivery and process management	Transaction capture, execution and maintenance	Miscommunication; data entry, maintenance or loading error; missed deadline or responsibility; model/system misoperation; accounting error/entity attribution error; other task misperformance; delivery failure; collateral management failure; reference data maintenance

Table 1.5 (continued)

Level 1	Level 2	Activity (level 3)
	Monitoring and reporting	Failed mandatory reporting obligation; inaccurate external report (loss incurred)
	Customer intake and documentation	Client permissions/disclaimers missing; legal documents missing/incomplete
	Customer/client account management	Unapproved access given to accounts; incorrect client records (loss incurred); negligent loss or damage of client assets
	Trade counterparties	Non-client counterparty misperformance; misc. non-client counterparty disputes
	Vendors and suppliers	Outsourcing; vendor disputes

1.4.1 2001 LDCE

The summary of this LDCE in BCBS [14] is based on individual operational risk loss data supplied by 30 banks from 11 countries in Europe, North America, Asia and Africa. This exercise collected 27,371 individual loss events for the 3 year period, 1998–2000. The majority of banks in the sample used minimum cut-off levels at or below Euro 10,000. Some of these cut-offs were different across business lines including, in some cases, cut-offs higher than Euro 10,000. The following observations on data clustering were made:

Frequency. The number of reported events appeared to be clustered in a few risk cells. In particular:

- Across business lines – The data were clustered in two of the eight business lines. "Retail Banking" BL(3) accounted for 67% of the total number of events; "Commercial Banking" BL(4) accounted for 13%.
- Across event types – A clustering is observed in two event types. "Execution, Delivery and Process Management" event type ET(7) accounted for 42% of the total number of events; "External Fraud" ET(2) accounted for 36%.
- Across business line/event type cells – Two most significant cells are BL(3)/ET(2) and BL(3)/ET(7), i.e. "External Fraud" ET(2) and "Execution, Delivery and Process Management" ET(7) in the "Retail Banking" BL(3). These cells accounted for over half of all individual loss events.

Aggregated loss. The total loss amount over all reported events was approximately Euro 2.6 billion. The largest losses in the sample ranged between Euro 50 million and Euro 100 million. The aggregate loss amounts appeared to be clustered in few risk cells too. In particular:

- Across business lines – "Retail Banking" BL(3) accounted for 39% of the total loss; "Commercial Banking" BL(4) accounted for 23%; "Trading and Sales" BL(2) was responsible for 19%.
- Across event types – "Execution, Delivery and Process Management" ET(7) accounted for 35%; "Clients, Products and Business Practices" ET(4) accounted for 28%; "External Fraud" ET(2) was responsible for 20%.

- Across business line/event type cells – Three risk cells accounted for approximately 40% of the total loss amount. These cells are: BL(2)/ET(7) – "Execution, Delivery and Process Management" in the "Trading and Sales" business line; BL(3)/ET(4) – "Clients, Products and Business Practices" in the "Retail Banking" business line; and BL(4)/ET(2) – "External Fraud" in the "Commercial Banking" business line.

1.4.2 2002 LDCE

The second LDCE was conducted by BCBS in 2002 across 89 banks from 19 countries in Europe, North and South America, Asia, and Australasia with the data summary provided in BCBS [15]. The data were submitted for losses occurred during 2001. Overall, the combined data for the 89 participating banks included more than 47,000 individual loss events. While the survey asked banks to report all events with gross loss amounts greater than or equal to Euro 10,000, in practice some banks used different minimum cut-off levels in reporting their data. The number of loss events and gross loss amounts per Business Line and Event Type reported in BCBS [15] are presented in Table 1.6.

Table 1.6 Number of loss events (%, top value in a cell) and total gross loss (%, bottom value in a cell) per business line and event type occurred in 2001 and reported in 2002 LDCE; see BCBS ([15], pp. 6–7). 100% corresponds to 47,269 events and Euro 7,795.5 million. Values exceeding 1% are indicated in bold. The original texts and data are available free of charge on the BIS website www.BIS.org/bcbs/publ.htm

	ET(1)	ET(2)	ET(3)	ET(4)	ET(5)	ET(6)	ET(7)	No ET	Total
BL(1)	0.04	0.04	0.15	0.15	0.03	0.02	0.45	0.00	0.89
	0.63	0.06	0.03	**2.03**	0.10	0.01	0.64	0.01	**3.51**
BL(2)	0.10	0.20	0.21	0.23	0.07	0.29	**9.74**	0.02	**10.86**
	0.76	0.52	0.83	**2.48**	**1.13**	0.23	**8.96**	0.1	**14.92**
BL(3)	**2.68**	**36.19**	**4.36**	**4.50**	**1.10**	0.34	**11.19**	0.73	**61.10**
	4.26	**10.10**	**4.36**	**3.26**	**1.12**	0.34	**5.45**	0.48	**29.36**
BL(4)	0.18	**3.81**	0.17	0.65	0.11	0.10	**2.14**	0.07	**7.22**
	0.27	**4.17**	0.26	**2.01**	**13.76**	0.23	**7.95**	0.30	**28.95**
BL(5)	0.05	0.68	0.11	0.05	0.02	0.17	**2.82**	0.01	**3.92**
	0.29	0.27	0.15	0.13	0.19	**1.01**	**1.20**	0.00	**3.25**
BL(6)	0.01	0.03	0.04	0.06	0.02	0.07	**2.92**	0.01	**3.15**
	0.00	0.05	0.10	0.06	**1.28**	0.51	**2.23**	0.01	**4.25**
BL(7)	0.06	0.09	0.08	0.28	0.01	0.03	**1.77**	0.02	**2.35**
	0.08	0.06	0.13	0.99	0.03	0.03	**1.45**	0.01	**2.78**
BL(8)	0.12	0.04	**1.68**	**1.14**	0.01	0.11	**3.75**	0.06	**6.91**
	0.79	0.02	0.65	**2.03**	**6.58**	0.36	**1.25**	0.04	**11.72**
No BL info	0.07	**1.31**	**1.70**	0.11	0.03	0.01	0.29	0.08	**3.59**
	0.13	0.30	0.24	0.15	0.09	0.01	0.29	0.05	**1.26**
Total	**3.31**	**42.39**	**8.52**	**7.17**	**1.40**	**1.14**	**35.07**	0.99	100.00
	7.23	**15.54**	**6.76**	**13.14**	**24.29**	**2.73**	**29.41**	0.91	100.00

Frequency. The following clustering can be observed for the number of events:

- Across business lines–The data are clustered into four of the eight business lines, with the highest concentration in "Retail Banking" BL(3) accounting for 61% of the individual observations. "Trading and Sales" BL(2) accounted for 11%; "Commercial Banking" BL(4)–7%; "Retail Brokerage" BL(8)–7%. Altogether, these four business lines accounted for 86% of all individual loss events reported.
- Across event types – 42% of the individual loss events were categorised as "External Fraud" ET(2), and 35% as "Execution, Delivery and Process Management" ET(7). "Employment Practices and Workplace Safety" ET(3) and "Clients, Products and Business Practices" ET(4) followed with 9 and 7% respectively. Altogether, these four event types accounted for 93% of the individual loss events.
- Across business line/event type cells – A considerable clustering is observed in the individual business line/event type cells. Just one cell, "External Fraud" in the "Retail Banking", BL(3)/ET(2), accounted for over 36% of the individual loss events. This was followed by BL(3)/ET(7) and BL(2)/ET(7) (i.e. "Execution, Delivery and Process Management" in "Retail Banking" and "Trading and Sales") with 11% and 10% respectively. Most of the cells (42 of the 56) accounted for less than 1% of the total events.

Aggregated loss. The total of gross operational risk loss amounts was just under Euro 7.8 billion. The aggregate gross loss amounts were distributed somewhat more evenly across business lines and Level 1 event types than the number of individual loss events. However, there was still evidence of clustering.

- Across business lines – "Retail Banking" BL(3) accounted for the largest share of gross loss amounts, slightly above 29% of the total. One can observe a lower percentage of loss amounts compared with loss numbers that reflects the dominance of smaller than average losses in this business line (recall that "Retail Banking" accounts for about 61% of the individual loss events). "Commercial Banking" BL(4) accounted for just under 29% of gross loss. Again, one can note a large difference between the share of gross losses accounted for by "Commercial Banking" (29%) and the share of the number of losses incurred by this business line (7%).
- Across event types – In terms of event types, gross loss amounts were concentrated in four categories: "Execution, Delivery and Process Management" ET(7), 29%; "Damage to Physical Assets" ET(5), 24%; "External Fraud" ET(2), 16%; and "Clients, Products and Business Practices" ET(4), 13%. Comparing the distribution of the number of losses by event types with the distribution of gross loss amounts, it is worth noting the difference in the "Damage to Physical Assets" ET(5). This event type accounted for less than 2% of the number of losses but over 24% of the gross losses. In contrast, "External Fraud" ET(2) accounted for over 42% of the number of operational losses but only 16% of the gross loss amounts.

- Across business line/event type cells – Looking at the individual cells of Table 1.6, two cells: BL(4)/ET(5) and BL(8)/ET(5) (i.e. "Damage to Physical Assets" in "Commercial Banking" and "Retail Brokerage") account for about 20% of gross losses. Three further cells: BL(3)/ET(2), BL(2)/ET(7) and BL(4)/ET(7) (i.e. "External Fraud" in the "Retail Banking" business line; and "Execution, Delivery and Process Management" in the "Trading and Sales" and "Commercial Banking" business lines) together account for a further 27% of the gross losses.

1.4.3 2004 LDCE

This survey was conducted by US Federal bank and Thrift Regulatory agencies in 2004 for US banks only to gain a better understanding of the potential effects of a Basel II-based regulatory capital regime on US institutions. Its results are summarised in the report from Federal Reserve System, Office of the Comptroller of the Currency, Office of Thrift Supervision and Federal Deposit Insurance Corporation [92]. Hereafter this document is referred to as FRS et al. [92]. Twenty three US banks provided LDCE data. In aggregate, approximately 1.5 million losses were submitted, totalling USD 25.9 billion. However, there was significant variation in the number of losses submitted by participating institutions. No specific loss threshold was required in the 2004 LDCE. Thresholds ranged from USD 0 to more than USD 10,000 across participating institutions.

Table 1.7 obtained from FRS et al. [92] provides the average annual number of losses and the average loss amount per year across all respondents by business line and event type for the events with the losses larger than USD 10,000.

Frequency. The reported events were clustered as follows:

- Across business lines – More than half of the losses (60%) occurred in "Retail Banking" BL(3). The majority of losses in this business line were attributed to two event types: "External Fraud" ET(2) and "Execution, Delivery and Process Management" ET(7). The business line with the second largest number of losses is the "Other" category with 8% of the total losses. Almost all respondents reported losses that fell within this category, suggesting that classification of losses affecting more than one business line remains an industry challenge.
- Across event types – With respect to event type, "External Fraud" ET(2) and "Execution, Delivery and Process Management" ET(7) had the largest number of losses per year with 39% and 35% of the reported losses respectively. ET(2) losses were primarily in "Retail Banking" BL(3), while ET(7) losses were spread across business lines more evenly with the largest contribution in "Retail Banking" BL(3).
- Across business line/event type cells – The largest risk cell was BL(3)/ET(2) accounting for 34% of the events. The next risk cell was BL(3)/ET(7) with 12% of the events.

Table 1.7 Number of loss events (%, top value in a cell) and total gross loss (%, bottom value in a cell) annualised per business line and event type reported by US banks in 2004 LDCE, FRS et al. ([92], tables 3 and 4). 100% corresponds to 18,371.1 events and USD 8,643.2 million. Losses ≥ USD 10,000 occurring during the period 1999–2004 in years when data capture was stable. Values exceeding 1% are indicated in bold

	ET(1)	ET(2)	ET(3)	ET(4)	ET(5)	ET(6)	ET(7)	Other	Fraud	Total
BL(1)	0.01	0.01	0.06	0.08	0.00		0.12	0.03	0.01	0.3
	0.14	0.00	0.03	0.30	0.00		0.05	0.01	0.00	0.5
BL(2)	0.02	0.01	0.17	0.19	0.03	0.24	**6.55**		0.05	**7.3**
	0.10	**1.17**	0.05	**4.29**	0.00	0.06	**2.76**		0.15	**8.6**
BL(3)	**2.29**	**33.85**	**3.76**	**4.41**	0.56	0.21	**12.28**	0.69	**2.10**	**60.1**
	0.42	**2.75**	0.87	**4.01**	0.1	0.21	**3.66**	0.06	0.26	**12.3**
BL(4)	0.05	**2.64**	0.17	0.36	0.01	0.03	**1.38**	0.02	0.44	**5.1**
	0.01	0.70	0.03	0.78	0.00	0.00	0.28	0.00	0.04	**1.8**
BL(5)	0.52	0.44	0.18	0.04	0.01	0.05	**2.99**	0.01	0.23	**4.5**
	0.08	0.13	0.02	0.01	0.00	0.02	0.28	0.00	0.05	0.6
BL(6)	0.01	0.03	0.04	0.31	0.01	0.14	**4.52**			**5.1**
	0.02	0.01	0.02	0.06	0.01	0.02	0.99			**1.1**
BL(7)	0.00	0.26	0.10	0.13	0.00	0.04	**1.82**		0.09	**2.4**
	0.00	0.02	0.02	**2.10**	0.00	0.01	0.38		0.01	**2.5**
BL(8)	0.06	0.10	**1.38**	**3.30**		0.01	**2.20**		0.20	**7.3**
	0.03	0.02	0.33	0.94		0.00	0.25		0.07	**1.6**
Other	0.42	**1.66**	**1.75**	0.40	0.12	0.02	**3.45**	0.07	0.08	**8.0**
	0.1	0.3	0.34	**67.34**	**1.28**	0.44	0.98	0.05	0.01	**70.8**
Total	**3.40**	**39.0**	**7.6**	**9.2**	0.7	0.7	**35.3**	**0.8**	**3.2**	100.0
	0.9	**5.1**	**1.7**	**79.8**	**1.4**	0.8	**9.6**	0.1	0.6	100.0

Aggregated loss. The following picture is observed for aggregated losses:

- Across business lines – The majority of the total loss amount (71%) was reported in the "Other" business line as losses that were not allocated to separate business lines. Note that these losses accounted for only 8.0% of annual loss frequency suggesting that the industry's loss experience is dominated by a small number of large losses spanning multiple business lines. Of the eight actual Basel business lines, "Retail Banking" BL(3) had the highest share (12%) of the annualised total loss though it was responsible for 60% of the number of losses.
- Across event types – 80% of the total loss amount per year was attributable to "Clients, Products and Business Practices" ET(4), with the largest portion of losses in the "Other" business line.
- Across business line/event type cells – Risk cells that appear to account for most of the total loss were: "Clients, Products and Business Practices" ET(4) in "Other" business line, 67%; "Clients, Products and Business Practices" ET(4) in the "Trading and Sales" BL(2) accounted for 4%.

1.4.4 2007 LDCE

The 2007 national LDCE was conducted in Japan jointly by the Financial Service Agency and the Bank of Japan and is referred to as 2007 LDCE. Summary results are reported in the document of Planning and Coordination Bureau, Financial Service Agency, Financial Systems and Bank Examination Department, Bank of Japan [193], hereafter referred to as PCB et al. [193]. Fourteen banks, including bank-holding companies, participated in this exercise providing 156,112 loss events which essentially occurred between 2002 and 2006. All of them provided data on individual losses of more than one yen. The number of losses and annualised gross amounts are given in Table 1.8. These data were extracted from PCB et al. [193]. Quick observations on data clustering are as follows.

Frequency. More than half of the losses occurred in "Retail Banking" BL(3). The business line with the second largest number of losses was "Commercial Banking" BL(4). With respect to event type – "Execution, Delivery and Process Management" ET(7) and "External Fraud" ET(2) accounted for the largest number of losses per year. The largest risk cell was BL(3)/ET(2) accounting for approximately 35% of events.

Table 1.8 Number of loss events (%, top value in a cell) and total gross loss (%, bottom value in a cell), annualised per business line and event type reported by banks in Japan for 2007 LDCE; see PCB et al. ([193], tables 3-3 and 3-4). 100% corresponds to 940.7 number of events and JPY 22,650 million. Based on stable data, greater than or equal to JPY 1 million. Values exceeding 1% are indicated in bold

	ET(1)	ET(2)	ET(3)	ET(4)	ET(5)	ET(6)	ET(7)	Total
BL(1)	0.00 0.00	0.01 0.00	0.01 0.00	0.11 0.09	0.00 0.00	0.07 0.04	0.28 0.18	0.5 0.3
BL(2)	0.01 0.00	0.00 0.00	0.04 0.04	0.22 0.09	0.00 0.00	0.26 0.00	**4.12** **25.03**	**4.7** **25.2**
BL(3)	0.54 **1.24**	**35.73** **6.36**	0.14 0.18	**3.62** **4.77**	**1.11** 0.26	**2.83** 0.26	**13.21** **8.43**	**57.2** **21.5**
BL(4)	0.29 0.13	0.64 **1.99**	**1.16** 0.71	**2.02** **18.32**	0.69 **4.19**	**5.87** **3.22**	**15.05** **16.87**	**25.7** **45.4**
BL(5)	0.00 0.00	0.00 0.00	0.00 0.00	0.00 0.00	0.00 0.00	0.38 0.04	0.23 0.00	0.6 0.1
BL(6)	0.00 0.00	0.00 0.00	0.01 0.00	0.92 0.62	0.01 0.00	**1.19** 0.22	**3.14** **3.00**	**5.3** **3.8**
BL(7)	0.00 0.00	0.00 0.00	0.1 0.04	0.48 0.49	0.00 0.00	0.05 0.00	**1.50** **1.02**	**2.1** **1.5**
BL(8)	0.84 **1.32**	0.00 0.00	0.01 0.00	**1.40** 0.53	0.00 0.00	0.18 0.04	**1.06** 0.13	**3.5** **2.0**
Other	0.09 0.18	0.11 0.00	0.02 0.00	0.00 0.00	0.13 0.00	0.09 0.00	0.02 0.00	0.4 0.2
Total	**1.8** **2.9**	**36.5** **8.3**	**1.5** **1.0**	**8.8** **24.8**	**1.9** **4.5**	**10.9** **3.9**	**38.6** **54.6**	100 100

Aggregated loss. Nearly half of the total loss amount was reported in "Commercial Banking" BL(4), followed by "Trading and Sales" BL(2). With respect to event type – "Execution, Delivery and Process Management" ET(7) and "Clients, Products and Business Practices" ET(4) accounted for more than three quarters of the total loss amount per year, with ET(7) accounting for more than half. Across risk cells, BL(2)/ET(7) produced the largest loss amount (25% of the total).

1.4.5 General Remarks

Even these very large databases almost certainly fail to provide a fully comprehensive sense of the range of potential operational risk loss events experienced by banks. Several general remarks can be made as follows:

- The data in all LDCEs exhibit considerable clustering around certain business lines and event types. In particular, there is considerable clustering in "Retail Banking" BL(3), which tends to have many but small operational risk events. There are also business line/event type combinations with few to no events reported. It is unclear whether the low reporting frequency in these areas reflects the low probability of event types occurring for certain business lines or the short data collection window or gaps in data collection.
- The number of large loss events (exceeding Euro 1 million) is comparatively small, representing just few percents of the observations. The findings reflect the evolution that was occurring in the data capture of operational risk losses in terms of methodologies and approaches for data collection in participating banks. Gaps in data collection almost certainly contributed to the considerable variation across banks in the number of events reported. It is also important to recognise that the findings discussed in the LDCE summaries reflect a short data collection window (one or few years), which even under the best of circumstances is unlikely to capture many large impact tail events.
- The data collected in the above LDCEs show some similarities and differences. In particular, the difference in results between national and international LDCEs may be attributed to variations between banking systems in different countries in terms of regulation, structure, scale, etc.

It is necessary to be cautious in using the summary data to draw conclusions about the extent of operational risk exposures. In considering the findings reported in Tables 1.6, 1.7, and 1.8, it would be inappropriate to conclude that business line and/or event types with a comparatively greater number or value of reported loss events are those representing the greatest sources of operational risk. To assess the extent of risk, it would be necessary to assess the extent of variability of both number and value of loss events around their expected values. Business lines or event types with large numbers of individual losses or with large aggregate losses could exhibit large or small variation over time, and therefore correspondingly large or small degrees of risk. A simple summary of the data does not supply significant insight in this regard. To gain such insight, it would be necessary to analyse the actual loss data.

Unfortunately the real operational risk datasets are not available for most of the researchers due to confidentiality issues. Though the author of this book had to deal with real data during consulting projects, it was not possible to get the real datasets to be used in the book. In this respect, two papers that will be referred to many times are of high importance: Moscadelli [166] analysing 2002 LDCE and Dutta and Perry [77] analysing 2004 LDCE .

1.5 Operational Risk Models

Many models have been suggested for modelling operational risk under the Basel II AMA. Excellent overviews of these can be found in Chernobai, Rachev and Fabozzi ([55], chapter 4), and Allen, Boudoukh and Saunders [9]. In brief, two conceptual approaches are the so-called *top-down* and *bottom-up* approach.

Top-down approach. Here, the data are typically analysed at the macro level (e.g. analysing overall bank losses) without attempting to model individual processes/risks types. Examples of the top-down models are:

- *Multifactor equity pricing models.* This approach assumes *market efficiency*, where the current asset price (stock price of the company) reflects all relevant information. Then the stock return process is assumed to be driven by many factors related to the market, credit and other non operational risks. The residual term of this regression is treated as due to operational risk. An example of such a study is Allen and Bali [8] that reported empirical evidence of the dependence between some operational risks and macroeconomic variables (such as GDP, unemployment, equity indices, interest rates, foreign exchange rates, regulatory environment variables and others).
- *Capital asset pricing model* (CAPM). Here, the asset risk premium is quantified, which is a difference between expected return and risk-free return. Then the contributions from credit and market risks are measured and the operational risk is treated as the residual. CAPM was introduced by Sharpe [213] for asset pricing. In the context of operational risk, it is discussed in van den Brink [38], and Hiwatashi and Ashida [122].
- *Income or expense based models.* These models are based on estimating the historical volatility of income or expense respectively subtracting the contributions from credit and market risks.
- *Risk indicator models.* These models link operational risk and exposure indicators such as gross income, volume of transactions, number of staff, etc. The Basel II Basic Indicator Approach (1.1) and Standardised Approach (1.2) are examples of a single indicator and multi-indicator models respectively.

Bottom-up approach. Broadly speaking, there are two bottom up approaches: process based models and loss distribution approach (LDA) models. The latter are often referred to as actuarial or statistical models.

- *Process based models.* Within this group of models, one can find *causal network models, multifactor causal models* (regression type models) and *reliability models.*

  - *Causal networks* are typically subjective models. These models are inherently linked to scorecard approaches. For each bank activity, a tree of events that may lead to operational risk loss is constructed. The probability of each event is specified by an expert. Typically, Bayesian belief networks are used to quantify the posterior probability of the loss. Various models of this type have been developed in the safety critical industries over many decades. For recent applications in air-transport safety, see Ale et al. [7], Neil, Malcolm and Shaw [173]. For application to operational risk, see Neil, Fenton and Tailor [171], Cruz ([65], section 9). Many examples can also be found in King ([134], chapters 8 and 9).

    Bayesian networks account for causal dependencies enabling linkage of the operational conditions to the probability and severity of the losses. There is a view that these models are certainly useful for risk management in finance but not as models for quantification of regulatory/economic capital. Nevertheless, for example, a dynamic Bayesian network recently developed in Neil, Häger and Andersen [172] allows quantification of the VaR of the total losses.
  - *Multifactor causal models* are based on regression of operational risk loss on a number of control factors (explanatory variables) such as number of staff, number of transactions, skill level, etc; see for example Cruz [65], Haubenstock [117]. Then these factors are used to predict future losses assuming that the factors are known for the next period of time.
  - *Reliability models* quantify the probability that a system will operate satisfactorily for a certain period of time. These are the models considered in operational research to study the trustworthiness of system elements. This is relevant to many processes in operational risk, for example, modelling the reliability of transaction processing systems; see Cruz ([65], section 7.7). For calculations of operational risk regulatory capital, it is not used as a standalone model but rather as a part of other models.

- **LDA models.** The LDA model is based on modelling frequency N and severities $X_1, X_2, \ldots$ of the operational risk events. Then, the annual loss is calculated by aggregation of the severities over a one-year period: $Z = X_1 + \ldots + X_N$. Both frequency and severity are modelled by random variables. The LDA is a focus of this book. Typically, the model is used to model operational risk within a business line/event type risk cell rather than at the process level.

The initial Basel II proposal for operational risk in 2001 suggested three approaches for AMA: the internal measurement approach, the loss distribution approach and the scorecard approach, see BCBS ([19], Annex 4). The latest Basel II document, BCBS [17], does not give any guidelines for the approaches and allows flexibility.

Hereafter, we consider the LDA model only.

Chapter 2
Loss Distribution Approach

Out of intense complexities intense simplicities emerge.
Sir Winston Churchill

Abstract This chapter introduces a basic model for the Loss Distribution Approach. We discuss the main aspects of the model and basic probabilistic concepts of risk quantification. The essentials of the frequentist and Bayesian statistical approaches are introduced. Basic Markov chain Monte Carlo methods that allow sampling from the posterior distribution, when the sampling cannot be done directly, are also described.

2.1 Loss Distribution Model

A popular method under the AMA is the loss distribution approach (LDA). Under the LDA, banks quantify distributions for frequency and severity of operational risk losses for each risk cell (business line/event type) over a 1-year time horizon. The banks can use their own risk cell structure but must be able to map the losses to the Basel II risk cells. Various quantitative aspects of LDA modelling are discussed in King [134]; Cruz [65, 66]; McNeil, Frey and Embrechts [157]; Panjer [181]; Chernobai, Rachev and Fabozzi [55]; Shevchenko [216]. The commonly used LDA model for the total annual loss Z_t in a bank can be formulated as

$$Z_t = \sum_{j=1}^{J} Z_t^{(j)}; \quad Z_t^{(j)} = \sum_{i=1}^{N_t^{(j)}} X_i^{(j)}(t). \tag{2.1}$$

Here:

- $t = 1, 2, \ldots$ is discrete time in annual units. If shorter time steps are used (e.g. quarterly steps to calibrate dependence structure between the risks), then extra summation over these steps can easily be added in (2.1).
- The annual loss $Z_t^{(j)}$ in risk cell j is modelled as a compound (*aggregate*) loss over one year with the *frequency* (annual number of events) $N_t^{(j)}$ implied by a counting process (e.g. Poisson process) and *severities* $X_i^{(j)}(t)$, $i = 1, \ldots, N_t^{(j)}$.
- Typically, the frequencies and severities are modelled by independent random variables.

P. Shevchenko, *Modelling Operational Risk Using Bayesian Inference*,
DOI 10.1007/978-3-642-15923-7_2,

Estimation of the annual loss distribution by modelling frequency and severity of losses is a well-known actuarial technique; see for example Klugman, Panjer and Willmot [136]. It is also used to model solvency requirements for the insurance industry; see Sandström [207] and Wüthrich and Merz [240]. Under model (2.1), the capital is defined as the 0.999 Value-at-Risk (VaR) which is the quantile of the distribution for the next year annual loss Z_{T+1}:

$$\mathrm{VaR}_q[Z_{T+1}] = \inf\{z \in \mathbb{R} : \Pr[Z_{T+1} > z] \leq 1 - q\} \tag{2.2}$$

at the level $q = 0.999$. Here, index $T + 1$ refers to the next year. The capital can be calculated as the difference between the 0.999 VaR and the expected loss if the bank can demonstrate that the expected loss is adequately captured through other provisions. If assumptions on correlations between some groups of risks (e.g. between business lines or between risk cells) cannot be validated then the capital should be calculated as the sum of the 0.999 VaRs over these groups. This is equivalent to the assumption of perfect positive dependence between annual losses of these groups.

Of course, instead of modelling frequency and severity to obtain the annual loss distribution, one can model aggregate loss per shorter time period (e.g. monthly total loss) and calculate the annual loss as a sum of these aggregate losses. However, the frequency/severity approach is more flexible and has good advantages, because some factors may affect frequency only while other factors may affect severity only. For example:

- As the business grows (e.g. volume of the transactions grows), the expected number of losses changes and this should be accounted for in forecasting the number of losses (frequency) over the next year.
- The general economic inflation affects the loss sizes (severity).
- The insurance for operational risk losses is more easily incorporated. This is because, typically, the insurance policies apply per event and affect the severity.

In this book, we focus on some statistical methods proposed in the literature for the LDA model (2.1). In particular we consider the problem of combining different data sources, modelling dependence and large losses, and accounting for parameter uncertainty.

2.2 Operational Risk Data

Basel II specifies the data that should be collected and used for AMA. In brief, a bank should have internal data, external data and expert opinion data. In addition, internal control indicators and factors affecting the businesses should be used. Development and maintenance of operational risk databases is a difficult and challenging task. Some of the main features of the required data are summarised as follows.

- *Internal data.* Internal data should be collected over a minimum five-year period to be used for capital charge calculations (when the bank starts the AMA, a three-year period is acceptable). Due to a short observation period, typically the internal data for many risk cells contain few low-frequency/high-severity losses or none. A bank must be able to map its historical internal loss data into the relevant Basel II risk cells; see Tables 1.1, 1.2 and 1.3. The data must capture all material activities and exposures from all appropriate sub-systems and geographic locations. A bank can have an appropriate low reporting threshold for internal loss data collection, typically of the order of EURO 10,000. Aside from information on gross loss amounts, a bank should collect information about the date of the event, any recoveries of gross loss amounts, as well as some descriptive information about the drivers or causes of the loss event.
- *External data.* A bank's operational risk measurement system must use relevant external data (either public data and/or pooled industry data). These external data should include data on actual loss amounts, information on the scale of business operations where the event occurred, and information on the causes and circumstances of the loss events. Industry data are available through external databases from vendors (e.g. Algo OpData provides publicly reported operational risk losses above USD 1million) and consortia of banks (e.g. ORX provides operational risk losses above EURO 20,000 reported by ORX members). External data are difficult to use directly due to different volumes and other factors. Moreover, the data have a survival bias as typically the data of all collapsed companies are not available. As discussed previously in Sect. 1.4, several Loss Data Collection Exercises (LDCE) for historical operational risk losses over many institutions were conducted and their analyses reported in the literature. In this respect, two papers are of high importance: Moscadelli [166] analysing 2002 LDCE and Dutta and Perry [77] analysing 2004 LDCE. In each case the data were mainly above EURO 10,000 and USD 10,000 respectively.
- *Scenario Analysis/expert opinion.* A bank must use scenario analysis in conjunction with external data to evaluate its exposure to high-severity events. Scenario analysis is a process undertaken by experienced business managers and risk management experts to identify risks, analyse past internal/external events, consider current and planned controls in the banks, etc. It may involve: workshops to identify weaknesses, strengths and other factors; opinions on the severity and frequency of losses; opinions on sample characteristics or distribution parameters of the potential losses. As a result some rough quantitative assessment of the risk frequency and severity distributions can be obtained. Scenario analysis is very subjective and should be combined with the actual loss data. In addition, it should be used for stress testing, for example to assess the impact of potential losses arising from multiple simultaneous loss events.
- *Business environment and internal control factors.* A bank's methodology must capture key business environment and internal control factors affecting operational risk. These factors should help to make forward-looking estimates, account for the quality of the controls and operating environments, and align capital assessments with risk management objectives.

Data important for modelling but often missing in external databases are risk exposure indicators and near-misses.

- *Exposure indicators.* The frequency and severity of operational risk events are influenced by indicators such as gross income, number of transactions, number of staff and asset values. For example, frequency of losses typically increases with increasing number of employees.
- *Near-miss losses.* These are losses that could occur but were prevented. Often these losses are included in internal datasets to estimate severity of losses but excluded in the estimation of frequency. For detailed discussion on management of near-misses, see Muermann and Oktem [167].

2.3 A Note on Data Sufficiency

Empirical estimation of the annual loss 0.999 quantile, using observed losses only, is impossible in practice. It is instructive to calculate the number of data points needed to estimate the 0.999 quantile empirically within the desired accuracy. Assume that independent data points $X_1, \ldots, X_n$ with common density $f(x)$ have been observed. Then the quantile q_α at confidence level α is estimated empirically as $\widehat{Q}_\alpha = \widetilde{X}_{\lfloor n\alpha \rfloor + 1}$, where $\widetilde{\mathbf{X}}$ is the data sample $\mathbf{X}$ sorted into the ascending order. The standard deviation of this empirical estimate is

$$\text{stdev}[\widehat{Q}_\alpha] = \frac{\sqrt{\alpha(1-\alpha)}}{f(q_\alpha)\sqrt{n}}; \tag{2.3}$$

see Glasserman ([108], section 9.1.2, p. 490). Thus, to calculate the quantile within relative error $\varepsilon = 2 \times \text{stdev}[\widehat{Q}_\alpha]/q_\alpha$, we need

$$n = \frac{4\alpha(1-\alpha)}{\varepsilon^2 (f(q_\alpha)q_\alpha)^2} \tag{2.4}$$

observations. Suppose that the data are from the lognormal distribution $\mathcal{LN}(\mu = 0, \sigma = 2)$. Then using formula (2.4), we obtain that $n = 140,986$ observations are required to achieve 10% accuracy ($\varepsilon = 0.1$) in the 0.999 quantile estimate. In the case of $n = 1,000$ data points, we get $\varepsilon = 1.18$, that is, the uncertainty is larger than the quantile we estimate.

Moreover, according to the regulatory requirements, the 0.999 quantile of the annual loss (rather than 0.999 quantile of the severity) should be estimated. As will be discussed many times in this book, operational risk losses are typically modelled by the so-called heavy-tailed distributions. In this case, the quantile at level q of the aggregate distributions can be approximated by the quantile of the severity distribution at level

$$p = 1 - \frac{1-q}{\mathrm{E}[N]};$$

see Sect. 6.7. Here, $\mathrm{E}[N]$ is the expected annual number of events. For example, if $\mathrm{E}[N] = 10$, then we obtain that the error of the annual loss 0.999 quantile is the same as the error of the severity quantile at the confidence level $p = 0.9999$. Again, using (2.4) we conclude that this would require $n \approx 10^6$ observed losses to achieve 10% accuracy. If we collect annual losses then $n/\mathrm{E}[N] \approx 10^5$ annual losses should be collected to achieve the same accuracy of 10%. These amounts of data are not available even from the largest external databases and extrapolation well beyond the data is needed. Thus parametric models must be used.

For an excellent discussion on data sufficiency in operational risk, see Cope, Antonini, Mignola and Ugoccioni [62].

2.4 Insurance

Some operational risks can be insured. If a loss occurs and it is covered by an insurance policy, then part of the loss will be recovered. Under the AMA, banks are allowed to recognise the risk mitigating impact of insurance on the regulatory capital charge. The reduction in the capital due to insurance is limited to 20%; see BCBS ([17], p. 155).

A typical policy will provide a recovery R for a loss X subject to the excess amount (deductible) D and top cover limit amount U as follows:

$$R = \begin{cases} 0, & \text{if } 0 \leq X < D, \\ X - D, & \text{if } D \leq X < U + D, \\ U, & \text{if } D + U \leq X. \end{cases} \tag{2.5}$$

That is, the recovery will take place if the loss is larger than the excess and the maximum recovery that can be obtained from the policy is U. Note that in (2.5), the time of the event is not involved and the top cover limit applies for a recovery per risk event, that is, for each event the obtained recovery is subject of the top cover limit. Including insurance into the LDA is simple; the loss severity in (2.1) should be reduced by the amount of recovery (2.5) and can be viewed as a simple transformation of the severity. However, there are several difficulties in practice, namely that

- policies may cover several different risks;
- different policies may cover the same risk;
- the top cover limit may apply for the aggregated recovery over many events of one or several risks (e.g. the policy will pay the recovery for losses until the top cover limit is reached by accumulated recovery).

These aspects and special restrictions on insurance recoveries required by Basel II make recovery dependent on time. Thus accurate accounting for insurance requires modelling the loss event times. For example, one can use a Poisson process to model the event times.

Remark 2.1 A convenient method to simulate event times from a Poisson process over a one-year time horizon is to simulate the annual number of events N from the Poisson distribution and then simulate the times of these N events as independent random variables from a uniform distribution $\mathcal{U}(0, 1)$.

It is not difficult to incorporate the insurance into an overall model if a Monte Carlo method[1] is used to quantify the annual loss distributions. The inclusion of the insurance will certainly reduce the capital charge, though the reduction is capped by 20% according to the Basel II requirement.

Finally, it is important to note that, incorporating insurance into the LDA is not only important for capital reduction but also beneficial for negotiating a fair premium with the insurer because the distribution of the recoveries and its characteristics can be estimated.

For implementation of insurance into the LDA, see Bazzarello, Crielaard, Piacenza and Soprano [22], Peters, Byrnes and Shevchenko [184]; also for guidelines on insurance within the AMA capital calculations, see Committee of European Banking Supervisors [59].

2.5 Basic Statistical Concepts

A concept of financial risk strongly relates to a notion of events that may occur and lead to financial consequences. Thus it is natural to model risks using probability theory. While a notion of randomness is very intuitive, it was only in 1933 that Kolmogorov [138] gave an axiomatic definition of randomness and probability. This theory gives a mathematical foundation to modern risk modelling. It is expected that the reader has a basic understanding of elementary statistics and probability. This section provides a description of essential concepts of probability theory used in the book and introduces relevant notation.

2.5.1 Random Variables and Distribution Functions

Hereafter, the following notation is used:

- Random variables are denoted by upper case symbols (capital letters) and their realisations are denoted by lower case symbols, e.g. random variable X and its realisation x.
- By convention, vectors are considered as column vectors and are written in bold, e.g. n-dimensional random vector $\mathbf{X} = (X_1, X_2, \dots, X_n)'$, where superscript '$'$' denotes transposition.
- The realisations of random variables considered in this book are real numbers, so that $\mathbf{x} = (x_1, x_2, \dots, x_n)'$ means a point in the n-dimensional Euclidean space of real numbers $\mathbb{R}^n$.

---

[1] Monte Carlo method is discussed in Sect. 3.2.

- To simplify notation, in general, the same symbol will be used to denote both a random variable and the space of its possible realisations. For example: Θ is a random variable; θ is realisation of Θ; and the space of all possible θ values is also denoted as Θ.
- Operators on random variables are written with square brackets, e.g. the variance of a random variable X is denoted as $\mathrm{Var}[X]$.
- Notationally, an *estimator* is a function of the sample while an *estimate* is the realised value of an estimator for a given realisation of the sample. For example, given a sample of random variables $X_1, X_2, \ldots, X_n$ the estimator is a function of $\mathbf{X}$ while the estimate is a function of the realisation $\mathbf{x}$.

A random variable has associated distribution function defined as follows.

Definition 2.1 (Univariate distribution function) The distribution function of a random variable X, denoted as $F_X(x)$, is defined as

$$F_X(x) = \Pr[X \leq x].$$

A corresponding *survival function* (*tail function*) is defined as

$$\overline{F}_X(x) = 1 - F_X(x) = \Pr[X > x].$$

Definition 2.2 (Multivariate distribution function) The multivariate distribution function of a random vector $\mathbf{X} = (X_1, X_2, \ldots, X_n)'$ is defined as

$$F_{\mathbf{X}}(x_1, x_2, \ldots, x_n) = \Pr[X_1 \leq x_1, X_2 \leq x_2, \ldots, X_n \leq x_n].$$

Often, for short notation we write $F_{\mathbf{X}}(\mathbf{x})$. A corresponding survival function is defined as

$$\overline{F}_{\mathbf{X}}(\mathbf{x}) = \Pr[\mathbf{X} > \mathbf{x}].$$

Remark 2.2

- Frequently used notation, $X \sim F_X(x)$, means a random variable X has a distribution function $F_X(x)$. Often, for simplicity of notation, we may drop the subscript and write $X \sim F(\cdot)$.
- All distributions used throughout the book are formally defined in Appendix A.

Random variables can be classified into different categories (*continuous, discrete* or *mixed*) according to their *support* (a set of all possible outcomes of a random variable). Precisely:

Definition 2.3 (Support of a random variable) The support of a random variable X with a distribution function $F_X(\cdot)$ is defined as a set of all points, where $F_X(\cdot)$ is strictly increasing.

Definition 2.4 (Continuous random variable) A continuous random variable X has its support on an interval, a union of intervals or real line (half-line). The distribution function of a continuous random variable can be written as

$$F_X(x) = \int_{-\infty}^{x} f_X(y)dy,$$

where $f_X(x)$ is called the continuous *probability density function.*

Definition 2.5 (Discrete random variable) A discrete random variable X has a finite or countable number of values $x_1, x_2, \ldots$. The distribution function of a discrete random variable has jump discontinuities at $x_1, x_2, \ldots$ and is constant between. The probability function (also called the *probability mass function*) of a discrete random variable is defined as

$$p_X(x_i) = \Pr[X = x_i], \quad i = 1, 2, \ldots$$
$$p_X(x) = 0 \text{ for } x \neq x_1, x_2, \ldots .$$

The corresponding probability density function can be written as

$$f_X(x) = \sum_{i \geq 1} p_X(x_i)\delta(x - x_i), \tag{2.6}$$

where $\delta(x)$ is the *Dirac δ-function* (also called the impulse δ-function) defined next.

Definition 2.6 (The Dirac δ-function) The Dirac δ-function is a function which is zero everywhere except from the origin where it is infinite and its integral over any arbitrary interval containing the origin is equal to one:

$$\delta(x) = 0 \text{ if } x \neq 0; \;\; \delta(0) = \infty,$$
$$\int_{-\epsilon}^{\epsilon} \delta(x)dx = 1 \text{ for any } \epsilon > 0.$$

Note that, this implies that for any function $g(x)$

$$\int_a^b g(x)\delta(x - x_0)dx = g(x_0) \;\text{ if }\; a < x_0 < b \tag{2.7}$$

and the integral is zero if (a, b) interval does not contain x_0. This definition of δ function is merely a heuristic definition but it is enough for the purposes of this book. The use and theory of the Dirac δ-function can be found in many books; see for example Pugachev ([196], section 9).

Definition 2.7 (Mixed random variable) Mixed random variable X is a continuous random variable with positive probability of occurrence on a countable set of exception points. Its distribution function F_X has jumps at these exception points and can be written as

$$F_X(x) = wF_X^{(d)}(x) + (1 - w)F_X^{(c)}(x)$$

where $0 \leq w \leq 1$, $F_X^{(c)}$ is a continuous distribution function and $F_X^{(d)}(x)$ is a discrete distribution function. The corresponding density function can be written as

$$f_X(x) = w \sum_{i \geq 1} p_X(x_i)\delta(x - x_i) + (1 - w) f_X^{(c)}(x), \tag{2.8}$$

where $f_X^{(c)}(x)$ is the continuous density function and $p_X(x_i)$ is a probability mass function of a discrete distribution.

Remark 2.3

- A mixed random variable is common in modelling financial risk and in operational risk in particular, when there is a probability of non-occurrence loss during a period of time (giving finite probability mass at zero) while the loss amount is a continuous random variable.
- In general, every distribution function may be represented as a mixture of three different types: discrete distribution function, continuous distribution function and singular continuous distribution function. The last is a continuous distribution function with points of increase on a set of zero Lebesgue measure. This type of random variable will not be considered in the book. The case of mixed random variables with two components (discrete and continuous) covers all situations encountered in operational risk practice.

2.5.2 *Quantiles and Moments*

We use the following standard definition of a generalised inverse function (also called *quantile function*) for a distribution function.

Definition 2.8 (Quantile function) Given a distribution function $F_X(x)$, the inverse function F_X^{-1} of F_X is

$$F_X^{-1}(\alpha) = \inf\{x \in \mathbb{R} : F_X(x) \geq \alpha\} = \sup\{x \in \mathbb{R} : F_X(x) < \alpha\},$$

where $0 < \alpha < 1$.

Given a probability level α, $F_X^{-1}(\alpha)$ is the α-th quantile of X (often, it is denoted as q_α). This generalised definition is needed to define a quantile for cases such as discrete and mixed random variables. If F_X is continuous, then the quantile function is the ordinary inverse function.

The expected value (*mean*) of a random variable X is denoted as $\mathrm{E}[X]$. A formal construction of the operator $\mathrm{E}[\cdot]$ is somewhat involved but for the purposes of this book we will use the following short definition.

Definition 2.9 (Expected value)

- If X is a continuous random variable with the density function $f_X(x)$, then

$$\mathrm{E}[X] = \int_{-\infty}^{\infty} x f_X(x) dx; \tag{2.9}$$

- If X is a discrete random variable with support $x_1, x_2, \ldots$ and probability mass function $p_X(x)$, then

$$\mathrm{E}[X] = \sum_{j \geq 1} x_j p_X(x_j);$$

- In the case of a mixed random variable X (see Definition 2.7), the expected value is

$$\mathrm{E}[X] = w \sum_{j \geq 1} x_j p_X(x_j) + (1 - w) \int_{-\infty}^{\infty} x f_X^{(c)}(x) dx.$$

Remark 2.4

- The expected value integral or sum may not converge to a finite value for some distributions. In this case it is said that the mean does not exist.
- The definition of the expected value (2.9) can also be used in the case of the discrete and mixed random variables if their density functions are defined as (2.6) and (2.8) respectively. This gives a unified notation for the expected value of the continuous, discrete and mixed random variables. Another way to introduce a unified notation is to use Riemann-Stieltjes integral

$$\mathrm{E}[X] = \int_{-\infty}^{\infty} x dF_X(x). \tag{2.10}$$

See Carter and Van Brunt [48] for a good introduction on this topic.

The expected value is the first moment about the origin (also called the first raw moment). There are two standard types of moments: the raw moments and central moments, defined as follows.

Definition 2.10 (Moments)

- The k-th moment about the origin (raw moment) of a random variable X is the expected value of X^k, i.e. $\mathrm{E}[X^k]$.
- The k-th central moment of a random variable X is the expected value of $(X - \mathrm{E}[X])^k$, i.e. $\mathrm{E}[(X - \mathrm{E}[X])^k]$.

Typically, k is nonnegative integer $k = 0, 1, 2, \ldots$. The expected value may not exist for some values of k; then it is said that the k-th moment does not exist. The first four moments are most frequently used and the relevant characteristics are:

- *Variance* – The variance of a random variable X is the second central moment

$$\mathrm{Var}[X] = \mathrm{E}[(X - \mathrm{E}[X])^2] = \mathrm{E}[X^2] - (\mathrm{E}[X])^2. \tag{2.11}$$

- *Standard deviation* – The standard deviation,

$$\mathrm{stdev}[X] = \sqrt{\mathrm{Var}[X]}, \tag{2.12}$$

is a measure of spread of the random variable around the mean. It is measured in the same units as the mean (i.e. the same units as the values of random variable).

- *Variational coefficient* – The *variational coefficient* (also called the *coefficient of variation*) is dimensionless quantity,

$$\text{Vco}[X] = \frac{\text{stdev}[X]}{\text{E}[X]}, \tag{2.13}$$

that measures the spread relative to the mean.
- *Skewness* – The skewness is a dimensionless quantity that measures an asymmetry of a random variable X and is defined as

$$\gamma_1 = \frac{\text{E}[(X - \text{E}[X])^3]}{(\text{stdev}[X])^3}. \tag{2.14}$$

For symmetric distributions, the skewness is zero.
- *Kurtosis* – The kurtosis is a dimensionless quantity that measures flatness of distribution relative to the normal distribution. It is defined as

$$\gamma_2 = \frac{\text{E}[(X - \text{E}[X])^4]}{(\text{stdev}[X])^4} - 3. \tag{2.15}$$

For the normal distribution, kurtosis is zero.

Again, for some distributions the above characteristics may not exist. Also, central moments can be expressed through the raw moments and vice-versa. Detailed discussion, definition and relationships for the above quantities can be found in virtually any statistical textbook. To conclude this section, we define the covariance and the linear correlation coefficient that measure the dependence between random variables.

Definition 2.11 (Covariance and linear correlation) The covariance of random variables X and Y is defined as

$$\text{Cov}[X, Y] = \text{E}[(X - \text{E}[X])(Y - \text{E}[Y])] = \text{E}[XY] - \text{E}[X]\text{E}[Y].$$

The linear correlation between X and Y is

$$\rho[X, Y] = \text{Cov}[X, Y]/\sqrt{\text{Var}[X]\text{Var}[Y]}.$$

These quantities are popular measures of the dependence between X and Y but, as will be discussed in Chap. 7, the linear correlation can be a bad indicator of dependence. Also, for some distributions these measures may not exist.

2.6 Risk Measures

Using economic reasoning, a list of axiomatic properties for a good (*coherent*) risk measure was suggested in the seminal paper by Artzner, Delbaen, Eber and Heath [10].

Definition 2.12 (A coherent risk measure) A coherent risk measure, $\varrho[X]$, is defined to have the following properties for any two random variables X and Y:

- Subadditivity: $\varrho[X+Y] \leq \varrho[X] + \varrho[Y]$;
- Monotonicity: if $X \leq Y$ for all possible outcomes, then $\varrho[X] \leq \varrho[Y]$;
- Positive homogeneity: for any positive constant c, $\varrho[cX] = c\varrho[X]$;
- Translation invariance: for any positive constant c, $\varrho[X+c] = \varrho[X] + c$.

For detailed discussions of this topic, see McNeil, Frey and Embrechts [157]. Two popular risk measures are the so-called *Value-at-Risk*(VaR) and *expected shortfall* defined and discussed below.

Definition 2.13 (Value-at-Risk) The VaR of a random variable $X \sim F_X(x)$ at the α-th probability level, $\mathrm{VaR}_\alpha(X)$, is defined as the α-th quantile of the distribution of X, i.e.

$$\mathrm{VaR}_\alpha[X] = F_X^{-1}(\alpha).$$

Remark 2.5 VaR is not a coherent measure. In general, VaR possesses all the properties of a coherent risk measure in Definition 2.12 except subadditivity. For some cases, such as a multivariate normal distribution, VaR is subadditive. However, in general, the VaR of a sum may be larger than the sum of VaRs. For examples and discussions, see McNeil, Frey and Embrechts [157]. This has a direct implication for measuring operational risk and will be discussed in Chap. 7.

A VaR at a specified probability level α does not provide any information about the fatness of the distribution upper tail. Often the management and regulators are concerned not only with probability of default but also with its severity. Therefore, other risk measures are often used. One of the most popular is *expected shortfall* (sometimes referred to as the tail Value-at-Risk), though, a formal Basel II regulatory requirement for operational risk capital charge refers to a VaR.

Definition 2.14 (Expected shortfall) The expected shortfall of a random variable $X \sim F_X(x)$ at the α-th probability level, $\mathrm{ES}_\alpha[X]$, is

$$\mathrm{ES}_\alpha[X] = \frac{1}{1-\alpha} \int_\alpha^1 \mathrm{VaR}_p[X] dp,$$

which is the "arithmetic average" of the VaRs of X from α to 1.

Remark 2.6 Expected shortfall is a coherent risk measure.

In the case of continuous distributions, it can be shown that $\mathrm{ES}_\alpha[X]$ is just expected loss given that the loss exceeds $\mathrm{VaR}_\alpha[X]$.

Proposition 2.1 *For a random variable* X *with a continuous distribution function* $F_X(x)$ *we have*

$$\mathrm{ES}_\alpha[X] = \mathrm{E}[X|X \geq \mathrm{VaR}_\alpha[X]],$$

which is the conditional expected loss given that the loss exceeds $\mathrm{VaR}_\alpha[X]$.

Proof Using Definition 2.14, the proof is trivial: simply change the integration variable to $x = F_X^{-1}(p)$. □

Remark 2.7 For a discontinuous distribution function $F_X(x)$, we have more general relation expression

$$\begin{aligned}\mathrm{ES}_\alpha[X] = \ & \mathrm{E}\left[X|X \geq \mathrm{VaR}_\alpha[X]\right] + \left(\frac{1}{1-\alpha} - \frac{1}{\overline{F}_X(\mathrm{VaR}_\alpha[X])}\right) \\ & \times \mathrm{E}\left[\max(X - \mathrm{VaR}_\alpha[X], 0)\right]. \end{aligned} \tag{2.16}$$

The quantity in brackets can be nonzero for some values of α, where there are jumps in distribution function. For a proof, see Proposition 3.2 in Acerbi and Tasche [4]

2.7 Capital Allocation

After the total capital is measured by $\varrho[\cdot]$, it is important to answer the question on how much a risk cell j contributes to the total capital. Calculation of the bank overall capital $\varrho[Z]$, where

$$Z = \sum_{j=1}^{J} Z^{(j)}$$

is the annual loss in a bank over next year as defined by (2.1),[2] should be followed by an important procedure of allocation of the capital into risk cells in such a way that

$$\varrho[Z] = \sum_{j}^{J} \mathrm{AC}_j. \tag{2.17}$$

[2] Here, for simplicity we drop the subscript indicating a year.

Here, AC_j denotes the capital allocated to the j-th risk cell. It can be used for performance measurement providing incentives for a business to improve its risk management practices. Naive choice $\mathrm{AC}_j = \varrho[Z^{(j)}]$ is certainly not appropriate because it disregards risk diversification. Also, the sum of $\varrho[Z^{(j)}]$ adds up to $\varrho[Z]$ only in the case of perfect positive dependence between risk cells.

Two popular methods, the *Euler principle* and *marginal contribution*, to allocate the capital are described below.

2.7.1 Euler Allocation

If risk measure ϱ is a positive homogeneous function (i.e. $\varrho[hX] = h\varrho[X], h > 0$) and differentiable, then by the *Euler principle*

$$\varrho[Z] = \sum_{j=1}^{J} \varrho_j^{Euler}, \tag{2.18}$$

where

$$\varrho_j^{Euler} = \left. \frac{\partial \varrho[Z + hZ^{(j)}]}{\partial h} \right|_{h=0}. \tag{2.19}$$

For a proof, see Problem 2.4. The Euler principle is used by many practitioners to calculate the allocated capitals as

$$\mathrm{AC}_j = \varrho_j^{Euler} = \left. \frac{\partial \varrho[Z + hZ^{(j)}]}{\partial h} \right|_{h=0}; \tag{2.20}$$

see Litterman [146], Tasche [232, 233] and McNeil, Frey and Embrechts ([157], section 6.3). These are called the Euler allocations and represent capital allocation per unit of exposure $Z^{(j)}$. Tasche [232] showed that it is the only allocation compatible with RORAC (return on risk adjusted capital, i.e expected return divided by risk capital) measure of performance in portfolio management. Another justification of the Euler allocations was given in Denaut [75] using game-theoretic considerations.

Standard deviation risk measure. In the case of standard deviation as a risk measure, $\varrho[Z] = \mathrm{stdev}[Z]$, it is easy to show that

$$\varrho_j^{Euler} = \frac{\mathrm{Cov}[Z^{(j)}, Z]}{\sqrt{\mathrm{Var}[Z]}}. \tag{2.21}$$

VaR and expected shortfall risk measures. For risk measures $\mathrm{VaR}_\alpha[\cdot]$ and $\mathrm{ES}_\alpha[\cdot]$, the derivatives in (2.20) can be calculated as

$$\left.\frac{\partial \mathrm{VaR}_\alpha[Z + hZ^{(j)}]}{\partial h}\right|_{h=0} = \mathrm{E}[Z^{(j)}|Z = \mathrm{VaR}_\alpha[Z]], \tag{2.22}$$

$$\left.\frac{\partial \mathrm{ES}_\alpha[Z + hZ^{(j)}]}{\partial h}\right|_{h=0} = \mathrm{E}[Z^{(j)}|Z \geq \mathrm{VaR}_\alpha[Z]]. \tag{2.23}$$

It is easy to verify that

$$\sum_{j=1}^{J} \mathrm{E}[Z^{(j)}|Z = \mathrm{VaR}_\alpha[Z]] = \mathrm{E}[Z|Z = \mathrm{VaR}_\alpha[Z]] = \mathrm{VaR}_\alpha[Z],$$

$$\sum_{j=1}^{J} \mathrm{E}[Z^{(j)}|Z \geq \mathrm{VaR}_\alpha[Z]] = \mathrm{E}[Z|Z \geq \mathrm{VaR}_\alpha[Z]] = \mathrm{ES}_\alpha[Z].$$

In general, the Euler allocations should be calculated numerically. Assume that the total capital is quantified using Monte Carlo methods. That is, a sample of independent and identically distributed annual losses $z_k^{(j)}$, $k = 1, \ldots, K$ is simulated for each risk cell j (here, the dependence between risk cells is allowed). Then, a sample $z_1, \ldots, z_K$, where $z_k = \sum_{j=1}^{J} z_k^{(j)}$, can be calculated and $\mathrm{VaR}_\alpha[Z]$ is estimated using the sample in the usual way. Denote this estimate by $\widehat{\mathrm{VaR}}_\alpha[Z]$. Then the Euler allocations in the case of expected shortfall (2.23) are

$$\mathrm{E}[Z^{(j)}|Z \geq \mathrm{VaR}_\alpha[Z]] \approx \frac{\sum_{k=1}^{K} z_k^{(j)} 1_{\{z_k \geq \widehat{\mathrm{VaR}}_\alpha[Z]\}}}{\sum_{k=1}^{K} 1_{\{z_k \geq \widehat{\mathrm{VaR}}_\alpha[Z]\}}}. \tag{2.24}$$

In the case of VaR, the Euler allocation can be difficult to estimate using the Monte Carlo sample, because $\Pr[Z = \mathrm{VaR}_\alpha[Z]] = 0$ in the case of continuous distributions. To handle this problem, the condition $Z = \mathrm{VaR}_\alpha[Z]$ can be replaced by $|Z - \mathrm{VaR}_\alpha[Z]| < \epsilon$ for some $\epsilon > 0$ large enough to have $\Pr[|Z - \mathrm{VaR}_\alpha[Z]| < \epsilon] > 0$. However, this condition will be satisfied by only a few Monte Carlo simulations and importance sampling techniques are needed to get an accurate estimation; see Glasserman [109]. For VaR, it can be somewhat easier to calculate the Euler allocations using the finite difference approximation

$$\left.\frac{\partial \varrho[Z + hZ^{(j)}]}{\partial h}\right|_{h=0} \approx \frac{\varrho[Z + \Delta Z^{(j)}] - \varrho[Z]}{\Delta} \tag{2.25}$$

with some small suitable $\Delta \neq 0$. Note that the choice of Δ depends on the numerical accuracy of the estimator for $\varrho[\cdot]$ and curvature of the $\varrho[\cdot]$ with respect to h. So, Δ should be neither very small nor too large. This is a typical problem with estimating derivatives via finite difference and details can be found in many books on numerical recipes; see for example Press, Teukolsky, Vetterling and Flannery ([195], section 5.7).

2.7.2 Allocation by Marginal Contributions

Another popular way to allocate capital is using marginal risk contribution

$$\varrho_j^{marg} = \varrho[Z] - \varrho[Z - Z^{(j)}], \tag{2.26}$$

which is the difference between total risk (across all risk cell) and total risk without risk cell j. This can be viewed as some crude approximation of Euler allocation derivatives (2.25) but of course differentiability is not required to calculate marginal contribution. The sum of marginal contributions may not add up to $\varrho[Z]$. In particular, in the case of subadditive risk measures, it can be shown that

$$\varrho_j^{marg} \le \varrho_j^{Euler}, \quad \sum_{j=1}^{J} \varrho_j^{marg} \le \varrho[Z]. \tag{2.27}$$

One can define

$$\mathrm{AC}_j = \frac{\varrho_j^{marg}}{\sum_{i=1}^{J} \varrho_i^{marg}} \varrho[Z], \tag{2.28}$$

to ensure that allocated capitals add up to $\varrho[Z]$.

Example 2.1 To illustrate, consider an example of three risk cells where the annual losses $Z^{(j)}$ are independent random variables from the lognormal distribution $\mathcal{LN}(0, \sigma_j)$ with $\sigma_1 = 1.5$, $\sigma_2 = 1.75$, and $\sigma_3 = 2$ respectively. Results based on 4×10^6 Monte Carlo simulations are given in Table 2.1. Here, we estimate VaR of the total loss, $\mathrm{VaR}_{0.999}[\sum_j Z^{(j)}] \approx 556$, and VaRs of individual risk cells $\mathrm{VaR}_{0.999}[Z^{(j)}]$, $j = 1, 2, 3$. The numerical error due to the finite number of simulations is of the order of 1%. $\widehat{\varrho}_j^{Euler}$ was estimated using finite difference approximation (2.25) with $\Delta = 0.02$. Due to this approximation, $\sum_j \widehat{\varrho}_j^{Euler} \approx 553$ is slightly different from $\mathrm{VaR}_{0.999}[\sum_j Z^{(j)}] \approx 556$, so the final estimate for capital allocations using Euler principle is

$$\mathrm{AC}_j^{Euler} = \frac{\widehat{\varrho}_j^{Euler}}{\sum_i \widehat{\varrho}_i^{Euler}} \mathrm{VaR}_{0.999}\left[\sum_i Z^{(i)}\right].$$

The total diversification

$$1 - \frac{\mathrm{VaR}_{0.999}[\sum_j Z^{(j)}]}{\sum_i \mathrm{VaR}_{0.999}[Z^{(i)}]} \tag{2.29}$$

is approximately 30%. It is easy to observe that, both marginal and Euler allocations AC_j are significantly less than corresponding $\mathrm{VaR}_{0.999}[Z^{(j)}]$.

Table 2.1 Allocation of capital $C = \mathrm{VaR}_{0.999}[\sum_j Z^{(j)}] \approx 556$ by marginal and Euler contributions. Here, $Z^{(j)} \sim \mathcal{LN}(0, \sigma_j)$. Estimated AC_j are given in absolute terms and as a percent of the total C. See Example 2.1 for details

j	σ_j	$\mathrm{VaR}_{0.999}[Z^{(j)}]$	ϱ_j^{marg}	AC_j^{marg}	ϱ_j^{Euler}	AC_j^{Euler}
1	1.5	103	9	13\2 %	20	20\4 %
2	1.75	221	58	84\15 %	102	103\18 %
3	2.0	490	314	459\83 %	431	433\78 %
Total		814	381	556\100 %	553	556\100 %

Also, $\widehat{\varrho}_j^{marg} < \widehat{\varrho}_j^{Euler}$. Finally, it is important to note that the relative importance of risk cells cannot be measured by simple ratios

$$\frac{\mathrm{VaR}_{0.999}[Z^{(j)}]}{\sum_i \mathrm{VaR}_{0.999}[Z^{(i)}]}, \quad j = 1, 2, 3,$$

which are, in this example, 13%, 27% and 60% respectively and very different from $\mathrm{AC}_j / \sum_i \mathrm{AC}_i$.

2.8 Model Fitting: Frequentist Approach

Estimation of the frequency and severity distributions is a challenging task, especially for low-frequency/high-severity losses, due to very limited data for these risks. The main tasks involved in fitting the frequency and severity distributions using data are:

- finding the best point estimates for the distribution parameters;
- quantification of the parameter uncertainties; and
- assessing the model quality (model error).

In general, these tasks can be accomplished by undertaking either the so-called frequentist or Bayesian approaches briefly discussed in this and the next section.

Fitting distribution parameters using data via the frequentist approach is a classical problem described in many textbooks. For the purposes of this book it is worth to mention several aspects and methods. Firstly, under the frequentist approach one says that the model parameters are fixed while their estimators have associated uncertainties that typically converge to zero when a sample size increases. Several popular methods to fit parameters (finding point estimators for the parameters) of the assumed distribution are:

- method of moments – finding the parameter estimators to match the observed moments;
- matching certain quantiles of the empirical distribution;
- maximum likelihood method – finding parameter values that maximise the joint density of observed data; and

- estimating parameters by minimising a certain distance between empirical and theoretical distributions, e.g. Anderson-Darling or other statistics; see Ergashev [89].

A *point estimator* is a function of a sample. Notationally, an *estimator* is a function of the sample while an *estimate* is the realised value of an estimator for a realisation of the sample. For example, given a vector of random variables $\mathbf{X} = (X_1, X_2, \ldots, X_K)'$, the estimator is a function of $\mathbf{X}$ while the estimate is a function of the realisation $\mathbf{x}$.

Given a sample $\mathbf{X} = (X_1, X_2, \ldots, X_K)'$ from a density $f(\mathbf{x}|\theta)$, we try to find a point estimator $\widehat{\Theta}$ for a parameter θ. In most cases different methods will lead to different point estimators. One of the standard ways to evaluate an estimator is to calculate its *mean squared error.*

Definition 2.15 (Mean squared error) The mean squared error (MSE) of an estimator $\widehat{\Theta}$ for a parameter θ is defined as

$$\mathrm{MSE}_{\widehat{\Theta}}(\theta) = \mathrm{E}[(\widehat{\Theta} - \theta)^2].$$

Any increasing function of $|\widehat{\Theta} - \theta|$ can be used as a measure of the accuracy of the estimator but MSE is the most popular due to tractability and good interpretation. In particular, it can be written as

$$\mathrm{MSE}_{\widehat{\Theta}}(\theta) = \mathrm{Var}[\widehat{\Theta}] + \left(\mathrm{E}[\widehat{\Theta}] - \theta\right)^2, \tag{2.30}$$

where the first term is due to the uncertainty (variability) of the estimator and the second term is due to the bias. The latter is defined as follows

Definition 2.16 (Bias of a point estimator) The *bias* of a point estimator $\widehat{\Theta}$ for a parameter θ is

$$\mathrm{Bias}_{\widehat{\Theta}}(\theta) = \mathrm{E}[\widehat{\Theta}] - \theta.$$

An estimator with zero bias, i.e. $\mathrm{E}[\widehat{\Theta}] = \theta$ is called *unbiased*. The MSE of an unbiased estimator is reduced to $\mathrm{MSE}_{\widehat{\Theta}}(\theta) = \mathrm{Var}[\widehat{\Theta}]$.

Example 2.2 Consider a sample of independent random variables $N_1, N_2, \ldots, N_M$ from $Poisson(\lambda)$, i.e. $\mathrm{E}[N_m] = \lambda$, and an estimator $\widehat{\Lambda} = \frac{1}{M}\sum_{m=1}^{M} N_m$ (in this case it is a maximum likelihood estimator; see Sect. 2.8.1 below). Then

$$\mathrm{E}[\widehat{\Lambda}] = \frac{1}{M}\mathrm{E}\left[\sum_{m=1}^{M} N_m\right] = \lambda.$$

Thus the estimator $\widehat{\Lambda}$ is an unbiased estimator of λ.

It is important for the point estimator of a parameter to be a *consistent* estimator, i.e. converge to the "true" value of the parameter in probability as the sample size increases. Formally, a property of consistency is defined for a sequence of estimators as follows.

Definition 2.17 (Consistent estimator) For a sample $X_1, X_2, \ldots$, a sequence of estimators

$$\widehat{\Theta}_n = \widehat{\Theta}_n(X_1, \ldots, X_n), \quad n = 1, 2, \ldots$$

for the parameter θ is a consistent sequence of estimators if for every $\epsilon > 0$

$$\lim_{n\to\infty} \Pr[|\widehat{\Theta}_n - \theta| < \epsilon] = 1.$$

A more informative estimation of the parameter (in comparison with the point estimator) is based on a confidence interval specifying the range of possible values.

Definition 2.18 (Confidence interval) Given a data realisation $\mathbf{X} = \mathbf{x}$, the $1 - \alpha$ *confidence interval* for a parameter θ is $[L(\mathbf{x}), U(\mathbf{x})]$ such that

$$\Pr[L(\mathbf{X}) \le \theta \le U(\mathbf{X})] \ge 1 - \alpha.$$

That is, the random interval $[L, U]$, where $L = L(\mathbf{X})$ and $U = U(\mathbf{X})$, contains the true value of parameter θ with at least probability $1 - \alpha$.

Typically, it is difficult to construct a confidence interval exactly. However, often it can be found approximately using Gaussian distribution approximation in the case of large data samples; see e.g. Sect. 2.8.1. Specifically, if a point estimator $\widehat{\Theta}$ is distributed from $\mathcal{N}(\theta, \sigma(\theta))$, then

$$\Pr\left[-F_N^{-1}(1-\alpha/2) \le \frac{\widehat{\Theta} - \theta}{\sigma(\theta)} \le F_N^{-1}(1-\alpha/2)\right] = 1 - \alpha,$$

where $F_N^{-1}(\cdot)$ is the inverse of the standard normal distribution $\mathcal{N}(0, 1)$. Note that $\sigma(\theta)$ depends on θ. For a given data realisation, typically $\sigma(\theta)$ is replaced by $\sigma(\widehat{\theta})$ to approximate a confidence interval by

$$\left[\widehat{\theta} - F_N^{-1}(1-\alpha/2)\sigma(\widehat{\theta}), \widehat{\theta} + F_N^{-1}(1-\alpha/2)\sigma(\widehat{\theta})\right]. \tag{2.31}$$

2.8.1 Maximum Likelihood Method

The most popular approach to fit the parameters of the assumed distribution is the maximum likelihood method. Given the model parameters $\boldsymbol{\theta} = (\theta_1, \theta_2, \ldots, \theta_K)'$, assume that the joint density of data $\mathbf{X} = (X_1, X_2, \ldots, X_n)'$ is $f(\mathbf{x}|\boldsymbol{\theta})$. Then the *likelihood function* is defined as the joint density $f(\mathbf{x}|\boldsymbol{\theta})$ considered as a function of parameters $\boldsymbol{\theta}$.

Definition 2.19 (Likelihood function) For a sample $\mathbf{X} = \mathbf{x}$ from the joint density $f(\mathbf{x}|\boldsymbol{\theta})$ with the parameter vector $\boldsymbol{\theta}$, the *likelihood function* is a function of $\boldsymbol{\theta}$:

$$\ell_{\mathbf{x}}(\boldsymbol{\theta}) = f(\mathbf{x}|\boldsymbol{\theta}). \tag{2.32}$$

The *log-likelihood function* is $\ln \ell_{\mathbf{x}}(\boldsymbol{\theta})$.

Often it is assumed that $X_1, X_2, \ldots, X_n$ are independent with a common density $f(x|\boldsymbol{\theta})$; then the likelihood function is $\ell_{\mathbf{x}}(\boldsymbol{\theta}) = \prod_{i=1}^{n} f(x_i|\boldsymbol{\theta})$.

The maximum likelihood estimators $\widehat{\boldsymbol{\Theta}}^{\text{MLE}} = \widehat{\boldsymbol{\Theta}}(\mathbf{X})$ of the parameters $\boldsymbol{\theta}$ are formally defined as follows.

Definition 2.20 (Maximum likelihood estimator) For a sample $\mathbf{X}$, $\widehat{\boldsymbol{\Theta}}(\mathbf{X})$ is the *maximum likelihood estimator* (MLE), if for each realisation $\mathbf{x}$, $\widehat{\boldsymbol{\theta}}(\mathbf{x})$ is a value of parameter $\boldsymbol{\theta}$ maximising the likelihood function $\ell_{\mathbf{x}}(\boldsymbol{\theta})$ or equivalently maximising the log-likelihood function $\ln \ell_{\mathbf{x}}(\boldsymbol{\theta})$.

An important property of MLEs is their convergence to the true value in probability as the sample size increases, i.e. MLEs are *consistent* estimators.

Theorem 2.1 *For a sample $X_1, X_2, \ldots, X_n$ of independent and identically distributed random variables from $f(x|\boldsymbol{\theta})$ and corresponding MLE $\widehat{\boldsymbol{\Theta}}_n$, under the suitable regularity conditions, as the sample size n increases,*

$$\lim_{n\to\infty} \Pr[|\widehat{\boldsymbol{\Theta}}_n - \boldsymbol{\theta}| \geq \epsilon] = 0 \quad \textit{for every} \quad \epsilon > 0. \tag{2.33}$$

The required regularity conditions are:

- The parameter is identifiable: $\boldsymbol{\theta} \neq \widetilde{\boldsymbol{\theta}} \Rightarrow f(x|\boldsymbol{\theta}) \neq f(x|\widetilde{\boldsymbol{\theta}})$.
- The true parameter should be an interior point of the parameter space.
- The support of $f(x|\boldsymbol{\theta})$ should not depend on $\boldsymbol{\theta}$.
- $f(x|\boldsymbol{\theta})$ should be differentiable in $\boldsymbol{\theta}$.

Asymptotically, for large sample size, under stronger conditions (that further require $f(x|\boldsymbol{\theta})$ to be differentiable three times with respect to $\boldsymbol{\theta}$ and to have continuous and bounded 3rd derivatives), the MLEs are distributed from the normal distribution:

Theorem 2.2 *Under the suitable regularity conditions, for a sample $X_1, X_2, \ldots, X_n$ of independent and identically distributed random variables from $f(x|\boldsymbol{\theta})$, $\boldsymbol{\theta} = (\theta_1, \theta_2, \ldots, \theta_K)'$, and corresponding MLE $\widehat{\boldsymbol{\Theta}}_n$:*

$$\sqrt{n}(\widehat{\boldsymbol{\Theta}}_n - \boldsymbol{\theta}) \to \mathcal{N}_K\left(0, [\mathbf{I}(\boldsymbol{\theta})]^{-1}\right), \tag{2.34}$$

as the sample size n increases. Here, $[\mathbf{I}(\boldsymbol{\theta})]^{-1}$ is the inverse matrix of the expected Fisher information matrix for one observation $\mathbf{I}(\boldsymbol{\theta})$, whose matrix elements are

$$\begin{aligned}\mathbf{I}(\boldsymbol{\theta})_{km} &= \mathrm{E}\left[\frac{\partial}{\partial\theta_k}\ln f(X_1|\boldsymbol{\theta})\frac{\partial}{\partial\theta_m}\ln f(X_1|\boldsymbol{\theta})\right]\\ &= -\mathrm{E}\left[\frac{\partial^2}{\partial\theta_k\partial\theta_m}\ln f(X_1|\boldsymbol{\theta})\right].\end{aligned} \tag{2.35}$$

That is, $\widehat{\boldsymbol{\Theta}}^{\mathrm{MLE}}$ converges to $\boldsymbol{\theta}$ as the sample size increases and asymptotically $\widehat{\boldsymbol{\Theta}}^{\mathrm{MLE}}$ is normally distributed with the mean $\boldsymbol{\theta}$ and covariance matrix $n^{-1}\mathbf{I}(\boldsymbol{\theta})^{-1}$. For precise details on regularity conditions and proofs, see Lehmann ([143], Theorem 6.2.1 and 6.2.3); these can also be found in many other books such as Casella and Berger ([49], p. 516), Stuart, Ord and Arnold ([225], chapter 18), Ferguson ([93], part 4) or Lehmann and Casella ([144], section 6.3).

In practice, this asymptotic result is often used even for small samples and for the cases that do not formally satisfy the regularity conditions. Note that the mean and covariances depend on the unknown parameters $\boldsymbol{\theta}$ and are usually estimated by replacing $\boldsymbol{\theta}$ with $\widehat{\boldsymbol{\theta}}^{\mathrm{MLE}}$ for a given realisation of data. Often in practice, the expected Fisher information matrix is approximated by the *observed information matrix*

$$\widehat{\mathbf{I}(\widehat{\boldsymbol{\theta}})}_{km} = -\frac{1}{n}\sum_{i=1}^{n}\left.\frac{\partial^2\ln f(x_i|\boldsymbol{\theta})}{\partial\theta_k\partial\theta_m}\right|_{\boldsymbol{\theta}=\widehat{\boldsymbol{\theta}}} = -\frac{1}{n}\left.\frac{\partial^2\ln\ell_{\mathbf{x}}(\boldsymbol{\theta})}{\partial\theta_k\partial\theta_m}\right|_{\boldsymbol{\theta}=\widehat{\boldsymbol{\theta}}} \tag{2.36}$$

for a given realisation of data. This should converge to the expected information matrix by the law of large numbers. It has been suggested in Efron and Hinkley [78], that the use of the observed information matrix leads to a better inference in comparison with the expected information matrix.

Though very useful and widely used, these asymptotic approximations are usually not accurate enough for small samples, that is the distribution of parameter errors can be materially different from normal and MLEs may have significant bias. Also, as for any asymptotic results, a priori, one cannot decide on a sample size that is large enough to use the asymptotic approximation.

To assess the quality of the fit, there are several popular goodness of fit tests including Kolmogorov-Smirnov, Anderson-Darling and Chi-square tests. Also, the likelihood ratio test and Akaike's information criterion are often used to compare models.

Usually maximisation of the likelihood (or minimisation of some distances in other methods) must be done numerically. Popular numerical optimisation algorithms include simplex method, Newton methods, expectation maximisation (EM) algorithm, and simulated annealing. It is worth mentioning that the last is attempting to find a global maximum while other methods find a local maximum. Also, EM is usually more stable and robust than the standard deterministic methods such as simplex or Newton methods.

Again, detailed descriptions of the above-mentioned methodologies can be found in many textbooks; for application in an operational risk context, see Panjer [181].

2.8.2 Bootstrap

Another popular method to estimate parameter uncertainties is the so-called *bootstrap*. This method is based on a simple idea: that we can learn about characteristics of a sample by taking resamples from the original sample and calculating the parameter estimates for each sample to asses the parameter variability. The bootstrap method was originally developed by Efron in the 1970s. For a good introduction to the method we refer the reader to Efron and Tibshirani [79]. Often the bootstrap estimators are reasonable and consistent. Two types of bootstrapping, *nonparametric bootstrap* and *parametric bootstrap*, are commonly used in practice.

Nonparametric bootstrap. Suppose we have a sample of independent and identically distributed random variables $\mathbf{X} = (X_1, X_2, \ldots, X_K)'$ and there is an estimator $\widehat{\Theta}(\mathbf{X})$. Then:

- Draw M independent samples

$$\mathbf{X}^{(m)} = (X_1^{(m)}, X_2^{(m)}, \ldots, X_K^{(m)})', \quad m = 1, \ldots, M$$

  with replacement from the original sample $\mathbf{X}$. That is $X_k^{(m)}$, $k = 1, \ldots, K$, $m = 1, \ldots, M$ are independent and identically distributed, and drawn from the empirical distribution of the original sample $\mathbf{X}$.
- Calculate estimator $\widehat{\Theta}^{(m)} = \widehat{\Theta}(\mathbf{X}^{(m)})$ for each resample $m = 1, \ldots, M$.
- Calculate

$$\widehat{\text{Var}}[\widehat{\Theta}] = \frac{1}{M-1}\sum_{m=1}^{M}\left(\widehat{\Theta}^{(m)} - \mu\right)^2, \text{ where } \mu = \frac{1}{M}\sum_{m=1}^{M}\widehat{\Theta}^{(m)}. \tag{2.37}$$

Parametric bootstrap. Suppose we have a sample of independent and identically distributed random variables $\mathbf{X} = (X_1, X_2, \ldots, X_K)'$ from $f(x|\theta)$ and we can calculate some estimator $\widehat{\Theta}(\mathbf{X})$ (e.g. MLE) for θ. Then:

- Draw M independent samples

$$\mathbf{X}^{(m)} = (X_1^{(m)}, X_2^{(m)}, \ldots, X_K^{(m)})', \quad m = 1, \ldots, M,$$

  where $X_k^{(m)}$, $k = 1, \ldots, K$, $m = 1, \ldots, M$ are independent and identically distributed from $f(x|\widehat{\theta})$.
- Calculate estimator $\widehat{\Theta}^{(m)} = \widehat{\Theta}(\mathbf{X}^{(m)})$ for each resample $m = 1, \ldots, M$.
- Calculate $\widehat{\text{Var}}[\widehat{\Theta}] = \frac{1}{M-1}\sum_{m=1}^{M}\left(\widehat{\Theta}^{(m)} - \mu\right)^2$, where $\mu = \frac{1}{M}\sum_{m=1}^{M}\widehat{\Theta}^{(m)}$.

The obtained $\widehat{\text{Var}}[\widehat{\Theta}]$ is used as an estimator for $\text{Var}[\widehat{\Theta}]$. Typically, for independent and identically distributed samples, this estimator is consistent, i.e.

$$\widehat{\text{Var}}[\widehat{\Theta}] \to \text{Var}[\widehat{\Theta}], \text{ as } M \to \infty \text{ and } K \to \infty, \tag{2.38}$$

though in more general situations it may not occur.

Remark 2.8 More accurate treatment of nonparametric bootstrap estimators involves an approximator

$$\widehat{\text{Var}}^*[\widehat{\Theta}] = \frac{1}{N-1}\sum_{m=1}^{N}\left(\widehat{\Theta}^{(m)} - \mu\right)^2, \quad \mu = \frac{1}{N}\sum_{m=1}^{N}\widehat{\Theta}^{(m)},$$

where $N = K^K$ is the total number of nondistinct resamples. N is very large even for small K, e.g. for $K = 10$, $N = 10^{10}$. Calculations of the variance estimators (2.37) with $M \ll N$ is considered as approximation for $\widehat{\text{Var}}^*$ variances. Then, convergence of bootstrap estimators is considered in two steps: $\widehat{\text{Var}}[\widehat{\Theta}] \to \text{Var}^*[\widehat{\Theta}]$ as $M \to \infty$; and $\text{Var}^*[\widehat{\Theta}] \to \text{Var}[\widehat{\Theta}]$ as $K \to \infty$.

2.9 Bayesian Inference Approach

There is a broad literature covering Bayesian inference and its applications for the insurance industry as well as other areas. For a good introduction to the Bayesian inference method, see Berger [27] and Robert [200]. This approach is well suited for operational risk and will be a central topic in this book. It is sketched below to introduce some notation and concepts, and then it will be discussed in detail in Chap. 4.

Consider a random vector of data $\mathbf{X} = (X_1, X_2, \ldots, X_n)'$ whose density, for a given vector of parameters $\boldsymbol{\theta} = (\theta_1, \theta_2, \ldots, \theta_K)'$, is $f_{\mathbf{X}|\boldsymbol{\Theta}}(\mathbf{x}|\boldsymbol{\theta})$. In the Bayesian approach, both data and parameters are considered to be random. A convenient interpretation is to think that parameter is a random variable with some distribution and the true value (which is deterministic but unknown) of the parameter is a realisation of this random variable. Then the joint density of the data and parameters is

$$f_{\mathbf{X},\boldsymbol{\Theta}}(\mathbf{x},\boldsymbol{\theta}) = f_{\mathbf{X}|\boldsymbol{\Theta}}(\mathbf{x}|\boldsymbol{\theta})\pi_{\boldsymbol{\Theta}}(\boldsymbol{\theta}) = \pi_{\boldsymbol{\Theta}|\mathbf{X}}(\boldsymbol{\theta}|\mathbf{x})f_{\mathbf{X}}(\mathbf{x}), \tag{2.39}$$

where

- $\pi_{\boldsymbol{\Theta}}(\boldsymbol{\theta})$ is the density of parameters (a so-called *prior density*);
- $\pi_{\boldsymbol{\Theta}|\mathbf{X}}(\boldsymbol{\theta}|\mathbf{x})$ is the density of parameters given data $\mathbf{X} = \mathbf{x}$ (a so-called *posterior density*);
- $f_{\mathbf{X},\boldsymbol{\Theta}}(\mathbf{x},\boldsymbol{\theta})$ is the joint density of the data and parameters;
- $f_{\mathbf{X}|\boldsymbol{\Theta}}(\mathbf{x}|\boldsymbol{\theta})$ is the density of the data given parameters $\boldsymbol{\Theta} = \boldsymbol{\theta}$. This is the same as a likelihood function, see (2.32), if considered as a function of $\boldsymbol{\theta}$ for a given $\mathbf{x}$, i.e. $\ell_{\mathbf{x}}(\boldsymbol{\theta}) = f_{\mathbf{X}|\boldsymbol{\Theta}}(\mathbf{x}|\boldsymbol{\theta})$;
- $f_{\mathbf{X}}(\mathbf{x})$ is the marginal density of $\mathbf{X}$. If $\pi_{\boldsymbol{\Theta}}(\boldsymbol{\theta})$ is continuous, then

$$f_{\mathbf{X}}(\mathbf{x}) = \int f_{\mathbf{X}|\boldsymbol{\Theta}}(\mathbf{x}|\boldsymbol{\theta})\pi_{\boldsymbol{\Theta}}(\boldsymbol{\theta})d\boldsymbol{\theta}$$

and if $\pi_{\Theta}(\boldsymbol{\theta})$ is a discrete, then the integration should be replaced by a corresponding summation.

Remark 2.9 Typically, $\pi_{\Theta}(\boldsymbol{\theta})$ depends on a set of further parameters, the so-called *hyper-parameters*, omitted here for simplicity of notation. The choice and estimation of the prior will be discussed later in Chap. 4.

Using (2.39), the well-known Bayes's theorem, Bayes [21], says that:

Theorem 2.3 (Bayes's theorem) *The posterior density can be calculated as*

$$\pi_{\Theta|\mathbf{X}}(\boldsymbol{\theta}|\mathbf{x}) = f_{\mathbf{X}|\Theta}(\mathbf{x}|\boldsymbol{\theta})\pi_{\Theta}(\boldsymbol{\theta})/f_{\mathbf{X}}(\mathbf{x}). \tag{2.40}$$

Here, $f_{\mathbf{X}}(\mathbf{x})$ plays the role of a normalisation constant and the posterior can be viewed as a combination of prior knowledge (contained in $\pi_{\Theta}(\boldsymbol{\theta})$) with information from the data (contained in $f_{\mathbf{X}|\Theta}(\mathbf{x}|\boldsymbol{\theta})$).

Given that $f_{\Theta}(\mathbf{x})$ is a normalisation constant, the posterior is often written as

$$\pi_{\Theta|\mathbf{X}}(\boldsymbol{\theta}|\mathbf{x}) \propto f_{\mathbf{X}|\Theta}(\mathbf{x}|\boldsymbol{\theta})\pi_{\Theta}(\boldsymbol{\theta}), \tag{2.41}$$

where "$\propto$" means "is proportional to" with a constant of proportionality independent of the parameter $\boldsymbol{\theta}$. Typically, in closed-form calculations, the right hand side of the equation is calculated as a function of $\boldsymbol{\theta}$ and then the normalisation constant is determined by integration over $\boldsymbol{\theta}$.

Using the posterior $\pi_{\Theta|\mathbf{X}}(\boldsymbol{\theta}|\mathbf{x})$, one can easily construct a probability interval for Θ, which is the analogue for confidence intervals (see Definition 2.18) under the frequentist approach.

Definition 2.21 (Credibility interval) Given a data realisation $\mathbf{X} = \mathbf{x}$, if $\pi_{\Theta|\mathbf{X}}(\theta|\mathbf{x})$ is the posterior density of Θ and

$$\Pr[a \le \Theta \le b|\mathbf{X} = \mathbf{x}] = \int_a^b \pi_{\Theta|\mathbf{X}}(\theta|\mathbf{x})d\theta \ge 1 \quad \alpha,$$

then the interval $[a, b]$ contains the true value of parameter θ with at least probability $1 - \alpha$. The interval $[a, b]$ is called a *credibility interval* (sometimes referred to as *predictive interval* or *credible interval*) for parameter θ.

Remark 2.10

- The inequality in the above definition is to cover the case of discrete posterior distributions.
- Typically, one chooses the smallest possible interval $[a, b]$. Also, one can consider one-sided intervals, e.g. $\Pr[\Theta \le b|\mathbf{X} = \mathbf{x}]$.
- Extension to the multivariate case, i.e. parameter vector $\boldsymbol{\theta}$, is trivial.
- Though the Bayesian credibility interval looks similar to the frequentist confidence interval (see Definition 2.18), these intervals are conceptually different. To determine a confidence (probability to contain the true value) the bounds of the frequentist confidence interval are considered to be random (functions of

random data) while bounds of the Bayesian credibility interval are functions of a data realisation. For some special cases the intervals are the same (for given data realisation) but in general they are different especially in the case of strong prior information.

If the data $X_1, X_2, \ldots$ are conditionally (given $\mathbf{\Theta} = \boldsymbol{\theta}$) independent then the posterior can be calculated iteratively, i.e. the posterior distribution calculated after k-1 observations can be treated as a prior distribution for the k-th observation. Thus the loss history over many years is not required, making the model easier to understand and manage, and allowing experts to adjust the priors at every step.

For simplicity of notation, the density and distribution subscripts indicating random variables will often be omitted, e.g. $\pi_{\mathbf{\Theta}}(\boldsymbol{\theta})$ will be written as $\pi(\boldsymbol{\theta})$.

2.9.1 Conjugate Prior Distributions

Sometimes the posterior density can be calculated in closed form, which is very useful in practice when Bayesian inference is applied. This is the case for the so-called conjugate prior distributions, where the prior and posterior distributions are of the same type.

Definition 2.22 (Conjugate prior) Let F denote a class of density functions $f(\mathbf{x}|\boldsymbol{\theta})$, indexed by $\boldsymbol{\theta}$. A class U of prior densities $\pi(\boldsymbol{\theta})$ is said to be a conjugate family for F and $F - U$ is called a conjugate pair, if the posterior density $\pi(\boldsymbol{\theta}|\mathbf{x}) = f(\mathbf{x}|\boldsymbol{\theta})\pi(\boldsymbol{\theta})/f(\mathbf{x})$, where $f(\mathbf{x}) = \int f(\mathbf{x}|\boldsymbol{\theta})\pi(\boldsymbol{\theta})d\boldsymbol{\theta}$, is in the class U for all $f \in F$ and $\pi \in U$.

Formally, if the family U contains all distribution functions then it is conjugate to any family F. However, to make a model useful in practice it is important that U should be as small as possible while containing realistic distributions. In Chap. 4, we present $F - U$ conjugate pairs (Poisson-gamma, lognormal-normal, Pareto-gamma) that are useful and illustrative examples of modelling frequencies and severities in operational risk. Several other pairs (binomial-beta, gamma-gamma, exponential-gamma) can be found for example in Bühlmann and Gisler [44]. In all these cases, the prior and posterior distributions have the same type and the posterior distribution parameters are easily calculated using the prior distribution parameters and observations (or recursively).

In general, if the posterior cannot be found in closed form or is difficult to evaluate, one can use Gaussian approximation or Markov chain Monte Carlo methods, discussed next.

2.9.2 Gaussian Approximation for Posterior

For a given data realisation $\mathbf{X} = \mathbf{x}$, denote the mode of the posterior $\pi(\boldsymbol{\theta}|\mathbf{x})$ by $\widehat{\boldsymbol{\theta}}$. If the prior is continuous at $\widehat{\boldsymbol{\theta}}$, then a Gaussian approximation for the posterior is obtained by a second-order Taylor series expansion around $\widehat{\boldsymbol{\theta}}$:

$$\ln \pi(\boldsymbol{\theta}|\mathbf{x}) \approx \ln \pi(\widehat{\boldsymbol{\theta}}|\mathbf{x}) + \frac{1}{2}\sum_{i,j} \left.\frac{\partial^2 \ln \pi(\boldsymbol{\theta}|\mathbf{x})}{\partial\theta_i \partial\theta_j}\right|_{\boldsymbol{\theta}=\widehat{\boldsymbol{\theta}}} (\theta_i - \widehat{\theta}_i)(\theta_j - \widehat{\theta}_j). \tag{2.42}$$

Under this approximation, $\pi(\boldsymbol{\theta}|\mathbf{x})$ is a multivariate normal distribution with the mean $\widehat{\boldsymbol{\theta}}$ and covariance matrix

$$\boldsymbol{\Sigma} = \mathbf{I}^{-1}, \; (\mathbf{I})_{ij} = -\left.\frac{\partial^2 \ln \pi(\boldsymbol{\theta}|\mathbf{x})}{\partial\theta_i \partial\theta_j}\right|_{\boldsymbol{\theta}=\widehat{\boldsymbol{\theta}}}. \tag{2.43}$$

Remark 2.11 In the case of improper constant priors, this approximation is comparable to the Gaussian approximation for the MLEs (2.34). Also, note that in the case of constant priors, the mode of the posterior and the MLE are the same. This is also true if the prior is uniform within a bounded region, provided that the MLE is within this region.

2.9.3 Posterior Point Estimators

Once the posterior density $\pi(\theta|\mathbf{x})$ is found, for given data $\mathbf{X}$, one can define point estimators of Θ. The mode and mean of the posterior are the most popular point estimators. These Bayesian estimators are typically referred to as the Maximum a Posteriori (MAP) estimator and the Minimum Mean Square Estimator (MMSE), formally defined as follows:

$$\text{MAP}: \quad \widehat{\Theta}^{\text{MAP}} = \arg\max_{\theta}\left[\pi(\boldsymbol{\theta} \mid \mathbf{X})\right], \tag{2.44}$$

$$\text{MMSE}: \quad \widehat{\Theta}^{\text{MMSE}} = \mathrm{E}\left[\Theta|\mathbf{X}\right]. \tag{2.45}$$

The median of the posterior is also often used as a point estimator for Θ. Also, note that if the prior $\pi(\theta)$ is constant and the parameter range includes the MLE, then the MAP of the posterior is the same as the MLE; see Remark 2.11.

More formally, the choice of point estimators is considered using a *loss function*, $l(\theta, \widehat{\theta})$, that measures the cost (loss) of a decision to use a particular point estimator $\widehat{\Theta}$. For example:

- quadratic loss: $l(\theta, \widehat{\theta}) = (\theta - \widehat{\theta})^2$;
- absolute loss: $l(\theta, \widehat{\theta}) = |\theta - \widehat{\theta}|$;
- all or nothing loss: $l(\theta, \widehat{\theta}) = 0$ if $\theta = \widehat{\theta}$ and $l(\theta, \widehat{\theta}) = 1$ otherwise;

- asymmetric loss function: e.g. $l(\theta, \widehat{\theta}) = \widehat{\theta} - \theta$ if $\widehat{\theta} > \theta$ and $l(\theta, \widehat{\theta}) = -2(\widehat{\theta} - \theta)$ otherwise.

Then the value of $\widehat{\Theta}$ that minimises $\mathrm{E}[l(\Theta, \widehat{\Theta})|\mathbf{X}]$ is called a Bayesian point estimator of Θ. Here, the expectation is calculated with respect to the posterior $\pi(\theta|\mathbf{X})$. In particular:

- The posterior mean is a Bayesian point estimator in the case of a quadratic loss function.
- In the case of an absolute loss function, the Bayesian point estimator is the median of the posterior.
- All or nothing loss function gives the mode of the posterior as the point estimator.

Remark 2.12 $\widehat{\Theta} = \widehat{\Theta}(\mathbf{X})$ is a function of data $\mathbf{X}$ and thus it is referred to as estimator. For a given data realisation $\mathbf{X} = \mathbf{x}$, we get $\widehat{\Theta} = \widehat{\theta}$ which is referred to as a point estimate.

Though the point estimators are useful, for quantification of operational risk annual loss distribution and capital we recommend the use of the whole posterior, as discussed in following chapters.

2.9.4 Restricted Parameters

In practice, it is not unusual to restrict parameters. In this case the posterior distribution will be a truncated version of the posterior distribution in the unrestricted case. That is, if $\boldsymbol{\theta}$ is restricted to some range $[\boldsymbol{\theta}_L, \boldsymbol{\theta}_H]$ then the posterior distribution will have the same type as in the unrestricted case but truncated outside this range.

For example, we choose the lognormal distribution, $\mathcal{LN}(\mu, \sigma)$ to model the data $\mathbf{X} = (X_1, \ldots, X_n)'$ and we choose a prior distribution for μ to be the normal distribution $\mathcal{N}(\mu_0, \sigma_0)$. This case will be considered in Sect. 4.3.4. However, if we know that μ cannot be negative, we restrict $\mathcal{N}(\mu_0, \sigma_0)$ to nonnegative values only.

Another example is the Pareto-gamma case, where the losses are modelled by $Pareto(\xi, L)$ and the prior distribution for the tail parameter ξ is $Gamma(\alpha, \beta)$; see Sect. 4.3.6. The prior is formally defined for $\xi > 0$. However, if we do not want to allow infinite mean predicted loss, then the parameter should be restricted to $\xi > 1$.

These cases can be easily handled by using the truncated versions of the prior-posterior distributions. Assume that $\pi(\theta)$ is the prior whose corresponding posterior density is $\pi(\theta|\mathbf{x}) = f(\mathbf{x}|\theta)\pi(\theta)/f(\mathbf{x})$, where θ is unrestricted. If the parameter is restricted to $a \leq \theta \leq b$, then we can consider the prior

$$\pi^{\mathrm{tr}}(\theta) = \frac{\pi(\theta)}{\Pr[a \leq \theta \leq b]} 1_{\{a \leq \theta \leq b\}}, \quad \Pr[a \leq \theta \leq b] = \int_a^b \pi(\theta) d\theta, \tag{2.46}$$

for some a and b with $\Pr[a \leq \theta \leq b] > 0$. $\Pr[a \leq \theta \leq b]$ plays the role of normalisation and thus the posterior density for this prior is simply

$$\pi^{\text{tr}}(\theta|\mathbf{x}) = \frac{\pi(\theta|\mathbf{x})}{\Pr[a \le \theta \le b|\mathbf{x}]} 1_{\{a \le \theta \le b\}}, \quad \Pr[a \le \theta \le b|\mathbf{x}] = \int_a^b \pi(\theta|\mathbf{x}) d\theta. \tag{2.47}$$

Remark 2.13 It is obvious that if $\pi(\theta)$ is a conjugate prior, then $\pi^{\text{tr}}(\theta)$ is a conjugate prior too.

2.9.5 Noninformative Prior

Sometimes there is no prior knowledge about the model parameter $\boldsymbol{\theta}$, or we would like to rely on data only and avoid an impact from any subjective information. In this case we need a *noninformative prior* (sometimes called *vague prior*) that attempts to represent a near-total absence of prior knowledge. A natural noninformative prior is the uniform density

$$\pi(\boldsymbol{\theta}) \propto \text{const} \quad \text{for all } \boldsymbol{\theta}. \tag{2.48}$$

If parameter $\boldsymbol{\theta}$ is restricted to a finite set, then this $\pi(\boldsymbol{\theta})$ corresponds to a proper uniform distribution. For example, the parameter p in a binomial distribution $Bin(n, p)$ is restricted to the interval [0, 1]. Then one can choose a noninformative constant prior which is the uniform distribution $\mathcal{U}(0, 1)$.

However, if the parameter $\boldsymbol{\theta}$ is not restricted, then a constant prior is not a proper density (since $\int f(\boldsymbol{\theta})d\boldsymbol{\theta} = \infty$). Such a prior is called an *improper prior*. For example, the parameter μ (mean) of the normal distribution $\mathcal{N}(\mu, \sigma)$ is defined on $(-\infty, \infty)$. Then, for any constant $c > 0$, $\pi(\mu) = c$ is not a proper density because $\int \pi(\mu)d\mu = \infty$. It is not a problem to use improper priors as long as the posterior is a proper distribution. Also, as noted in previous sections, if the prior $\pi(\boldsymbol{\theta})$ is constant and the parameter range includes the MLE, then the mode of the posterior is the same as the MLE; see Remark 2.11.

A constant prior is often used as a noninformative prior, though it can be criticised for a lack of invariance under transformation. For example, if a constant prior is used for parameter θ and model is reparameterised in terms of $\widetilde{\theta} = \exp(\theta)$, then the prior density for $\widetilde{\theta}$ is proportional to $1/\widetilde{\theta}$. Thus we cannot choose a constant prior for both θ and $\widetilde{\theta}$. In this case, one typically argues that some chosen parameterisation is the most intuitively reasonable and absence of prior information corresponds to a constant prior in this parameterisation. One can propose noninformative priors through consideration of problem transformations. This has been considered in many studies starting with Jeffreys [127]. For discussion on this topic, see Berger ([27], section 3.3). Here, we just mention that for a scale densities of the form $\theta^{-1} f(x/\theta)$, the recommended noninformative prior for a scale parameter $\theta > 0$ is

$$\pi(\theta) \propto \frac{1}{\theta}, \tag{2.49}$$

which is an improper prior because $\int_0^\infty \pi(\theta)d\theta = \infty$.

2.10 Mean Square Error of Prediction

To illustrate the difference between the frequentist and Bayesian approaches, consider the so-called (conditional) mean square error of prediction (MSEP) which is often used for prediction of uncertainty.

Consider a sample $X_1, X_2, \ldots, X_n, \ldots$ and assume that, given data

$$\mathbf{X} = (X_1, X_2, \ldots, X_n)',$$

we are interested in prediction of a random variable R which is a some function of $X_{n+1}, X_{n+2}, \ldots$. Assume that $\widehat{R}$ is a predictor for R and an estimator for $\mathrm{E}[R|\mathbf{X}]$. Then, the conditional MSEP is defined by

$$\mathrm{MSEP}_{R|\mathbf{X}}\left(\widehat{R}\right) = \mathrm{E}[(R - \widehat{R})^2|\mathbf{X}]. \tag{2.50}$$

It allows for a good interpretation if decoupled into *process variance* and *estimation error* as

$$\begin{aligned}\mathrm{MSEP}_{R|\mathbf{X}}\left(\widehat{R}\right) &= \mathrm{Var}[R|\mathbf{X}] + \left(\mathrm{E}[R|\mathbf{X}] - \widehat{R}\right)^2 \\ &= \text{process variance} + \text{estimation error}.\end{aligned} \tag{2.51}$$

It is clear that the estimator $\widehat{R}$ that minimises conditional MSEP is $\widehat{R} = \mathrm{E}[R|\mathbf{X}]$. Assume that the model is parameterised by the parameter vector $\boldsymbol{\theta} = (\theta_1, \ldots, \theta_K)'$. Then under the frequentist and Bayesian approaches we get the following estimators of MSEP.

Frequentist approach. Unfortunately, in frequentist approach $\mathrm{E}[R|\mathbf{X}]$ is unknown and the second term in (2.51) is often estimated by $\mathrm{Var}[\widehat{R}]$; see Wüthrich and Merz ([240], section 6.4.3). Under the frequentist approach, $\mathrm{Var}[R|\mathbf{X}]$ and $\mathrm{E}[R|\mathbf{X}]$ are functions of parameter $\boldsymbol{\theta}$ and can be denoted as $\mathrm{Var}_{\boldsymbol{\theta}}[R|\mathbf{X}]$ and $\mathrm{E}_{\boldsymbol{\theta}}[R|\mathbf{X}]$ respectively. Typically these are estimated as $\widehat{\mathrm{Var}}_{\boldsymbol{\theta}}[R|\mathbf{X}] = \mathrm{Var}_{\widehat{\boldsymbol{\Theta}}}[R|\mathbf{X}]$ and $\widehat{\mathrm{E}}_{\boldsymbol{\theta}}[R|\mathbf{X}] = \mathrm{E}_{\widehat{\boldsymbol{\Theta}}}[R|\mathbf{X}]$, where $\widehat{\boldsymbol{\Theta}}$ is a point estimator of $\boldsymbol{\theta}$ obtained by maximum likelihood or other methods. Also, typically one chooses $\widehat{R} = \mathrm{E}_{\widehat{\boldsymbol{\Theta}}}[R|\mathbf{X}]$, so that now $\widehat{R}$ is a function of $\widehat{\boldsymbol{\Theta}}$, that we denote as $\widehat{R}(\widehat{\boldsymbol{\Theta}})$. The parameter uncertainty term $\mathrm{Var}_{\boldsymbol{\theta}}[\widehat{R}]$ is usually estimated using the first-order Taylor expansion of $\widehat{R}(\widehat{\boldsymbol{\Theta}})$ around $\boldsymbol{\theta}$

$$\widehat{R}(\widehat{\boldsymbol{\theta}}) \approx \widehat{R}(\boldsymbol{\theta}) + \sum_i \left.\frac{\partial \widehat{R}(\widehat{\boldsymbol{\theta}})}{\partial \widehat{\theta}_i}\right|_{\widehat{\boldsymbol{\theta}}=\boldsymbol{\theta}} (\widehat{\theta}_i - \theta_i)$$

leading to

$$\mathrm{Var}_{\boldsymbol{\theta}}[\widehat{R}(\widehat{\boldsymbol{\Theta}})] \approx \sum_{i,j} \left.\frac{\partial \widehat{R}}{\partial \widehat{\theta}_i}\right|_{\widehat{\boldsymbol{\theta}}=\boldsymbol{\theta}} \left.\frac{\partial \widehat{R}}{\partial \widehat{\theta}_j}\right|_{\widehat{\boldsymbol{\theta}}=\boldsymbol{\theta}} \mathrm{Cov}[\widehat{\Theta}_i, \widehat{\Theta}_j].$$

Estimating $\boldsymbol{\theta}$ by $\widehat{\boldsymbol{\Theta}}$ gives the final estimator

$$\widehat{\mathrm{Var}}_{\boldsymbol{\theta}}[\widehat{R}(\widehat{\boldsymbol{\Theta}})] = \mathrm{Var}_{\widehat{\boldsymbol{\Theta}}}[\widehat{R}(\widehat{\boldsymbol{\Theta}})].$$

Note that if the point estimators are unbiased, i.e. $\mathrm{E}[\widehat{\Theta}_i - \theta_i] = 0$ then $\mathrm{E}[\widehat{R}(\widehat{\boldsymbol{\theta}})] \approx \widehat{R}(\boldsymbol{\theta})$. Finally, the estimator for conditional MSEP is

$$\begin{aligned} \widehat{\mathrm{MSEP}}_{R|\mathbf{X}}\left[\widehat{R}\right] &= \widehat{\mathrm{Var}}[R|\mathbf{X}] + \widehat{\mathrm{Var}}[\widehat{R}] \\ &= \text{process variance} + \text{estimation error}. \end{aligned} \tag{2.52}$$

The above estimators are typically consistent and unbiased in the limit of large sample size.

Bayesian approach. Under the Bayesian inference approach, where the unknown parameters $\boldsymbol{\theta}$ are modelled as random variables $\boldsymbol{\Theta}$, $\mathrm{Var}[R|\mathbf{X}]$ can be decomposed as

$$\begin{aligned} \mathrm{Var}[R|\mathbf{X}] &= \mathrm{E}[\mathrm{Var}[R|\boldsymbol{\Theta},\mathbf{X}]|\mathbf{X}] + \mathrm{Var}[\mathrm{E}[R|\boldsymbol{\Theta},\mathbf{X}]|\mathbf{X}] \\ &= \text{average process variance } + \text{parameter estimation error} \end{aligned} \tag{2.53}$$

that equals $\mathrm{MSEP}_{R|\mathbf{X}}\left[\widehat{R}\right]$ if we choose $\widehat{R} = \mathrm{E}[R|\mathbf{X}]$. Estimation of the terms involved requires knowledge of the posterior distribution for $\boldsymbol{\Theta}$ that can be obtained either analytically or approximated accurately using Markov chain Monte Carlo methods discussed in the next section.

2.11 Markov Chain Monte Carlo Methods

As has already been mentioned, the posterior distribution is often not known in closed form. Thus, typically, estimation of the posterior empirically by direct simulation is also problematic. Then, in general, Markov chain Monte Carlo methods (hereafter referred to as MCMC methods) can be used. These are described below.

Simulation from the known density function can be accomplished using well-known generic methods such as the inverse transform, or accept-reject methods; see Glasserman ([108], section 2.2).

Corollary 2.1 (The inverse transform) *If* $U \sim \mathcal{U}(0,1)$, *then the distribution of the random variable* $X = F^{-1}(U)$ *is* $F(x)$.

Remark 2.14 That is, to simulate X from the distribution $F(x)$ using the inverse transform, generate $U \sim \mathcal{U}(0,1)$ and calculate $X = F^{-1}(U)$.

Corollary 2.2 *Simulating* X *from the density* $f(x)$ *is equivalent to simulating* (X, U) *from the uniform distribution on* (x, u), *where* $0 \le u \le f(x)$.

Remark 2.15 This means that to simulate X from the density $f(x)$, generate (X, U) from the uniform distribution under the curve of $f(x)$. The latter is typically done through accept-reject algorithm (or sometimes called as rejection sampling).

Corollary 2.3 (Accept-reject method) *Assume that the density $f(x)$ is bounded by M (i.e. $f(x) \leq M$) and defined on the support $a \leq x \leq b$. Then, to simulate X with the density $f(x)$:*

- *draw $X \sim \mathcal{U}(a, b)$ and $U \sim \mathcal{U}(0, M)$;*
- *accept the sample of X if $U \leq f(X)$, otherwise repeat the above steps.*

If another density $g(x)$ such that $Mg(x) \geq f(x)$ can be found for constant M, then to simulate X with the density $f(x)$:

- *draw X from $g(x)$ and $U \sim \mathcal{U}(0, Mg(X))$;*
- *accept the sample of X if $U \leq f(X)$, otherwise repeat the above steps.*

The inverse method cannot be used if the normalisation constant is unknown, and the above accept-reject method cannot be used if you cannot easily find the bounds for the density. These difficulties are typical for the posterior densities. In general, estimation (sampling) of the posterior $\pi(\boldsymbol{\theta}|\mathbf{x})$ numerically can be accomplished using MCMC methods; for a good introduction see Robert and Casella [201]. MCMC has almost unlimited applicability though its performance depends on the problem particulars. The idea of MCMC methods is based on a simple observation that to obtain an acceptable approximation to some integrals depending on a distribution of interest $\pi(\boldsymbol{\theta}|\mathbf{x})$, it is enough to sample a sequence (Markov chain) $\{\boldsymbol{\theta}^{(1)}, \boldsymbol{\theta}^{(2)}, \dots\}$, whose limiting density is the density of interest $\pi(\boldsymbol{\theta}|\mathbf{x})$. This idea appeared as early as the original Monte Carlo method but became very popular and practical in the last few decades only when fast computing platforms became available.

A Markov chain is a sequence of random variables defined as follows:

Definition 2.23 (Markov chain) A sequence of random variables

$$\{\boldsymbol{\Theta}^{(0)}, \boldsymbol{\Theta}^{(1)}, \dots, \boldsymbol{\Theta}^{(l)}, \dots\}$$

is a *Markov chain* if, for any l, the conditional distribution of $\boldsymbol{\Theta}^{(l+1)}$ given $\boldsymbol{\Theta}^{(i)}$, $i = 0, 1, \dots, l$ is the same as the conditional distribution of $\boldsymbol{\Theta}^{(l+1)}$ given $\boldsymbol{\Theta}^{(l)}$. A conditional probability density of $\boldsymbol{\Theta}^{(l+1)}$ given $\boldsymbol{\Theta}^{(l)}$ is called *transition kernel* of the chain and is usually denoted as $K(\boldsymbol{\Theta}^{(l)}, \boldsymbol{\Theta}^{(l+1)})$.

The MCMC approach produces an *ergodic Markov chain* with a *stationary distribution* (which is also a *limiting distribution*). These chains are also *recurrent* and *irreducible*. The precise definitions of these properties are somewhat involved and can be found for example in Robert and Casella [201]. For the purposes of this book we remark as follows:

Remark 2.16

- We are interested in the case when the chain stationary distribution corresponds to the posterior density $\pi(\boldsymbol{\theta}|\mathbf{x})$.

- The *ergodic* property means that the distribution of $\boldsymbol{\Theta}^{(l)}$ converges to a *limiting distribution* $\pi(\boldsymbol{\theta}|\mathbf{x})$ for almost any starting value of $\boldsymbol{\Theta}^{(0)}$. Therefore for large l, $\boldsymbol{\Theta}^{(l)}$ is approximately distributed from $\pi(\boldsymbol{\theta}|\mathbf{x})$ regardless of the starting point. Of course the problem is to decide what is large l. This can formally be accomplished by running diagnostic tests on the stationarity of the chain.
- A Markov chain is said to have a *stationary distribution* if there is a distribution $\pi(\boldsymbol{\theta}|\mathbf{x})$ such that if $\boldsymbol{\Theta}^{(l)}$ is distributed from $\pi(\boldsymbol{\theta}|\mathbf{x})$ then $\boldsymbol{\Theta}^{(l+1)}$ is distributed from $\pi(\boldsymbol{\theta}|\mathbf{x})$ too.
- A Markov chain is *irreducible* if it is guaranteed to visit any set $\mathcal{A}$ of the support of $\pi(\boldsymbol{\theta}|\mathbf{x})$. This property implies that the chain is *recurrent*, i.e. that the average number of visits to an arbitrary set $\mathcal{A}$ is infinite and even *Harris recurrent*. The latter means that the chain has the same limiting behaviour for *every* starting value rather than *almost every* starting value.
- Markov chains considered in MCMC algorithms are almost always *homogeneous*, i.e. the distribution of $\boldsymbol{\Theta}^{(l_0+1)}, \boldsymbol{\Theta}^{(l_0+2)}, \ldots, \boldsymbol{\Theta}^{(l_0+k)}$ given $\boldsymbol{\Theta}^{(l_0)}$ is the same as the distribution of $\boldsymbol{\Theta}^{(1)}, \boldsymbol{\Theta}^{(2)}, \ldots, \boldsymbol{\Theta}^{(k)}$ given $\boldsymbol{\Theta}^{(0)}$ for any $l_0 \geq 0$ and $k > 0$.
- Another important stability property is called *reversibility* that means that the direction of the chain does not matter. That is, the distribution of $\boldsymbol{\Theta}^{(l+1)}$ conditional on $\boldsymbol{\Theta}^{(l+2)} = \boldsymbol{\theta}$ is the same as the distribution of $\boldsymbol{\Theta}^{(l+1)}$ conditional on $\boldsymbol{\Theta}^{(l)} = \boldsymbol{\theta}$. The chain is *reversible* if the transition kernel satisfies the *detailed balance condition*:

$$K(\boldsymbol{\theta}, \boldsymbol{\theta}')\pi(\boldsymbol{\theta}|\mathbf{x}) = K(\boldsymbol{\theta}', \boldsymbol{\theta})\pi(\boldsymbol{\theta}'|\mathbf{x}). \tag{2.54}$$

  The detailed balance condition is not necessary but sufficient condition for $\pi(\boldsymbol{\theta}|\mathbf{x})$ to be stationary density associated with the transitional kernel $K(\cdot\ ,\ \cdot)$ that usually can easily be checked for MCMC algorithms.

Of course, the samples $\boldsymbol{\Theta}^{(1)}, \boldsymbol{\Theta}^{(2)}, \ldots$ are not independent. However, the independence is not required if we have to calculate some functionals of $\pi(\boldsymbol{\theta}|\mathbf{x})$, because the Ergodic Theorem implies that for large L, the average

$$\frac{1}{L}\sum_{l=1}^{L} g(\boldsymbol{\Theta}^{(l)}) \tag{2.55}$$

converges to $\mathrm{E}[g(\boldsymbol{\Theta})|X = \mathbf{x}]$ (if this expectation is finite), where expectation is calculated with respect to $\pi(\boldsymbol{\theta}|\mathbf{x})$.

2.11.1 Metropolis-Hastings Algorithm

Metropolis-Hastings algorithm is almost a universal algorithm used to generate a Markov chain with a stationary distribution $\pi(\boldsymbol{\theta}|\mathbf{x})$. It has been developed by Metropolis et al. [161] in mechanical physics and generalised by Hastings [116] in a statistical setting. It can be applied to a variety of problems since it requires the knowledge of the distribution of interest up to a constant only. Given a density $\pi(\boldsymbol{\theta}|\mathbf{x})$, known up to a normalisation constant, and a conditional density $q(\boldsymbol{\theta}^*|\boldsymbol{\theta})$, the method generates the chain $\{\boldsymbol{\theta}^{(1)}, \boldsymbol{\theta}^{(2)}, \dots\}$ using the following algorithm:

Algorithm 2.1 (Metropolis-Hastings algorithm)

1. Initialise $\boldsymbol{\theta}^{(l=0)}$ with any value within a support of $\pi(\boldsymbol{\theta}|\mathbf{x})$;
2. For $l = 1, \dots, L$
   a. Set $\boldsymbol{\theta}^{(l)} = \boldsymbol{\theta}^{(l-1)}$;
   b. Generate a proposal $\boldsymbol{\theta}^*$ from $q(\boldsymbol{\theta}^*|\boldsymbol{\theta}^{(l)})$;
   c. Accept proposal with the acceptance probability

$$p(\boldsymbol{\theta}^{(l)}, \boldsymbol{\theta}^*) = \min\left\{1, \frac{\pi(\boldsymbol{\theta}^*|\mathbf{x})q(\boldsymbol{\theta}^{(l)}|\boldsymbol{\theta}^*)}{\pi(\boldsymbol{\theta}^{(l)}|\mathbf{x})q(\boldsymbol{\theta}^*|\boldsymbol{\theta}^{(l)})}\right\}, \tag{2.56}$$

   i.e. simulate U from the uniform distribution function $\mathcal{U}(0, 1)$ and set $\boldsymbol{\theta}^{(l)} = \boldsymbol{\theta}^*$ if $U < p(\boldsymbol{\theta}^{(l)}, \boldsymbol{\theta}^*)$. Note that the normalisation constant of the posterior does not contribute here;
3. Next l (i.e. do an increment, $l = l + 1$, and return to step 2).

Remark 2.17

- The density $\pi(\boldsymbol{\theta}|\mathbf{x})$ is called the *target* or *objective density*.
- $q(\boldsymbol{\theta}^*|\boldsymbol{\theta})$ is called the *proposal density* and will be discussed shortly.

2.11.2 Gibbs Sampler

The Gibbs sampler is a technique for generating random variables from a distribution indirectly, without having to calculate the density. The method takes its name from the Gibbs random fields in image-processing models starting with the paper of Geman and Geman [101]. Its roots can be traced back to the 1950s; see Robert and Casella [201] for a brief summary of the early history.

To illustrate the idea of the Gibbs sampler, consider the case of two random variables X and Y that have a joint bivariate density $f(x, y)$. Assume that simulation of X from $f(x)$ cannot be done directly but we can easily sample X from $f(x|y)$ and Y from $f(y|x)$. Then, the Gibbs sampler generates samples as follows:

Algorithm 2.2 (Gibbs sampler, bivariate case)

1. Initialise $y^{(l=0)}$ with an arbitrary value within a support of Y.
2. For $l = 1, \dots, L$
   a. simulate $x^{(l)}$ from $f\left(x|y^{(l-1)}\right)$;

b. simulate $y^{(l)}$ from $f\left(y|x^{(l)}\right)$;

3. Next l (i.e. do an increment, $l = l + 1$, and return to step 2).

Under quite general conditions $f(x, y)$ is a stationary distribution of the chain $\{(x^{(l)}, y^{(l)}), l = 1, 2, \ldots\}$; and the chain is ergodic with a limiting distribution $f(x, y)$, that is the distribution of $x^{(l)}$ converges to $f(x)$ for large l.

Gibbs sampling can be thought of as a practical implementation of the fact that knowledge of the conditional distributions is sufficient to determine a joint distribution (if it exists!).

The generalisation of the Gibbs sampling to a multidimensional case is as follows. Consider a random vector $\mathbf{X}$ with a joint density $f(\mathbf{x})$. Denote full conditionals $f_i(x_i|\mathbf{x}_{-i}) = f(x_i|x_1, \ldots, x_{i-1}, x_{i+1}, \ldots, x_N)$. Then, do the following steps:

Algorithm 2.3 (Gibbs sampler, multivariate case)

- Initialise $x_2^{(l=0)}, \ldots, x_N^{(l=0)}$ with an arbitrary value.
- For $l = 1, \ldots, L$

  1) simulate $x_1^{(l)}$ from $f_1\left(x_1|x_2^{(l-1)}, \ldots, x_N^{(l-1)}\right)$;

  2) simulate $x_2^{(l)}$ from $f_2\left(x_2|x_1^{(l)}, x_3^{(l-1)}, \ldots, x_N^{(l-1)}\right)$;

  $\vdots$

  N) simulate $x_N^{(l)}$ from $f_N\left(x_N|x_1^{(l)}, \ldots, x_{N-1}^{(l-1)}\right)$;

- Next l.

Again, under general conditions the joint density $f(\mathbf{x})$ is a stationary distribution of the generated chain $\{\mathbf{x}^{(l)}, l = 1, 2, \ldots\}$; and the chain is ergodic, that is $f(\mathbf{x})$ is a limiting distribution of the chain.

2.11.3 Random Walk Metropolis-Hastings Within Gibbs

The *Random Walk Metropolis-Hastings* (RW-MH) *within Gibbs* algorithm is easy to implement and often efficient if the likelihood function can be easily evaluated. It is referred to as *single-component Metropolis-Hastings* in Gilks, Richardson and Spiegelhalter ([106], section 1.4). The algorithm is not well known among operational risk practitioners and we would like to mention its main features; see Shevchenko and Temnov [217] for application in the context of operational risk and Peters, Shevchenko and Wüthrich [186] for application in the context of a similar problem in the insurance.

The RW-MH within Gibbs algorithm creates a reversible Markov chain with a stationary distribution corresponding to our target posterior distribution. Denote by $\boldsymbol{\theta}^{(l)}$ the state of the chain at iteration l. The algorithm proceeds by proposing to move the i-th parameter from the current state $\theta_i^{(l-1)}$ to a new proposed state θ_i^* sampled from the MCMC proposal transition kernel. Typically the parameters are restricted by simple ranges, $\theta_i \in [a_i, b_i]$, and proposals are sampled from the normal distribution. Then, the logical steps of the algorithm are as follows.

Algorithm 2.4 (RW-MH within Gibbs)

1. Initialise $\theta_i^{(l=0)}, i = 1, \ldots, I$ by e.g. using MLEs.
2. For $l = 1, \ldots, L$
   a. Set $\boldsymbol{\theta}^{(l)} = \boldsymbol{\theta}^{(l-1)}$.
   b. For $i = 1, \ldots, I$
      i. Sample proposal θ_i^* from the transition kernel, e.g. from the truncated normal density

$$f_N^{\mathrm{tr}}(\theta_i^* | \theta_i^{(l)}, \sigma_i) = \frac{f_N(\theta_i^* | \theta_i^{(l)}, \sigma_i)}{F_N(b_i | \theta_i^{(l)}, \sigma_i) - F_N(a_i | \theta_i^{(l)}, \sigma_i)}, \quad (2.57)$$

      where $f_N(x|\mu, \sigma)$ and $F_N(x|\mu, \sigma)$ are the normal density and its distribution with mean μ and standard deviation σ.
      ii. Accept proposal with the acceptance probability

$$p(\boldsymbol{\theta}^{(l)}, \boldsymbol{\theta}^*) = \min\left\{1, \frac{\pi(\boldsymbol{\theta}^* | \mathbf{x}) f_N^{\mathrm{tr}}(\theta_i^{(l)} | \theta_i^*, \sigma_i)}{\pi(\boldsymbol{\theta}^{(l)} | \mathbf{x}) f_N^{\mathrm{tr}}(\theta_i^* | \theta_i^{(l)}, \sigma_i)}\right\}, \quad (2.58)$$

      where $\boldsymbol{\theta}^* = (\theta_1^{(l)}, \ldots, \theta_{i-1}^{(l)}, \theta_i^*, \theta_{i+1}^{(l-1)}, \ldots)$, i.e. simulate U from the uniform $\mathcal{U}(0, 1)$ and set $\theta_i^{(l)} = \theta_i^*$ if $U < p(\boldsymbol{\theta}^{(l)}, \boldsymbol{\theta}^*)$. Note that the normalisation constant of the posterior does not contribute here.
   c. Next i
3. Next l.

This procedure builds a set of correlated samples from the target posterior distribution. One of the most useful asymptotic properties is the convergence of ergodic averages constructed using the Markov chain samples to the averages obtained under the posterior distribution. The chain has to be run until it has sufficiently converged to the stationary distribution (the posterior distribution) and then one obtains samples from the posterior distribution. General properties of this algorithm, including convergence results, can be found in Robert and Casella ([201], sections 6–10).

The RW-MH algorithm is simple in nature and easy to implement. However, for a bad choice of the proposal distribution, the algorithm gives a very slow convergence to the stationary distribution. There have been several recent studies regarding the optimal scaling of the proposal distributions to ensure optimal convergence rates; see Bedard and Rosenthal [24]. The suggested asymptotic acceptance rate optimising the efficiency of the process is 0.234. Usually, it is recommended that the σ_i in (2.57) are chosen to ensure that the acceptance probability is roughly close to 0.234. This requires some tuning of the σ_i prior to the final simulations.

2.11.4 ABC Methods

The standard MCMC described above assumes that the likelihood of the data for given model parameters can easily be evaluated. If this is not the case, but synthetic data are easily simulated from the model for given parameters, then the so-called *approximate Bayesian computation* (ABC) methods can be utilised to estimate the model. For example, this is the case when the severity is modelled by the α-stable or g-and-h distributions that can easily be simulated but the density is not available in closed form. ABC methods are relatively recent developments in computational statistics; see Beaumont, Zhang and Balding [23] and Tavaré, Marjoram, Molitor and Plagnol [234]. For applications in the context of operational risk and insurance; see Peters and Sisson [188], and Peters, Wüthrich and Shevchenko [190].

Consider the data $\mathbf{X}$ and denote the model parameters by $\boldsymbol{\theta}$. Then the posterior from which we wish to draw samples is $\pi(\boldsymbol{\theta}|\mathbf{x}) \propto f(\mathbf{x}|\boldsymbol{\theta})\pi(\boldsymbol{\theta})$. The purpose of ABC is to sample from the posterior $\pi(\boldsymbol{\theta}|\mathbf{x})$ without evaluating computationally intractable $f(\mathbf{x}|\boldsymbol{\theta})$. The logical steps of the simplest ABC algorithm are as follows.

Algorithm 2.5 (Rejection Sampling ABC)

1. Choose a small tolerance level ϵ.
2. For $l = 1, 2, \ldots$
   a. Draw $\boldsymbol{\theta}^*$ from the prior $\pi(\cdot)$.
   b. Simulate a synthetic dataset $\mathbf{x}^*$ from the model given parameters $\boldsymbol{\theta}^*$, i.e. simulate from $f(\cdot|\boldsymbol{\theta}^*)$.
   c. Rejection condition: calculate a distance metric $\rho(\mathbf{x}, \mathbf{x}^*)$ that measures a difference between $\mathbf{x}$ and $\mathbf{x}^*$. Accept the sample, i.e. set $\boldsymbol{\theta}^{(l)} = \boldsymbol{\theta}^*$ if $\rho(\mathbf{x}, \mathbf{x}^*) \leq \epsilon$, otherwise return to step a).
3. Next l.

It is easy to show that, if the support of the distributions on $\mathbf{x}$ is discrete and the rejection condition $\rho(\mathbf{x}, \mathbf{x}^*) \leq \epsilon$ is a simplest condition accepting the proposal only if $\mathbf{x}^* = \mathbf{x}$, then the obtained $\boldsymbol{\theta}^{(1)}, \boldsymbol{\theta}^{(2)}, \ldots$ are samples from $\pi(\boldsymbol{\theta}|\mathbf{x})$. For general case, the obtained samples $\boldsymbol{\theta}^{(l)}$, are from

$$\pi_{ABC}(\boldsymbol{\theta}|\mathbf{x},\epsilon) \propto \int \pi(\boldsymbol{\theta})\pi(\mathbf{x}^*|\boldsymbol{\theta})g_\epsilon(\mathbf{x}|\mathbf{x}^*)d\mathbf{x}^*, \tag{2.59}$$

where the weighting function

$$g_\epsilon(\mathbf{x}|\mathbf{x}^*) \propto \begin{cases} 1, \text{ if } & \rho(\mathbf{x},\mathbf{x}^*) \le \epsilon, \\ 0, \text{ otherwise.} \end{cases} \tag{2.60}$$

As $\epsilon \to 0$, for appropriate choices of distance $\rho(\cdot,\cdot)$,

$$\pi_{ABC}(\boldsymbol{\theta}|\mathbf{x},\epsilon) \to \pi(\boldsymbol{\theta}|\mathbf{x}).$$

Of course, for a finite ϵ we obtain an approximation for $\pi(\boldsymbol{\theta}|\mathbf{x})$.

To improve the efficiency, $\rho(\mathbf{x},\mathbf{x}^*)$ is often replaced by $\rho(S(\mathbf{x}),S(\mathbf{x}^*))$, where $S(\mathbf{x})$ is a summary statistic of the data sample. Other weighting functions can be used. In general, the procedure is simple: given a realisation of the model parameters, a synthetic dataset $\mathbf{x}^*$ is simulated and compared to the original dataset $\mathbf{x}$. Then the summary statistic $S(\mathbf{x}^*)$ is calculated for simulated dataset $\mathbf{x}^*$ and compared to the summary statistic of the observed data $S(\mathbf{x})$; and a distance $\rho(S(\mathbf{x}),S(\mathbf{x}^*))$ is calculated. Finally, a greater weight is given to the parameter values producing $S(\mathbf{x}^*)$ close to $S(\mathbf{x})$ according to the weighting function $g_\epsilon(\mathbf{x}|\mathbf{x}^*)$. The obtained sample is from $\pi_{ABC}(\boldsymbol{\theta}|\mathbf{x},\epsilon)$ that converges to the target posterior $\pi(\boldsymbol{\theta}|\mathbf{x})$ as $\epsilon \to 0$, assuming that $S(\mathbf{x})$ is a *sufficient statistic*[3] and the weighting function converges to a point mass on $S(\mathbf{x})$. The tolerance, ϵ is typically set as small as possible for a given computational budget. One can calculate the results for subsequently reduced values of ϵ until the further reduction does not make material difference for the model outputs. The described ABC can be viewed as a general augmented model

$$\pi(\boldsymbol{\theta},\mathbf{x},\mathbf{x}^*) = \pi(\mathbf{x}|\mathbf{x}^*,\boldsymbol{\theta})\pi(\mathbf{x}^*|\boldsymbol{\theta})\pi(\boldsymbol{\theta}),$$

where $\pi(\mathbf{x}|\mathbf{x}^*,\boldsymbol{\theta})$ is replaced by $g(\mathbf{x}|\mathbf{x}^*)$.

To improve the performance of ABC algorithm, it can be combined with MCMC producing the stationary distribution $\pi_{ABC}(\boldsymbol{\theta}|\mathbf{x},\epsilon)$. For example, the MCMC-ABC can be implemented as follows.

Algorithm 2.6 (MCMC-ABC)

1. Initialise $\boldsymbol{\theta}^{(l=0)}$.
2. For $l = 1,\ldots,L$
   a. Draw proposal $\boldsymbol{\theta}^*$ from the proposal density $q(\cdot|\boldsymbol{\theta}^{(l-1)})$.

[3] A sufficient statistic is a function of the dataset $\mathbf{x}$ which summarises all the available sample information about $\boldsymbol{\theta}$; for a formal definition, see Berger ([27], section 1.7).

b. Simulate a synthetic dataset $\mathbf{x}^*$ from the model given parameters $\boldsymbol{\theta}^*$.
c. Accept the proposal with the acceptance probability

$$p(\boldsymbol{\theta}^{(l-1)}, \boldsymbol{\theta}^*) = \min\left\{1, \frac{\pi(\boldsymbol{\theta}^*)q(\boldsymbol{\theta}^{(l-1)}|\boldsymbol{\theta}^*)}{\pi(\boldsymbol{\theta}^{(l-1)})q(\boldsymbol{\theta}^*|\boldsymbol{\theta}^{(l-1)})} 1_{\{\rho(S(\mathbf{x}),S(\mathbf{x}^*))\leq\epsilon\}}\right\},$$

i.e. simulate U from the uniform (0,1) and set $\boldsymbol{\theta}^{(l)} = \boldsymbol{\theta}^*$ if $U \leq p(\boldsymbol{\theta}^{(l-1)}, \boldsymbol{\theta}^*)$, otherwise set $\boldsymbol{\theta}^{(l)} = \boldsymbol{\theta}^{(l-1)}$. Here, $1_{\{.\}}$ is a standard indicator function.

3. Next l.

Various summary statistics of the dataset $x_1, \ldots, x_N$ are used in practice. For example, the statistic $S(\mathbf{x})$ can be defined as the following vectors:

- $\mathbf{S} = (\widetilde{\mu}, \widetilde{\sigma})$, where $\widetilde{\mu}$ and $\widetilde{\sigma}$ are empirical mean and standard deviation of the dataset $\mathbf{x}$ respectively;
- $\mathbf{S} = (x_1, \ldots, x_N)$, i.e. all data points in the dataset.

Popular choices for the distance metrics, $\rho(\mathbf{S}, \mathbf{S}^*)$, include:

- Euclidean distance: $\rho(\mathbf{S}, \mathbf{S}^*) = \sum_{l=1}^{L}(S_l - S_l^*)^2$;
- $\mathcal{L}^1$-distance $\rho(\mathbf{S}, \mathbf{S}^*) = \sum_{l=1}^{L}|S_l - S_l^*|$.

2.11.5 Slice Sampling

Often, the full conditional distributions in Gibbs sampler do not take standard explicit closed forms and typically the normalising constants are not known in closed form. Therefore this will exclude straightforward simulation using the inversion method (see Corollary 2.1) or basic rejection sampling (see Corollaries 2.2 and 2.3). In this case, for sampling, one may adopt a Metropolis-Hastings within Gibbs algorithm (described in Sect. 2.11.3). This typically requires tuning of the proposal for a given target distribution that becomes computationally expensive, especially for high dimensional problems. To overcome this problem one may use an adaptive Metropolis-Hastings within Gibbs sampling algorithm; see Atchade and Rosenthal [11] and Rosenthal [205]. An alternative approach, which is more efficient in some cases, is known as a univariate *slice sampler*; see Neal [170]. The latter was developed with the intention of providing a "black box" approach for sampling from a target distribution which may not have a simple form.

A single iteration of the slice sampler algorithm for a toy example is presented in Fig. 2.1. The intuition behind the slice sampling arises from the fact that sampling

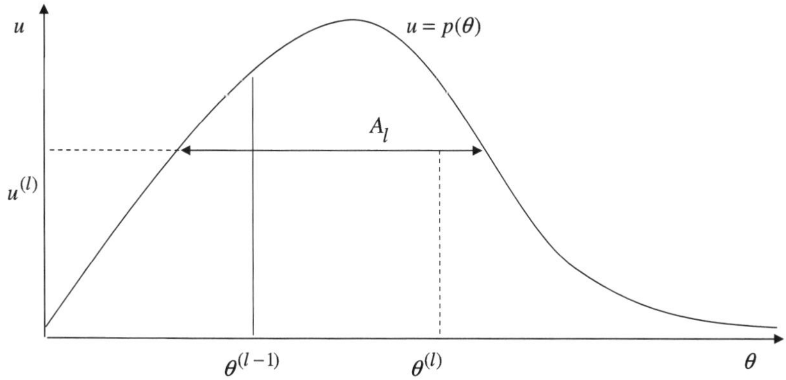

Fig. 2.1 Markov chain created for Θ and auxiliary random variable U, $\left(u^{(1)}, \theta^{(1)}\right), \ldots, \left(u^{(l-1)}, \theta^{(l-1)}\right), \left(u^{(l)}, \theta^{(l)}\right), \ldots$ has a stationary distribution with the desired marginal density $p(\theta)$

from a univariate density $p(\theta)$ can always be achieved by sampling uniformly from the region under the density $p(\theta)$.

Algorithm 2.7 (Univariate slice sampler)

1. Initialise $\theta^{(0)}$ by any value within the support of $p(\theta)$.
2. For $l = 1, 2, \ldots$
   a. Sample a value $u^{(l)} \sim \mathcal{U}\left(0, p\left(\theta^{(l-1)}\right)\right)$.
   b. Sample a value $\theta^{(l)}$ uniformly from the level set $A_l = \left\{\theta : p(\theta) > u^{(l)}\right\}$, i.e. $\theta^{(l)} \sim \mathcal{U}(A_l)$.
3. Next l.

By discarding the auxiliary variable sample $u^{(l)}$, one obtains correlated samples $\theta^{(l)}$ from $p(\cdot)$. Neal [170], demonstrates that a Markov chain (U, Θ) constructed in this way will have a stationary distribution defined by a uniform distribution under $p(\theta)$ and the marginal of Θ has desired stationary density $p(\theta)$. Additionally, Mira and Tierney [165] proved that the slice sampler algorithm, assuming a bounded target density $p(\theta)$ with bounded support, is uniformly ergodic.

There are many approaches that could be used in the determination of the level sets A_l for the density $p(\cdot)$; see Neal ([170], section 4). For example, one can use a stepping out and a shrinkage procedure; see Neal ([170], p. 713, Figure 1). The basic idea is that given a sampled vertical level $u^{(l)}$, the level sets A_l can be found by positioning an interval of width w randomly around $\theta^{(l-1)}$. This interval is expanded in step sizes of width w until both ends are outside the slice. Then a new state is obtained by sampling uniformly from the interval until a point in the slice A_l is obtained. Points that fail can be used to shrink the interval.

Additionally, it is important to note that we only need to know the target full conditional posterior up to normalisation; see Neal ([170], p. 710). To make more precise the intuitive description of the slice sampler presented above, we briefly detail the argument made by Neal on this point. Suppose we wish to sample a random vector $\boldsymbol{\Theta}$ whose density $p(\boldsymbol{\theta})$ is proportional to some function $f(\boldsymbol{\theta})$. This can be achieved by sampling uniformly from the $(n+1)$-dimensional region that lies under the plot of $f(\boldsymbol{\theta})$. This is formalised by introducing the auxiliary random variable U and defining a joint distribution over $\boldsymbol{\Theta}$ and U (which is uniform over the region $\{(\boldsymbol{\Theta}, U) : 0 < u < f(\boldsymbol{\theta})\}$ below the surface defined by $f(\boldsymbol{\theta})$) given by

$$p(\boldsymbol{\theta}, u) = \begin{cases} 1/Z, & \text{if} \quad 0 < u < f(\boldsymbol{\theta}), \\ 0, & \text{otherwise}, \end{cases} \tag{2.61}$$

where $Z = \int f(\boldsymbol{\theta})\, d\boldsymbol{\theta}$. Then the target marginal density for $\boldsymbol{\Theta}$ is given by

$$p(\boldsymbol{\theta}) = \int_0^{f(\boldsymbol{\theta})} \frac{1}{Z} du = \frac{f(\boldsymbol{\theta})}{Z}, \tag{2.62}$$

as required.

The simplest way to apply the slice sampler in a multivariate case is by applying the univariate slice sampler for each full conditional distribution within the Gibbs sampler, as in the example in Sect. 7.13.1.

2.12 MCMC Implementation Issues

There are several numerical issues when implementing MCMC. In practice, a MCMC run consists of three stages: *tuning*, *burn-in* and *sampling* stages. Also, it is important to assess the numerical errors of the obtained estimators due to finite number of MCMC iterations.

2.12.1 Tuning, Burn-in and Sampling Stages

Tuning. The use of MCMC samples can be very inefficient for an arbitrary chosen proposal distribution. Typically, parameters of a chosen proposal distribution are adjusted to achieve a reasonable acceptance rate for each component. There have been several studies regarding the optimal scaling of proposal distributions to ensure optimal convergence rates. Gelman, Gilks and Roberts [100], Bedard and Rosenthal [24] and Roberts and Rosenthal [202] were the first authors to publish theoretical results for the optimal scaling problem in RW-MH algorithms with Gaussian proposals. For the d-dimensional target distributions with independent and identically distributed components, the asymptotic acceptance rate optimising the efficiency of the process is 0.234 independent of the target density. Though for most problems the posterior parameters are not independent Gaussian, it provides a practical guide.

There is no need to be very precise in this stage. In practice, the chains with acceptance rate between 0.2 and 0.8 work well. Typically, tuning is easy. In an ad-hoc procedure, one can initialise the proposal distribution parameters with the values corresponding to the proposal with a very small variability; and start the chain. This will lead to a very high acceptance rate. Then run the chain and gradually change the parameters towards the values that correspond to the proposal with a large uncertainty. This will gradually decrease the acceptance rate. Continue this procedure until the acceptance rate is within 0.2–0.8 range. For example, for Gaussian proposal choose a very small standard deviation parameter. Then increase the standard deviation in small steps and measure the average acceptance rate over the completed iterations until the rate is within 0.2–0.8 range. One can apply a reverse procedure, that is start with parameter values corresponding to a very uncertain proposal resulting in a very low acceptance rate. Then gradually change the parameters towards the values corresponding to the proposal with small variability. Many other alternative ways can be used in this spirit.

Gaussian proposals are often useful with the covariance matrix given by (2.43), that is using Gaussian approximation for the posterior, or just MLE observed information matrix (2.36) in the case of constant prior. An alternative approach is to utilise a new class of Adaptive MCMC algorithms recently proposed in the literature; see Atchade and Rosenthal [11], and Rosenthal [204].

Burn-in stage. Subject to regularity conditions, the chain converges to the stationary target distribution. The number of iterations required for the chain to converge should be discarded and called *burn-in* iterations. Again, we do not need to identify this quantity precisely. Rough approximations of the order of magnitude work well. Visual inspections of the chain plot is the most commonly used method. If the chain is run for long enough then the impact of these *burn-in* iterations on the final estimates is not material. There are many formal *convergence diagnostics* that can be used to determine the length of *burn-in*; for a review, see Cowles and Carlin [63].

Sampling stage. Consider the chain $\{\boldsymbol{\theta}^{(1)}, \boldsymbol{\theta}^{(2)}, \ldots, \boldsymbol{\theta}^{(L)}\}$ and the number of *burn-in* iterations is L_b. Then, $\boldsymbol{\theta}^{(L_b+1)}, \boldsymbol{\theta}^{(L_b+2)}, \ldots, \boldsymbol{\theta}^{(L)}$ are considered as dependent samples from the target distribution $\pi(\boldsymbol{\theta}|\mathbf{x})$ and used for estimation purposes. For example, $\mathrm{E}[g(\boldsymbol{\Theta})|\mathbf{X} = \mathbf{x}]$ is estimated as

$$\mathrm{E}[g(\boldsymbol{\Theta})|\mathbf{X} = \mathbf{x}] = \int g(\boldsymbol{\theta})\pi(\boldsymbol{\theta}|\mathbf{x})d\boldsymbol{\theta} \approx \frac{1}{L - L_b} \sum_{l=L_b+1}^{L} g(\boldsymbol{\theta}^{(l)}). \tag{2.63}$$

Typically, when we calculate the posterior characteristics using MCMC samples, we assume that the samples are taken after burn-in and L_b is dropped in corresponding formulas to simplify notation.

In addition to visual inspection of MCMC, checking that after the burn-in period the samples are mixing well over the support of the posterior distribution, it is useful to monitor the serial correlation of the MCMC samples. For a given chain sample $\theta_i^{(1)}, \ldots, \theta_i^{(L)}$, the autocorrelation at lag k is estimated as

$$\widehat{\mathrm{ACF}}[\theta_i, k] = \frac{1}{(L-k)\widehat{s}^2} \sum_{l=1}^{L-k} (\theta_i^{(l)} - \widehat{\mu})(\theta_i^{(l+k)} - \widehat{\mu}), \tag{2.64}$$

where $\widehat{\mu}$ and $\widehat{s}^2$ are the mean and variance of a sample $\theta_i^{(1)}, \ldots, \theta_i^{(L)}$. In well mixed MCMC samples, the autocorrelation falls to near zero quickly and stays near zero at larger lags. It is useful to find a lag $k^{\max}$ where the autocorrelations seem to have *"died out"*, that is fallen to near zero (for some interesting discussion on this issue, see for example Kass, Carlin, Gelman and Neal [133]). It is not unusual to choose a $k_i^{\max}$ for each component such that the autocorrelation at lag $k_i^{\max}$ has reduced to less than 0.01.

Example 2.3 To illustrate the above described stages, consider a dataset of the annual counts $\mathbf{n} = (9, 12, 7, 9)$ simulated from $Poisson(10)$. Then, we obtain the chain $\lambda^{(0)}, \lambda^{(1)}, \ldots$ using RW-MH algorithm with the Gaussian proposal distribution for the $Poisson(\lambda)$ model and constant prior on a very wide range $[0.1, 100]$. Figure 2.2 shows the chains in the case of different starting values $\lambda^{(0)}$ and different standard deviations σ_{RW} of the Gaussian proposal. One can see that after the burn-in stage indicated by the vertical broken line, the chain looks like stationary. Figure 2.2a, b were obtained when $\sigma_{RW} = \widehat{\mathrm{stdev}}[\widehat{\lambda}^{\mathrm{MLE}}] \approx 1.521$ leading to the acceptance probability approximately 0.7, while Fig. 2.2c, d were obtained when $\sigma_{RW} = 0.4$ and $\sigma_{RW} = 30$ leading to the acceptance probability about 0.91 and 0.10 respectively. The MLE was calculated in the usual way as $\widehat{\mathrm{stdev}}[\widehat{\lambda}^{\mathrm{MLE}}] = (\sum_{i=1}^{m} n_i/m)^{1/2}/\sqrt{m}$, where $m = 4$. The impact of the value of σ_{RW} is easy to see: the chains on Fig. 2.2c, d are *mixing* slowly (moves slowly around the support of the posterior) while the chains on Fig. 2.2a, b are mixing rapidly. Slow mixing means that much longer chain should be run to get good estimates.

2.12.2 Numerical Error

Due to the finite number of iterations, MCMC estimates have numerical error that reduces as the chain length increases. Consider the estimator

$$\widehat{\Omega} = \widehat{\mathrm{E}}[g(\boldsymbol{\Theta})|\mathbf{X} = \mathbf{x}] = \frac{1}{L} \sum_{l=1}^{L} g(\boldsymbol{\Theta}^{(l)}). \tag{2.65}$$

If the samples $\boldsymbol{\Theta}^{(1)}, \ldots, \boldsymbol{\Theta}^{(L)}$ are independent and identically distributed then the standard error of $\widehat{\Omega}$ (due to the finite L) is estimated using

$$\mathrm{stdev}[\widehat{\Omega}] = \mathrm{stdev}[g(\boldsymbol{\Theta})|\mathbf{X} = \mathbf{x}]/\sqrt{L},$$

where $\mathrm{stdev}[g(\boldsymbol{\Theta})|\mathbf{X}]$ is estimated by the standard deviation of the sample $g(\boldsymbol{\Theta}^{(l)})$, $l = 1, \ldots, L$. This formula does not work for MCMC samples due to serial

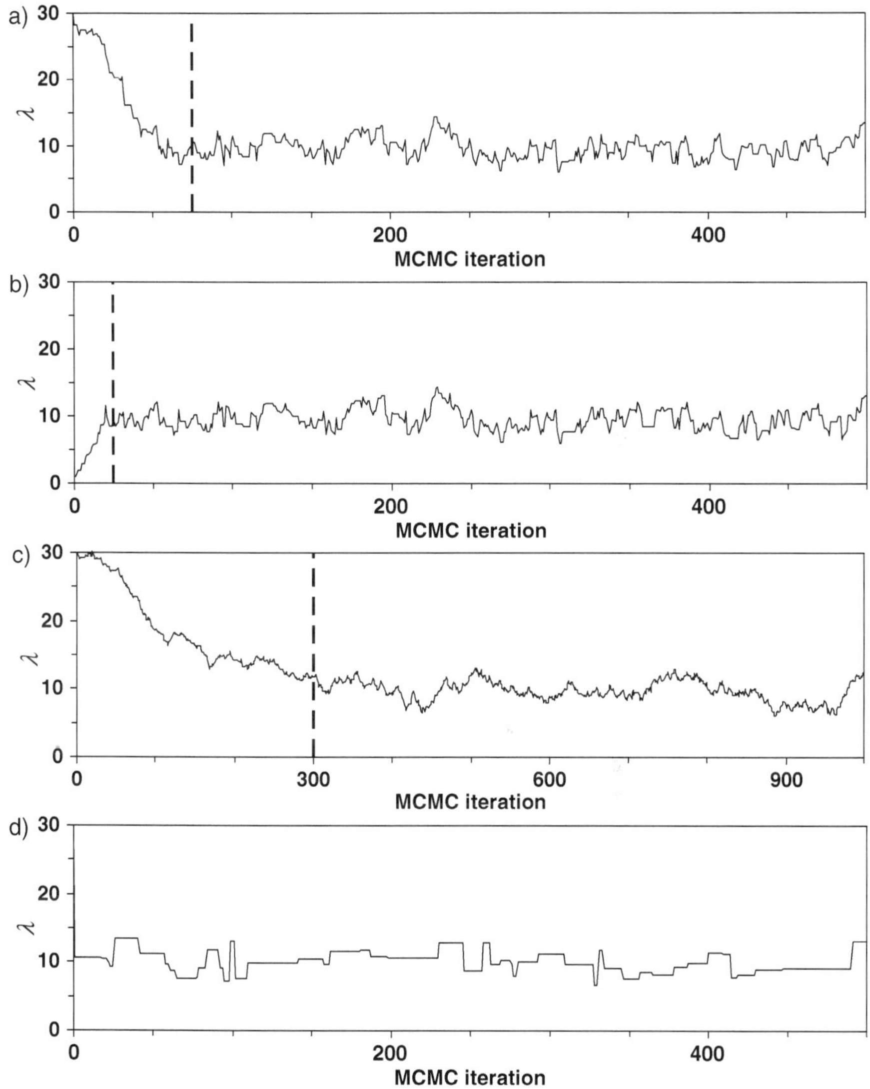

Fig. 2.2 MCMC chains of λ parameter of $Poisson(\lambda)$ model in the case of different starting points $\lambda^{(0)}$ and different standard deviations of the Gaussian proposal distribution: (**a**) starting point $\lambda^{(0)} = 30$ and $\sigma_{RW} = 1.521$; (**b**) $\lambda^{(0)} = 1$ and $\sigma_{RW} = 1.521$; (**c**) $\lambda^{(0)} = 30$ and $\sigma_{RW} = 0.4$; (**d**) $\lambda^{(0)} = 30$ and $\sigma_{RW} = 30$. The *burn-in* stage is to the *left* of the vertical broken line. The dataset consisting of the annual number of events (9, 12, 7, 9) over 4 years was simulated from $Poisson(10)$

correlations between the samples. Of course one can keep every $k_{\max}$-th sample from the chain to get approximately independent samples, but it is always a suboptimal approach; see MacEachern and Berliner [152].

Effective sample size. If there is only one parameter θ, then one of the popular approaches is to calculate *effective sample size*, $T_{eff} = T/\tau$, where τ is autocorrelation time

$$\tau = 1 + 2\sum_{k=1}^{\infty} \mathrm{ACF}[\theta, k]. \tag{2.66}$$

To estimate τ, it is necessary to cut off the sum in (2.66) at a value of $k = k^{\max}$, where the autocorrelations seem to have fallen to near zero. Then the standard error of the $\widehat{\Omega}$ (2.65) is estimated using

$$\mathrm{stdev}[\widehat{\Omega}] = \frac{\mathrm{stdev}[g(\Theta)]}{\sqrt{L/\tau}};$$

see Ripley [199], Neal [168].

Batch sampling. Probably the most popular approach to estimate the numerical error of the MCMC posterior averages is a so-called *batch sampling*; see section 3.4.1 in Gilks, Richardson and Spiegelhalter [106]. Consider MCMC posterior samples $\boldsymbol{\Theta}^{(1)}, \ldots, \boldsymbol{\Theta}^{(L)}$ of $\boldsymbol{\Theta}$ with the length $L = K \times N$, and an estimator $\widehat{\Omega} = \sum_{l=1}^{L} g(\boldsymbol{\Theta}^{(l)})$ of $\mathrm{E}[g(\boldsymbol{\Theta})]$. If N is sufficiently large, the means

$$\widehat{\Omega}_j = \frac{1}{N} \sum_{i=(j-1)N+1}^{j \times N} g(\boldsymbol{\Theta}^{(i)}), \quad j = 1, \ldots, K \tag{2.67}$$

are approximately independent and identically distributed. Then the overall estimator and its variance are

$$\widehat{\Omega} = \frac{1}{K}(\widehat{\Omega}_1 + \cdots + \widehat{\Omega}_K),$$
$$\mathrm{Var}[\widehat{\Omega}] = \frac{1}{K^2}(\mathrm{Var}[\widehat{\Omega}_1] + \cdots + \mathrm{Var}[\widehat{\Omega}_K]) = \frac{\sigma^2}{K},$$

where $\sigma^2 = \mathrm{Var}[\widehat{\Omega}_1] = \cdots = \mathrm{Var}[\widehat{\Omega}_K]$. In the limit of large K, by the central limit theorem (i.e. we also assume that σ^2 is finite), the distribution of $\widehat{\Omega}$ is normal with the standard deviation $\sigma/\sqrt{K}$. The latter is referred to as the standard error of $\widehat{\Omega}$. Finally, σ^2 can be estimated using sample variance

$$\widehat{\sigma}^2 = \frac{1}{K-1}\sum_{j=1}^{K}(\widehat{\Omega}_j - \widehat{\Omega})^2. \tag{2.68}$$

Note that K is the number of quasi-independent bins, and $N = L/K$ is the size of each bin or batch. Typically, in practice $K \geq 20$ and $N \geq 100k^{\max}$, where $k^{\max} = \max(k_1^{\max}, k_2^{\max}, \ldots)$ is the maximum of the cut-off lags over components. In general, we would like to run the chain until the numerical error is not material. So, one can set N using $k^{\max}$ identified during tuning and burning stages, e.g. set $N = 100k^{\max}$, then run the chain in batches until the numerical error of the estimates is less than the desired accuracy.

2.12.3 MCMC Extensions

Sometimes, in the developed Bayesian models, there is a strong correlation between the model parameters in the posterior. In extreme cases, this can cause slow rates of convergence in the Markov chain to reach the ergodic regime, translating into longer Markov chain simulations. In such a situation several approaches can be tried to overcome this problem.

The first involves the use of a mixture transition kernel combining local and global moves. For example, one can perform local moves via a univariate slice sampler and global moves via an independent Metropolis-Hastings sampler with adaptive learning of its covariance structure. Such an approach is known as a hybrid sampler; see comparisons in Brewer, Aitken and Talbot [36]. Alternatively, for the global move, if determination of level sets in multiple dimensions is not problematic (for the model under consideration), then some of the multivariate slice sampler approaches designed to account for correlation between parameters can be incorporated; see Neal [170] for details.

Another approach to break correlation between parameters in the posterior is via the transformation of the parameter space. If the transformation is effective this will reduce correlation between parameters of the transformed target posterior. Sampling can then proceed in the transformed space, and then samples can be transformed back to the original space. It is not always straightforward to find such transformations.

A third alternative is based on *simulated tempering*, introduced by Marinari and Parisi [153] and discussed extensively in Geyer and Thompson [103]. In particular a special version of simulated tempering, first introduced by Neal [169], can be utilised in which one considers a sequence of target distributions $\{\pi_l\}$ constructed such that they correspond to the objective posterior in the following way,

$$\pi_l = (\pi(\boldsymbol{\theta}|\mathbf{x}))^{\gamma_l} \tag{2.69}$$

with sequence $\{\gamma_l\}$. Then one can use the standard MCMC algorithms (e.g. slice sampler), where π is replaced with π_l.

Running a Markov chain such that at each iteration l we target the posterior π_l and then only keeping samples from the Markov chain corresponding to situations in which $\gamma_l = 1$ can result in a significant improvement in exploration around the posterior support. This can overcome slow mixing arising from a univariate sampling regime. The intuition for this is that for values of $\gamma_l << 1$ the target posterior is almost uniform over the space, resulting in large moves being possible around the support of the posterior. Then as γ_l returns to a value of 1, several iterations later, it will be in potentially new unexplored regions of the posterior support.

For example, one can utilise a sine function,

$$\gamma_l = \min\left(\sin\left(\frac{2\pi}{K}l\right) + 1, 1\right)$$

with large K (e.g. $K = 1{,}000$), which has its amplitude truncated to ensure it ranges between 0 and 1. That is the function is truncated at $\gamma_l = 1$ for extended iteration periods for our simulation index l to ensure the sampler spends significant time sampling from the actual posterior distribution.

In the application of tempering one must discard many simulated states of the Markov chain, whenever $\gamma_l \neq 1$. There is, however, a computational way to avoid discarding these samples; see Gramacy, Samworth and King [111].

Finally, we note that there are several alternatives to a Metropolis-Hastings within Gibbs sampler such as a basic Gibbs sampler combined with *adaptive rejection sampling* (ARS), Gilks and Wild [107]. Note that ARS requires distributions to be log-concave. Alternatively an adaptive version of this known as the adaptive Metropolis rejection sampler could be used; see Gilks, Best and Tan [105].

2.13 Bayesian Model Selection

Consider a model M with parameter vector $\boldsymbol{\theta}$. The model likelihood with data $\mathbf{x}$ can be found by integrating out the parameter $\boldsymbol{\theta}$

$$\pi(\mathbf{x}|M) = \int \pi(\mathbf{x}|\boldsymbol{\theta}, M)\pi(\boldsymbol{\theta}|M)d\boldsymbol{\theta}, \tag{2.70}$$

where $\pi(\boldsymbol{\theta}|M)$ is the prior density of $\boldsymbol{\theta}$ in M. Given a set of K competing models $(M_1, \ldots, M_K)$ with parameters $\boldsymbol{\theta}_{[1]}, \ldots, \boldsymbol{\theta}_{[K]}$ respectively, the Bayesian alternative to traditional hypothesis testing is to evaluate and compare the posterior probability ratio between the models. Assuming we have some prior knowledge about the model probability $\pi(M_i)$, we can compute the posterior probabilities for all models using the model likelihoods

$$\pi(M_i|\mathbf{x}) = \frac{\pi(\mathbf{x}|M_i)\,\pi(M_i)}{\sum_{k=1}^{K}\pi(\mathbf{x}|M_k)\,\pi(M_k)}. \tag{2.71}$$

Consider two competing models M_1 and M_2, parameterised by $\boldsymbol{\theta}_{[1]}$ and $\boldsymbol{\theta}_{[2]}$ respectively. The choice between the two models can be based on the posterior model probability ratio, given by

$$\frac{\pi(M_1|\mathbf{x})}{\pi(M_2|\mathbf{x})} = \frac{\pi(\mathbf{x}|M_1)\,\pi(M_1)}{\pi(\mathbf{y}|M_2)\,\pi(M_2)} = \frac{\pi(M_1)}{\pi(M_2)} B_{12}, \tag{2.72}$$

where $B_{12} = \pi(\mathbf{x}|M_1)/\pi(\mathbf{x}|M_2)$ is the Bayes factor, the ratio of the posterior odds of model M_1 to that of model M_2. As shown by Lavin and Scherrish [142], an accurate interpretation of the Bayes factor is that the ratio B_{12} captures the change of the odds in favour of model M_1 as we move from the prior to the posterior. Jeffreys [127] recommended a scale of evidence for interpreting the Bayes factors, which was later modified by Wasserman [238]. A Bayes factor $B_{12} > 10$ is considered strong evidence in favour of M_1. Kass and Raftery [131] give a detailed review of the Bayes factors.

Typically, the integral (2.70) required by the Bayes factor is not analytically tractable, and sampling based methods must be used to obtain estimates of the model likelihoods. There are quite a few methods in the literature for direct computation of the Bayes factor or indirect construction of the Bayesian model selection criterion, both based on MCMC outputs. The popular methods are direct estimation of the model likelihood thus the Bayes factor; indirect calculation of an asymptotic approximation as the model selection criterion; and direct computation of the posterior model probabilities, as discussed below.

Popular model selection criteria, based on simplifying approximations, include the Deviance information criterion (DIC) and Bayesian information criterion (BIC); see e.g. Robert ([200], chapter 7).

In general, given a set of possible models $(M_1, \ldots, M_K)$, the model uncertainty can be incorporated in Bayesian framework via considering the joint posterior for the model and the model parameters $\pi(M_k, \boldsymbol{\theta}_{[k]}|\mathbf{x})$, where $\boldsymbol{\theta}_{[k]}$ is a vector of parameters for model k. Subsequently calculated posterior model probabilities $\pi(M_k|\mathbf{x})$ can be used to select an optimal model as the model with the largest probability or average over possible models according to the full joint posterior.

Accurate estimation of the required posterior distributions usually involves development of a Reversible Jump MCMC framework. This type of Markov chain sampler is complicated to develop and analyse. It goes beyond the scope of this book but interested reader can find details in Green [112]. In the case of small number of models, Congdon [60] suggests to run a standard MCMC (e.g. RW-MH) for each model separately and use the obtained MCMC samples to estimate $\pi(M_k|\mathbf{x})$. It was adopted in Peters, Shevchenko and Wüthrich [186] for modelling claims reserving problem in the insurance. Using the Markov chain results for each model, in the case of equiprobable nested models, this procedure calculates the posterior model probabilities $\pi(M_i|\mathbf{x})$ as

$$\pi(M_i|\mathbf{x}) = \frac{1}{L}\sum_{l=1}^{L} \frac{f\left(\mathbf{x}|M_i, \boldsymbol{\theta}_{[i]}^{(l)}\right)}{\sum_{j=1}^{K} f\left(\mathbf{x}|M_j, \boldsymbol{\theta}_{[j]}^{(l)}\right)}, \tag{2.73}$$

where $\boldsymbol{\theta}_{[i]}^{(l)}$ is the MCMC posterior sample at Markov chain step l for model M_i, $f(\mathbf{x}|M_i, \boldsymbol{\theta}_{[i]}^{(l)})$ is the joint density of the data $\mathbf{x}$ given the parameter vector $\boldsymbol{\theta}_{[i]}^{(l)}$ for model M_i, and L is the total number of MCMC steps after burn-in period.

2.13.1 Reciprocal Importance Sampling Estimator

Given MCMC samples $\boldsymbol{\theta}^{(l)}, l = 1, \ldots, L$ from the posterior distribution obtained through MCMC, Gelfand and Dey [99] proposed the *reciprocal importance sampling estimator* (RISE) to approximate the model likelihood

$$\widehat{p}_{RI}(\mathbf{x}) = \left[\frac{1}{L} \sum_{l=1}^{L} \frac{h(\boldsymbol{\theta}^{(l)})}{\pi(\mathbf{x}|\boldsymbol{\theta}^{(l)})\, \pi(\boldsymbol{\theta}^{(l)})} \right]^{-1}, \tag{2.74}$$

where h plays the role of an importance sampling density roughly matching the posterior. Gelfand and Dey [99] suggested the multivariate normal or t distribution density with mean and covariance fitted to the posterior sample.

The RISE estimator can be regarded as a generalisation of the *harmonic mean estimator* suggested by Newton and Raftery [175]. The latter is obtained from the RISE estimator by setting $h = 1$. Other estimators include the *bridge sampling* proposed by Meng and Wong [159], and the *Chib's candidate's estimator* in Chib [56]. In a recent comparison study by Miazhynskaia and Dorffner [162], these estimators were employed as competing methods for Bayesian model selection on GARCH-type models, along with the reversible jump MCMC. It was demonstrated that the RISE estimator (either with normal or t importance sampling density), the bridge sampling method, and the Chib's algorithm gave statistically equal performance in model selection. Also, the performance more or less matched the much more involved reversible jump MCMC.

2.13.2 Deviance Information Criterion

For a dataset $\mathbf{X} = \mathbf{x}$ generated by the model with the posterior density $\pi(\boldsymbol{\theta}|\mathbf{x})$, define the deviance

$$D(\boldsymbol{\theta}) = -2 \ln \pi(\mathbf{x}|\boldsymbol{\theta}) + C, \tag{2.75}$$

where the constant C is common to all candidate models. Then the *deviance information criterion* (DIC) is calculated as

$$\begin{aligned} DIC &= 2\mathrm{E}[D(\boldsymbol{\Theta})|\mathbf{X} = \mathbf{x}] - D(\mathrm{E}[\boldsymbol{\Theta}|\mathbf{X} = \mathbf{x}]) \\ &= \mathrm{E}[D(\boldsymbol{\Theta})|\mathbf{X} = \mathbf{x}] + (\mathrm{E}[D(\boldsymbol{\Theta})|\mathbf{X} = \mathbf{x}] - D(\mathrm{E}[\boldsymbol{\Theta}|\mathbf{X} = \mathbf{x}])), \end{aligned} \tag{2.76}$$

where

- $\mathrm{E}[\cdot|\mathbf{X} = \mathbf{x}]$ is the expectation with respect to the posterior density of $\boldsymbol{\Theta}$.
- The expectation $\mathrm{E}[D(\boldsymbol{\Theta})|\mathbf{X} = \mathbf{x}]$ is a measure of how well the model fits the data; the smaller this is, the better the fit.
- The difference $\mathrm{E}[D(\boldsymbol{\Theta})|\mathbf{X} = \mathbf{x}] - D(\mathrm{E}[\boldsymbol{\Theta}|\mathbf{X} = \mathbf{x}])$ can be regarded as the effective number of parameters. The larger this difference, the easier it is for the model to fit the data.

The DIC criterion favours the model with a better fit but at the same time penalises the model with more parameters. Under this setting the model with the smallest DIC value is the preferred model.

DIC is a Bayesian alternative to BIC (*Schwarz's criterion* and also called the *Bayesian information criterion*, Schwarz [209]) and AIC (*Akaike's information criterion*, Akaike [5]). For more details on the above-mentioned criteria, see e.g. Robert ([200], chapter 7).

Problems[4]

2.1 ($\star$) Given independent and identically distributed data $N_1, N_2, \ldots, N_m$ from $Poisson(\lambda)$, find the maximum likelihood estimator $\widehat{\Lambda}^{\text{MLE}}$ (for parameter λ) and its variance. Show that this variance is the same as the one obtained from a large sample size normal approximation for MLE.

2.2 ($\star\star\star$) Suppose there are independent and identically distributed data $\mathbf{N} = (N_1, \ldots, N_m)'$ from $Poisson(\lambda)$.

- Find in closed form the mean and variance of the posterior $\pi(\lambda|\mathbf{N})$. Compare these with the MLE and its variance calculated in Problem 2.1.
- Simulate Markov chain $\{\lambda^{(1)}, \lambda^{(2)}, \ldots, \lambda^{(L)}\}$ for parameter λ using RW-MH MCMC and dataset $\mathbf{N}$ as in Example 2.3. Estimate the mean and variance of the chain samples and compare with the above calculated closed form posterior mean and variance. Assume that $L = 1000$.

2.3 ($\star\star\star$) For a Markov chain $\{\lambda^{(1)}, \lambda^{(2)}, \ldots, \lambda^{(L)}\}$, $L = 1000$, simulated in Problem 2.2, estimate the numerical error of the posterior mean that was estimated using the chain samples. Repeat calculations for $L = 4 \times 10^3$, $L = 16 \times 10^3$ and compare results.

2.4 ($\star\star$) Consider random variables $L_1, \ldots, L_J$ and $L = L_1 + \cdots + L_J$. If risk measure $\varrho[L]$ is positively homogeneous, i.e. $\varrho[hZ] = h\varrho[Z]$ for $h > 0$ and differentiable, show that

$$\varrho[L] = \sum_{j=1}^{J} \left. \frac{\partial \varrho[L + hL_j]}{\partial h} \right|_{h=0}. \tag{2.77}$$

2.5 ($\star\star$) Given three independent risks, $Z_i \sim Gamma(\alpha_i, \beta)$, with $\alpha_1 = 0.5, \alpha_2 = 1, \alpha_3 = 1.5$ respectively and the scale parameter $\beta = 1$, find:

- the 0.999 VaR for each risk, $\text{VaR}_0.999[Z_i]$, $i = 1, 2, 3$;
- the 0.999 VaR of the total risk, $\text{VaR}_{0.999}[Z_1 + Z_2 + Z_3]$; and
- diversification

---

[4] Problem difficulty is indicated by asterisks: ($\star$) – low; ($\star\star$) – medium, ($\star\star\star$) – high.

$$1 - \mathrm{VaR}_{0.999}\left[\sum_j Z_j\right] \Big/ \sum_j \mathrm{VaR}_{0.999}[Z_j].$$

Hint: use the fact that the sum of two independent random variables, $X_1 \sim Gamma(\alpha_1, \beta)$ and $X_2 \sim Gamma(\alpha_2, \beta)$, is distributed from $Gamma(\alpha_1+\alpha_2, \beta)$.

2.6 ($\star$) Show that expected shortfall of a continuous random variable X (see Definition 2.14) can be calculated as

$$\mathrm{ES}_\alpha[X] = \mathrm{E}[X|X \geq \mathrm{VaR}_\alpha[X]].$$

That is, prove Proposition 2.1.

2.7 ($\star$) Calculate mean, variance and 0.9 quantile of a random variable X that has:

- a finite mass at zero, $\Pr[X = 0] = 0.5$; and
- density $\frac{1}{2} f^{(c)}(x)$ for $x > 0$, where $f^{(c)}(x)$ is the density of the lognormal distribution $\mathcal{LN}(\mu, \sigma)$ with $\mu = 0$ and $\sigma = 1$.

Compare the results with the case when $X \sim \mathcal{LN}(0, 1)$.

2.8 ($\star$) Calculate mean, variance, skewness, mode, median and 0.9 quantile of a random variable $X \sim Pareto(\xi = 3, x_0 = 1)$.

2.9 ($\star$) Suppose $X \sim Pareto(\xi, x_0)$. Given two quantiles q_1 and q_2 of random variable X at the confidence levels α_1 and α_2 respectively ($\alpha_1 \neq \alpha_2$), find the distribution parameters ξ and x_0.

Chapter 3
Calculation of Compound Distribution

Science never solves a problem without creating ten more.
Bernard Shaw

Abstract Estimation of the capital under the LDA requires evaluation of compound loss distributions. Closed-form solutions are not available for the distributions typically used in operational risk and numerical evaluation is required. This chapter describes numerical algorithms that can be successfully used for this problem. In particular Monte Carlo, Panjer recursion and Fourier transformation methods are presented. Also, several closed-form approximations are reviewed.

3.1 Introduction

The LDA model (2.1) requires calculation of the distribution for the aggregate (compound) loss $X_1 + \cdots + X_N$, where the frequency N is a discrete random variable. This is one of the classical problems in risk theory. Before the era of personal computers, it was calculated using approximations such as that based on the asymptotic central limit theory or on ad-hoc reasoning using, for example, shifted gamma approximation. With modern computer processing power, these distributions can be calculated virtually *exactly* using numerical algorithms. The easiest to implement is the Monte Carlo method. However, because it is typically slow, Panjer recursion and Fourier inversion techniques are widely used. Both have a long history, but their applications to computing very high quantiles of the compound distribution functions with high frequencies and heavy tails are only recent developments and various pitfalls exist. The methods described in this chapter are based on the following model assumptions.

Model Assumptions 3.1 *The annual loss in a risk cell is modelled by a compound random variable*

$$Z = \sum_{i=1}^{N} X_i, \tag{3.1}$$

P. Shevchenko, *Modelling Operational Risk Using Bayesian Inference*,
DOI 10.1007/978-3-642-15923-7_3,

where

- *N is the number of events (frequency) over one year modelled as a discrete random variable with probability mass function $p_k = \Pr[N = k]$, $k = 0, 1, \ldots$;*
- *X_i, $i \geq 1$ are positive severities of the events (loss amounts) modelled as independent and identically distributed random variables from a continuous distribution function $F(x)$ with $x \geq 0$ and $F(0) = 0$. The corresponding density function is denoted as $f(x)$;*
- *N and X_i are independent for all i, i.e. the frequencies and severities are independent;*
- *The distribution and density functions of the annual loss Z are denoted as $H(z)$ and $h(z)$ respectively;*
- *All model parameters (parameters of the frequency and severity distributions) are assumed to be known.*

Remark 3.1

- Only one risk cell and one time period are considered, so the indices indicating the time period and risk cell in the LDA model (2.1) are dropped in this section.
- Typically, the calculation of the annual loss distribution is required for the next year, i.e. year $T + 1$ in (2.1).
- In this chapter, the model parameters are assumed to be known. However, in reality, the model parameters are unknown and estimated using past data over T years. Estimation of the parameters and implications for the annual loss distribution are the topics of the chapters that follow.
- Note that there is a finite probability of no loss occurring over the considered time period if $N = 0$ is allowed, i.e. $\Pr[Z = 0] = \Pr[N = 0]$.
- The methods described in this chapter can be used to calculate the distribution of compound loss over any time period. For simplicity, only the most relevant case of a one-year time period is considered here. Extension to the case of other time periods is trivial.

In general, there are two types of analytic solutions for calculating the compound distribution $H(z)$. These are based on convolutions and method of characteristic functions. Typically, the analytic solutions do not have closed form and numerical methods (such as Monte Carlo, Panjer recursion, Fast Fourier Transform (FFT) or direct integration) are required. These solutions and methods are described in the following sections.[1]

3.1.1 Analytic Solution via Convolutions

It is well known that the density and distribution functions of the sum of two independent continuous random variables $Y_1 \sim F_1(\cdot)$ and $Y_2 \sim F_2(\cdot)$, with the densities

[1] Computing time quoted in this chapter is for a standard Dell laptop Latitude D820 with Intel(R) CPU T2600 @ 2.16 GHz and 3.25GB of RAM.

$f_1(\cdot)$ and $f_2(\cdot)$ respectively, can be calculated via convolution as

$$f_{Y_1+Y_2}(y) = (f_1 * f_2)(y) = \int f_2(y - y_1) f_1(y_1) dy_1 \tag{3.2}$$

and

$$F_{Y_1+Y_2}(y) = (F_1 * F_2)(y) = \int F_2(y - y_1) f_1(y_1) dy_1 \tag{3.3}$$

respectively. Hereafter, notation $f_1 * f_2$ denotes convolution of f_1 and f_2 functions as defined above. Thus the distribution of the annual loss (3.1) can be calculated via convolutions as

$$\begin{aligned} H(z) = \Pr[Z \le z] &= \sum_{k=0}^{\infty} \Pr[Z \le z | N = k] \Pr[N = k] \\ &= \sum_{k=0}^{\infty} p_k F^{(k)*}(z). \end{aligned} \tag{3.4}$$

Here, $F^{(k)*}(z) = \Pr[X_1 + \cdots + X_k \le z]$ is the k-th convolution of $F(\cdot)$ calculated recursively as

$$F^{(k)*}(z) = \int_0^z F^{(k-1)*}(z - x) f(x) dx$$

with

$$F^{(0)*}(z) = \begin{cases} 1, & z \ge 0, \\ 0, & z < 0. \end{cases}$$

Note that the integration limits are 0 and z because the considered severities are non-negative. Though the obtained formula is analytic, its direct calculation is difficult because, in general, the convolution powers are not available in closed form. Panjer recursion and FFT, discussed in Sects. 3.3 and 3.4, are very efficient numerical methods to calculate these convolutions.

3.1.2 Analytic Solution via Characteristic Functions

The method of characteristic functions for computing probability distributions is a powerful tool in mathematical finance. It is explained in many textbooks on probability theory such as Pugachev ([196], chapter 4). In particular, it is used for calculating aggregate loss distributions in the insurance, operational risk and credit risk. Typically, the frequency-severity compound distributions cannot be found in closed form but can be conveniently expressed through the inverse transform of the characteristic functions. The characteristic function of the severity density $f(x)$ is formally defined as

$$\varphi(t) = \int_{-\infty}^{\infty} f(x) e^{itx} dx, \tag{3.5}$$

where $i = \sqrt{-1}$ is a unit imaginary number. Also, the *probability generating function* of a frequency distribution with probability mass function $p_k = \Pr[N = k]$ is

$$\psi(s) = \sum_{k=0}^{\infty} s^k p_k. \tag{3.6}$$

Then, the characteristic function of the compound loss Z in model (3.1), denoted by $\chi(t)$, can be expressed through the probability generating function of the frequency distribution and characteristic function of the severity distribution as

$$\chi(t) = \sum_{k=0}^{\infty} (\varphi(t))^k p_k = \psi(\varphi(t)). \tag{3.7}$$

In particular:

- If frequency N is distributed from $Poisson(\lambda)$, then

$$\chi(t) = \sum_{k=0}^{\infty} (\varphi(t))^k \frac{e^{-\lambda}\lambda^k}{k!} = \exp(\lambda\varphi(t) - \lambda); \tag{3.8}$$

- If N is from negative binomial distribution $NegBin(m, q)$, then

$$\begin{aligned} \chi(t) &= \sum_{k=0}^{\infty} (\varphi(t))^k \binom{k+m-1}{k} (1-q)^k q^m \\ &= \left(\frac{q}{1-(1-q)\varphi(t)}\right)^m. \end{aligned} \tag{3.9}$$

Given characteristic function, the density of the annual loss Z can be calculated via the inverse Fourier transform as

$$h(z) = \frac{1}{2\pi} \int_{-\infty}^{\infty} \chi(t) \exp(-itz) dt, \quad z \geq 0. \tag{3.10}$$

In the case of nonnegative severities, the density and distribution functions of the compound loss can be calculated using the following lemma.

Lemma 3.1 *For a nonnegative random variable Z with a characteristic function $\chi(t)$, the density $h(z)$ and distribution $H(z)$ functions, $z \geq 0$, are*

$$h(z) = \frac{2}{\pi} \int_0^\infty \text{Re}[\chi(t)] \cos(tz) dt, \quad z \geq 0; \tag{3.11}$$

$$H(z) = \frac{2}{\pi} \int_0^\infty \text{Re}[\chi(t)] \frac{\sin(tz)}{t} dt, \quad z \geq 0. \tag{3.12}$$

Proof The characteristic function of a non-negative random variable Z with the density $h(z)$, $z \geq 0$ can be written as

$$\chi(t) = \int_{-\infty}^{\infty} h(z) e^{itz} dz = \text{Re}[\chi(t)] + i\,\text{Im}[\chi(t)],$$

where

$$\text{Re}[\chi(t)] = \int_0^\infty h(z) \cos(tz) dz, \quad \text{Im}[\chi(t)] = \int_0^\infty h(z) \sin(tz) dz.$$

Define a function $\widetilde{h}(z)$ such that $\widetilde{h}(z) = h(z)$ if $z \geq 0$ and $\widetilde{h}(z) = h(-z)$ if $z < 0$. Using symmetry property, the characteristic function for this extended function is

$$\widetilde{\chi}(t) = \int_{-\infty}^{\infty} \widetilde{h}(z) e^{itz} dz = 2 \int_0^\infty h(z) \cos(tz) dz = 2\text{Re}[\chi(t)], \quad \widetilde{\chi}(t) = \widetilde{\chi}(-t).$$

Thus the density $h(z) = \widetilde{h}(z)$, $z \geq 0$ can be retrieved as

$$h(z) = \frac{1}{2\pi} \int_{-\infty}^{\infty} \widetilde{\chi}(t) e^{-itz} dt = \frac{1}{\pi} \int_0^\infty \widetilde{\chi}(t) \cos(tz) dt = \frac{2}{\pi} \int_0^\infty \text{Re}[\chi(t)] \cos(tz) dt$$

and the distribution can be calculated as

$$H(z) = \int_0^z h(y) dy = \int_0^z \frac{2}{\pi} dy \int_0^\infty \text{Re}[\chi(t)] \cos(ty) dt = \frac{2}{\pi} \int_0^\infty \text{Re}[\chi(t)] \frac{\sin(tz)}{t} dt.$$

This completes the proof. □

Remark 3.2 Changing variable $x = t \times z$, the formula (3.12) can be rewritten as

$$H(z) = \frac{2}{\pi}\int_{0}^{\infty} \mathrm{Re}[\chi(x/z)]\frac{\sin(x)}{x}dx,$$

which is often a useful representation to study limiting properties. In particular, in the limit $z \to 0$, it gives

$$H(z \to 0) = \frac{2}{\pi}\mathrm{Re}[\chi(\infty)]\int_{0}^{\infty} \frac{\sin(x)}{x}dx = \mathrm{Re}[\chi(\infty)].$$

This leads to a correct limit $H(0) = \Pr[N = 0]$, because the severity characteristic function $\varphi(\infty) \to 0$ (in the case of continuous severity distribution function). For example, $H(0) = \exp(-\lambda)$ in the case of $N \sim Poisson(\lambda)$, and $H(0) = q^m$ for $N \sim NegBin(m, q)$.

FFT and direct integration methods to calculate the above Fourier transforms are discussed in details in the following sections.

3.1.3 Compound Distribution Moments

In general, the compound distribution cannot be found in closed form. However, its moments can be expressed through the moments of the frequency and severity. It is convenient to calculate the moments via characteristic function. In particular, one can calculate the moments as

$$\mathrm{E}[Z^k] = (-i)^k \left.\frac{d^k \chi(t)}{dt^k}\right|_{t=0}, \quad k = 1, 2, \ldots . \tag{3.13}$$

Similarly, the central moments can be found as

$$\begin{aligned}\mu_k &= \mathrm{E}[(Z - \mathrm{E}[Z])^k] \\ &= (-i)^k \left.\frac{d^k \chi(t)\exp(-it\mathrm{E}[Z])}{dt^k}\right|_{t=0}, \quad k = 1, 2, \ldots . \end{aligned} \tag{3.14}$$

Here, for compound distribution, $\chi(t)$ is given by (3.7). Then, one can derive the explicit expressions for all moments of compound distribution via the moments of frequency and severity noting that $\varphi(0) = 1$ and using relations

$$\left.\frac{d^k \psi(s)}{ds^k}\right|_{s=1} = \mathrm{E}[N(N-1)\cdots(N-k+1)], \tag{3.15}$$

$$(-i)^k \left.\frac{d^k \varphi(t)}{dt^k}\right|_{t=0} = \mathrm{E}[X_1^k], \tag{3.16}$$

that follow from the definitions of the probability generating and characteristic functions (3.6) and (3.5) respectively, though the expression is lengthy for high moments. Sometimes, it is easier to work with the so-called cumulants (or semi-invariants)

$$\kappa_k = (-i)^k \left.\frac{d^k \ln \chi(t)}{dt^k}\right|_{t=0}, \tag{3.17}$$

which are closely related to the moments. The moments can be calculated via the cumulants and vice versa. In application, only the first four moments are most often used with the following relations

$$\mu_2 = \kappa_2 \equiv \mathrm{Var}[Z]; \quad \mu_3 = \kappa_3; \quad \mu_4 = \kappa_4 + 3\kappa_2^2. \tag{3.18}$$

Then, closely related distribution characteristics, skewness and kurtosis, are

$$\text{skewness} = \frac{\mu_3}{(\mu_2)^{3/2}}, \tag{3.19}$$

$$\text{kurtosis} = \frac{\mu_4}{(\mu_2)^2} - 3. \tag{3.20}$$

The above formulas relating characteristic function and moments can be found in many textbooks on probability theory such as Pugachev ([196], section 27). The explicit expressions for the first four moments are given by the following proposition.

Proposition 3.1 (Moments of compound distribution) *The first four moments of the compound random variable $Z = X_1 + \cdots + X_N$, where $X_1, \ldots, X_N$ are independent and identically distributed, and independent of N, are given by*

$$\begin{aligned}
\mathrm{E}[Z] &= \mathrm{E}[N]\mathrm{E}[X_1],\\
\mathrm{Var}[Z] &= \mathrm{E}[N]\mathrm{Var}[X_1] + \mathrm{Var}[N](\mathrm{E}[X_1])^2,\\
\mathrm{E}[(Z-\mathrm{E}[Z])^3] &= \mathrm{E}[N]\mathrm{E}[(X_1-\mathrm{E}[X_1])^3] + 3\mathrm{Var}[N]\mathrm{Var}[X_1]\mathrm{E}[X_1]\\
&\quad + \mathrm{E}[(N-\mathrm{E}[N])^3](\mathrm{E}[X_1])^3,\\
\mathrm{E}[(Z-\mathrm{E}[Z])^4] &= \mathrm{E}[N]\mathrm{E}[(X_1-\mathrm{E}[X_1])^4] + 4\mathrm{Var}[N]\mathrm{E}[(X_1-\mathrm{E}[X_1])^3]\mathrm{E}[X_1]\\
&\quad + 3\left(\mathrm{Var}[N] + \mathrm{E}[N](\mathrm{E}[N]-1)\right)(\mathrm{Var}[X_1])^2\\
&\quad + 6\Big(\mathrm{E}[(N-\mathrm{E}[N])^3] + \mathrm{E}[N]\mathrm{Var}[N]\Big)(\mathrm{E}[X_1])^2\mathrm{Var}[X_1]\\
&\quad + \mathrm{E}[(N-\mathrm{E}[N])^4](\mathrm{E}[X_1])^4.
\end{aligned}$$

Here, it is assumed that the required moments of severity and frequency exist.

Proof This follows from the expression for characteristic function of the compound distribution (3.7) and formulas (3.15). The calculus is simple but lengthy and is left for the reader as Problem 3.9. □

Example 3.1 If frequencies are Poisson distributed, $N \sim Poisson(\lambda)$, then

$$\begin{aligned} &\mathrm{E}[N] = \mathrm{Var}[N] = \mathrm{E}[(N - \mathrm{E}[N])^3] = \lambda, \\ &\mathrm{E}[(N - \mathrm{E}[N])^4] = \lambda(1 + 3\lambda), \end{aligned}$$

and compound loss moments calculated using Proposition 3.1 are

$$\begin{aligned} &\mathrm{E}[Z] = \lambda\mathrm{E}[X_1], \quad \mathrm{Var}[Z] = \lambda\mathrm{E}[X_1^2], \quad \mathrm{E}[(Z - \mathrm{E}[Z])^3] = \lambda\mathrm{E}[X_1^3], \\ &\mathrm{E}[(Z - \mathrm{E}[Z])^4] = \lambda\mathrm{E}[X_1^4] + 3\lambda^2(\mathrm{E}[X_1^2])^2 \end{aligned} \tag{3.21}$$

Moreover, if the severities are lognormally distributed, $X_1 \sim \mathcal{LN}(\mu, \sigma)$, then

$$\mathrm{E}[X_1^k] = \exp(k\mu + k^2\sigma^2/2). \tag{3.22}$$

It is illustrative to see that in the case of compound Poisson, the moments can easily be derived using the following proposition

Proposition 3.2 (Cumulants of compound Poisson) *The cumulants of the compound random variable $Z = X_1 + \cdots + X_N$, where $X_1, \ldots, X_N$ are independent and identically distributed, and independent of N, are given by*

$$\kappa_k = \lambda\mathrm{E}[X_1^k], \quad k = 1, 2, \ldots$$

Proof Using the definition of cumulants (3.17) and the characteristic function for compound Poisson (3.8), calculate

$$\kappa_k = (-i)^k \left.\frac{d^k \ln \chi(t)}{dt^k}\right|_{t=0} = \lambda(-i)^k \left.\frac{d^k \varphi(t)}{dt^k}\right|_{t=0} = \lambda\mathrm{E}[X_i^k], \quad k = 1, 2, \ldots$$

□

3.1.4 Value-at-Risk and Expected Shortfall

Having calculated the compound loss distribution, the risk measures such as VaR and expected shortfall should be evaluated. Analytically, VaR of the compound loss is calculated as the inverse of the compound distribution

$$\mathrm{VaR}_\alpha[Z] = H^{-1}(\alpha) \tag{3.23}$$

and the expected shortfall of the compound loss above the quantile $q_\alpha = \mathrm{VaR}_\alpha[Z]$, assuming that $q_\alpha > 0$, is

$$\mathrm{ES}_\alpha[Z] = \mathrm{E}[Z|Z \geq q_\alpha] = \frac{1}{1-H(q_\alpha)} \int_{q_\alpha}^{\infty} z h(z) dz$$
$$= \frac{\mathrm{E}[Z]}{1-H(q_\alpha)} - \frac{1}{1-H(q_\alpha)} \int_{0}^{q_\alpha} z h(z) dz, \qquad (3.24)$$

where $\mathrm{E}[Z] = \mathrm{E}[N]\mathrm{E}[X_1]$ is the mean of compound loss Z. Note that $\mathrm{ES}_\alpha[Z]$ is defined for a given quantile q_α, that is, the quantile $H^{-1}(\alpha)$ has to be computed first. It is easy to show (see Exercise 3.1) that in the case of nonnegative severities, the above integral can be calculated via characteristic function as

$$\mathrm{ES}_\alpha[Z] = \frac{1}{1-H(q_\alpha)} \times \left[\mathrm{E}[Z] - H(q_\alpha) q_\alpha + \frac{2q_\alpha}{\pi} \int_0^\infty \mathrm{Re}\left[\chi(x/q_\alpha)\right] \frac{1-\cos x}{x^2} dx \right]. \qquad (3.25)$$

Remark 3.3

- Strictly speaking, in the above formulas (3.24) and (3.25), we assumed that the quantile is positive, $q_\alpha > 0$, i.e. $\alpha > \Pr[Z = 0]$ and we do not have complications due to discontinuity at zero. The case of $q_\alpha = 0$ is not really important to operational risk practice, but can easily be treated if required; see Remark 2.7 and formula (2.16).
- In the above formulas (3.24) and (3.25), $H(q_\alpha)$ can be replaced by α. We kept $H(q_\alpha)$, so that the formulas can easily be modified if expected exceedance $\mathrm{E}[Z|Z \geq L]$ should be calculated. In this case, q_α should be replaced by L in these formulas.

3.2 Monte Carlo Method

The easiest numerical method to calculate the compound loss distribution is Monte Carlo with the following logical steps.

Algorithm 3.1 (Monte Carlo for compound loss distribution)

1. For $k = 1, \ldots, K$
   a. Simulate the annual number of events N from the frequency distribution.
   b. Simulate independent severities $X_1, \ldots, X_N$ from the severity distribution.
   c. Calculate $Z_k = \sum_{i=1}^{N} X_i$.
2. Next k (i.e. do an increment $k = k + 1$ and return to step 1).

All random numbers simulated in the above are independent.

Obtained $Z_1, \ldots, Z_K$ are samples from a compound distribution $H(\cdot)$. Distribution characteristics can be estimated using the simulated samples in the usual way described in many textbooks. Here, we just mention the quantile and expected shortfall which are of primary importance for operational risk.

3.2.1 Quantile Estimate

Denote samples $Z_1, \ldots, Z_K$ sorted into the ascending order as $\widetilde{Z}_1 \leq \ldots \leq \widetilde{Z}_K$, then a standard estimator of the quantile $q_\alpha = H^{-1}(\alpha)$ is

$$\widehat{Q}_\alpha = \widetilde{Z}_{\lfloor K\alpha \rfloor + 1}. \tag{3.26}$$

Here, $\lfloor . \rfloor$ denotes rounding downward. Then, for a given realisation of the sample $\mathbf{Z} = \mathbf{z}$, the quantile estimate is $\widehat{q}_\alpha = \widetilde{z}_{\lfloor K\alpha \rfloor + 1}$. It is important to estimate numerical error (due to the finite number of simulations K) in the quantile estimator. Formally, it can be assessed using the following asymptotic result

$$\frac{h(q_\alpha)\sqrt{K}}{\sqrt{\alpha(1-\alpha)}}(\widehat{Q}_\alpha - q_\alpha) \to \mathcal{N}(0,1), \quad \text{as} \quad K \to \infty; \tag{3.27}$$

see e.g. Stuart and Ord ([224], pp. 356–358) and Glasserman ([108], p. 490). This means that the quantile estimator $\widehat{Q}_\alpha$ converges to the true value q_α as the sample size K increases and asymptotically $\widehat{Q}_\alpha$ is normally distributed with the mean q_α and standard deviation

$$\text{stdev}[\widehat{Q}_\alpha] = \frac{\sqrt{\alpha(1-\alpha)}}{h(q_\alpha)\sqrt{K}}. \tag{3.28}$$

However, the density $h(q_\alpha)$ is not known and the use of the above formula is difficult. In practice, the error of the quantile estimator is calculated using a nonparametric statistic by forming a conservative confidence interval $[\widetilde{Z}^{(r)}, \widetilde{Z}^{(s)}]$ to contain the true quantile value q_α with the probability at least γ:

$$\Pr[\widetilde{Z}_r \leq q_\alpha \leq \widetilde{Z}_s] \geq \gamma, \quad 1 \leq r < s \leq K. \tag{3.29}$$

Indices r and s can be found by utilising the fact that the true quantile q_α is located between $\widetilde{Z}_M$ and $\widetilde{Z}_{M+1}$ for some M. The number of losses M not exceeding the quantile q_α has a binomial distribution, $Bin(K, \alpha)$, because it is the number of successes from K independent and identical attempts with success probability α.

Thus the probability that the interval $[\widetilde{Z}_r, \widetilde{Z}_s]$ contains the true quantile is simply

$$\Pr[r \leq M \leq s-1] = \sum_{i=r}^{s-1} \binom{K}{i} \alpha^i (1-\alpha)^{K-i}. \tag{3.30}$$

One typically tries to choose r and s that are symmetric around and closest to the index $\lfloor K\alpha \rfloor + 1$, and such that the probability (3.30) is not less than the desired confidence level γ. The mean and variance of the binomial distribution are $K\alpha$ and $K\alpha(1-\alpha)$ respectively. For large K, approximating the binomial by the normal distribution with these mean and variance leads to a simple approximation for the conservative confidence interval bounds:

$$\begin{aligned} r &= \lfloor l \rfloor, \quad l = K\alpha - F_N^{-1}((1+\gamma)/2)\sqrt{K\alpha(1-\alpha)}, \\ s &= \lceil u \rceil, \quad u = K\alpha + F_N^{-1}((1+\gamma)/2)\sqrt{K\alpha(1-\alpha)}, \end{aligned} \tag{3.31}$$

where $\lceil . \rceil$ denotes rounding upwards and $F_N^{-1}(\cdot)$ is the inverse of the standard normal distribution $\mathcal{N}(0,1)$. The above formula works very well for $K\alpha(1-\alpha) \geq 50$ approximately.

Remark 3.4

- A large number of simulations, typically $K \geq 10^5$, should be used to achieve a good numerical accuracy for the 0.999 quantile. However, a priori, the number of simulations required to achieve a specific accuracy is not known. One of the approaches is to continue simulations until a desired numerical accuracy is achieved.
- If the number of simulations to get acceptable accuracy is very large (e.g. $K > 10^7$) then you might not be able to store the whole array of samples $Z_1, \ldots, Z_K$ when implementing the algorithm, due to computer memory limitations. However, if you need to calculate just the high quantiles then you need to save only $\lfloor K\alpha \rfloor + 1$ largest samples to estimate the quantile (3.26). This can be done by using the sorting *on the fly* algorithms, where you keep a specified number of largest samples as you generate the new samples; see Press, Teukolsky, Vetterling and Flannery ([195], section 8.5). Moments (mean, variance, etc) can also be easily calculated *on the fly* without saving all samples into the computer memory.
- To use (3.31) for estimation of the quantile numerical error, it is important that MC samples $Z_1, \ldots, Z_K$ are independent and identically distributed. If the samples are correlated, for example generated by MCMC, then (3.31) can significantly underestimate the error. In this case, one can use *batch sampling* or *effective sample size* methods described in Sect. 2.12.2.

Example 3.2 Assume that $K = 5 \times 10^4$ independent samples were drawn from $\mathcal{LN}(0,2)$. Suppose that we would like to construct a conservative confidence interval to contain the 0.999 quantile with probability at least $\gamma = 0.95$. Then, sort the samples in ascending order and using (3.31) calculate $F_N^{-1}((1+\gamma)/2) \approx 1.96$, $r = 49,936$ and $s = 49,964$ and $\lfloor K\alpha \rfloor + 1 = 49,951$.

3.2.2 Expected Shortfall Estimate

Given independent samples $Z_1, \ldots, Z_K$ from the same distribution and the estimator $\widehat{Q}_\alpha$ of $\mathrm{VaR}_\alpha[Z]$, a typical estimator for expected shortfall $\omega_\alpha = \mathrm{E}[Z|Z \geq \mathrm{VaR}_\alpha[Z]]$ is

$$\widehat{\Omega}_\alpha = \frac{\sum_{k=1}^K Z_k \mathbf{1}_{\{Z_k \geq \widehat{Q}_\alpha\}}}{\sum_{k=1}^K \mathbf{1}_{\{Z_k \geq \widehat{Q}_\alpha\}}} = \frac{\sum_{k=1}^K Z_k \mathbf{1}_{\{Z_k \geq \widehat{Q}_\alpha\}}}{K - \lfloor K\alpha \rfloor}. \tag{3.32}$$

It gives an expected shortfall estimate $\widehat{\omega}_\alpha$ for a given sample realisation, $\mathbf{Z} = \mathbf{z}$. From the strong law of large numbers applied to the numerator and denominator and the convergence of the quantile estimator (3.27), it is clear that

$$\widehat{\Omega}_\alpha \to \omega_\alpha \tag{3.33}$$

with probability 1, as the sample size increases. If we assume that the quantile q_α is known, then in the limit $K \to \infty$, the central limit theorem gives

$$\frac{\sqrt{K}}{\sigma}(\widehat{\Omega}_\alpha - \omega_\alpha) \to \mathcal{N}(0, 1), \tag{3.34}$$

where σ, for a given realisation $\mathbf{Z} = \mathbf{z}$, can be estimated as

$$\widehat{\sigma}^2 = K \frac{\sum_{k=1}^K (z_k - \widehat{\omega}_\alpha)^2 \mathbf{1}_{z_k \geq q_\alpha}}{\left(\sum_{k=1}^K \mathbf{1}_{z_k \geq q_\alpha}\right)^2}.$$

Then, the standard deviation of $\widehat{\Omega}_\alpha$ is estimated by $\widehat{\sigma}/\sqrt{K}$; see Glasserman [109]. However, it will underestimate the error in expected shortfall estimate because the quantile q_α is not known and estimated itself by $\widehat{q}_\alpha$. Approximation for asymptotic standard deviation of expected shortfall estimate can be found in Yamai and Yoshiba ([242], Appendix 1). In general, the standard deviation of the MC estimates can always be evaluated by simulating K samples many times; see the batch sampling method described in Sect. 2.12.2. For heavy-tailed distributions and high quantiles, it is typically observed that the error in quantile estimate is much smaller than the error in expected shortfall estimate.

Remark 3.5 Expected shortfall does not exist for distributions with infinite mean. Such distributions were reported in the analysis of operational risk losses; see Moscadelli [166]. This will be discussed more in Chap. 6.

3.3 Panjer Recursion

It appears that, for some class of frequency distributions, the compound distribution calculation via the convolution (3.4) can be reduced to a simple recursion introduced by Panjer [180] and referred to as Panjer recursion. A good introduction of this method in the context of operational risk can be found in Panjer ([181], sections 5 and 6). Also, a detailed treatment of Panjer recursion and its extensions are given in a recently published book Sundt and Vernic [229]. Below we summarise the method and discuss implementation issues.

Firstly, Panjer recursion is designed for discrete severities. Thus, to apply the method for operational risk, where severities are typically continuous, the continuous severity should be replaced with the discrete one. For example, one can round all amounts to the nearest multiple of monetary unit δ, e.g. to the nearest USD 1000. Define

$$f_k = \Pr[X_i = k\delta], \quad p_k = \Pr[N = k], \quad h_k = \Pr[Z = k\delta], \tag{3.35}$$

with $f_0 = 0$ and $k = 0, 1, \ldots$. Then, the discrete version of (3.4) is

$$\begin{aligned} h_n &= \sum_{k=1}^{n} p_k f_n^{(k)*}, \quad n \geq 1, \\ h_0 &= \Pr[Z = 0] = \Pr[N = 0] = p_0, \end{aligned} \tag{3.36}$$

where $f_n^{(k)*} = \sum_{i=0}^{n} f_{n-i}^{(k-1)*} f_i$ with $f_0^{(0)*} = 1$ and $f_n^{(0)*} = 0$ if $n \geq 1$.

Remark 3.6

- Note that the condition $f_0 = \Pr[X = 0] = 0$ implies that $f_n^{(k)*} = 0$ for $k > n$ and thus the above summation is up to n only.
- If $f_0 > 0$, then $f_n^{(k)*} > 0$ for all n and k; and the upper limit in summation (3.36) should be replaced by infinity.
- The number of operations to calculate $h_0, h_1, \ldots, h_n$ using (3.36) explicitly is of the order of n^3.

If the maximum value for which the compound distribution should be calculated is large, the number of computations become prohibitive due to $O(n^3)$ operations. Fortunately, if the frequency N belongs to the so-called Panjer classes, (3.36) is reduced to a simple recursion introduced by Panjer [180] and referred to as Panjer recursion.

Theorem 3.1 (Panjer recursion) *If the frequency probability mass function p_n, $n = 0, 1, \ldots$ satisfies*

$$p_n = \left(a + \frac{b}{n}\right) p_{n-1}, \quad \text{for} \quad n \geq 1 \quad \text{and } a, b \in \mathbb{R}, \tag{3.37}$$

then it is said to be in Panjer class $(a, b, 0)$ *and the compound distribution (3.36) satisfies the recursion*

$$h_n = \frac{1}{1-af_0}\sum_{j=1}^{n}\left(a+\frac{bj}{n}\right)f_j h_{n-j}, \quad n \geq 1,$$

$$h_0 = \sum_{k=0}^{\infty}(f_0)^k p_k. \tag{3.38}$$

The initial condition in (3.38) is simply a probability generating function of N at f_0, i.e. $h_0 = \psi(f_0)$, see (3.6). If $f_0 = 0$, then it simplifies to $h_0 = p_0$. It was shown in Sundt and Jewell [228], that (3.37) is satisfied for the Poisson, negative binomial and binomial distributions. The parameters (a, b) and starting values h_0 are listed in Table 3.1; also see Appendix A.1 for definition of the distributions.

Remark 3.7

- If severity is restricted by a value of the largest possible loss m, then the upper limit in the recursion (3.38) should be replaced by $\min(m, n)$.
- The Panjer recursion requires $O(n^2)$ operations to calculate $h_0, \ldots, h_n$ in comparison with asymptotic $O(n^3)$ of explicit convolution.
- Strong stability of Panjer recursion was established for the Poisson and negative binomial cases; see Panjer and Wang [182]. The accumulated rounding error of the recursion increases linearly in n with a slope not exceeding one. Serious numerical problems may occur for the case of binomial distribution. Typically, instabilities in the recursion appear for significantly underdispersed frequencies of severities with a large negative skewness which are not typical in operational risk.
- In the case of severities from a phase-type distribution (distribution with a rational probability generating function), the recursion (3.38) is reduced to $O(n)$ operations; see Hipp [121]. Typically, the severity distributions are not phase-type distributions and approximation is required. This is useful for modelling small losses but not suitable for heavy-tailed distributions because the phase-type distributions are light tailed; see Bladt [28] for a review.

Table 3.1 Panjer recursion starting values h_0 and (a, b) parameters for Poisson, binomial and negative binomial distributions

	a	b	h_0
$Poisson(\lambda)$	0	λ	$\exp(\lambda(f_0 - 1))$
$NegBin(r, q)$	$1-q$	$(1-q)(r-1)$	$\left(1+(1-f_0)\frac{1-q}{q}\right)^{-r}$
$Bin(m, q)$	$-\frac{q}{1-q}$	$\frac{q(m+1)}{1-q}$	$(1+q(f_0-1))^m$

The Panjer recursion can be implemented as follows:

Algorithm 3.2 (Panjer recursion)

1. Initialisation: calculate f_0 and h_0, see Table 3.1, and set $H_0 = h_0$.
2. For $n = 1, 2, \ldots$
   a. Calculate f_n. If severity distribution is continuous, then f_n can be found as described in Sect. 3.3.1.
   b. Calculate $h_n = \frac{1}{1-af_0}\sum_{j=1}^{n}\left(a+\frac{bj}{n}\right) f_j h_{n-j}$.
   c. Calculate $H_n = H_{n-1} + h_n$.
   d. Interrupt the procedure if H_n is larger than the required quantile level α, e.g. $\alpha = 0.999$. Then the estimate of the quantile q_α is $n \times \delta$.
3. Next n (i.e. do an increment $n = n + 1$ and return to step 2).

3.3.1 Discretisation

Typically, severity distributions are continuous and thus discretisation is required. To concentrate severity, whose continuous distribution is $F(x)$, on $\{0, \delta, 2\delta, \ldots\}$, one can choose $\delta > 0$ and use the central difference approximation

$$\begin{aligned} f_0 &= F(\delta/2), \\ f_n &= F(n\delta + \delta/2) - F(n\delta - \delta/2), \quad n = 1, 2, \ldots . \end{aligned} \tag{3.39}$$

Then the compound discrete density h_n is calculated using Panjer recursion and compound distribution is calculated as $H_n = \sum_{i=0}^{n} h_i$. As an example, Table 3.2 gives results of calculation of the $Poisson(100) - \mathcal{LN}(0, 2)$ compound distribution up to the 0.999 quantile in the case of step $\delta =$ USD 1. Of course the accuracy of the result depends on the step size as shown by the results for the 0.999 quantile vs δ, see Table 3.3 and Fig. 3.1. It is, however, important to note that the error of the result is due to discretisation only and there is no truncation error (i.e. the severity is not truncated by some large value).

Table 3.2 Example of Panjer recursion calculating the $Poisson(100) - \mathcal{LN}(0, 2)$ compound distributions using central difference discretisation with the step $\delta = 1$

n	f_n	h_n	H_n
0	0.364455845	2.50419×10^{-28}	2.50419×10^{-28}
1	0.215872117	5.40586×10^{-27}	5.65628×10^{-27}
2	0.096248034	6.07589×10^{-26}	6.64152×10^{-26}
$\vdots$	$\vdots$	$\vdots$	$\vdots$
5847	2.81060×10^{-9}	4.44337×10^{-7}	0.998999329
5848	2.80907×10^{-9}	4.44061×10^{-7}	0.998999773
5849	2.80755×10^{-9}	4.43785×10^{-7}	0.999000217

Table 3.3 Convergence of Panjer recursion estimate, $\widehat{q}_{0.999}$, of the 0.999 quantile for the $Poisson(100) - \mathcal{LN}(0,2)$ compound distributions using central difference discretisation vs the step size δ. Here, $N = \widehat{q}_{0.999}/\delta$ is the number of steps required

δ	N	$\widehat{q}_{0.999}$	time (sec)
16	360	5760	0.19
8	725	5800	0.20
4	1457	5828	0.28
2	2921	5842	0.55
1	5849	5849	1.59
0.5	11703	5851.5	5.77
0.25	23411	5852.75	22.47
0.125	46824	5853	89.14
0.0625	93649	5853.0625	357.03

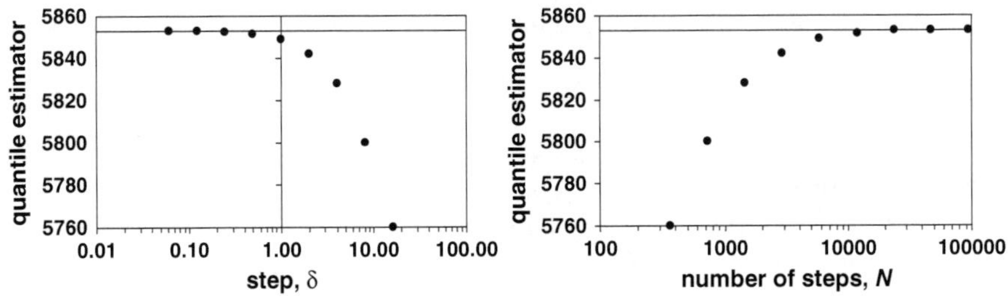

Fig. 3.1 Panjer recursion estimate, $\widehat{q}_{0.999}$, of the 0.999 quantile for the $Poisson(100) - \mathcal{LN}(0,2)$ compound distribution vs the step size δ (*left figure*) and vs the number of steps $N = \widehat{q}_{0.999}/\delta$ (*right figure*)

Discretisation can also be done via the forward and backward differences:

$$f_n^U = F(n\delta + \delta) - F(n\delta);$$
$$f_n^L = F(n\delta) - F(n\delta - \delta). \tag{3.40}$$

These allow for calculation of the upper and lower bounds for the compound distribution:

$$H_n^U = \sum_{i=0}^{n} h_i^U;$$
$$H_n^L = \sum_{i=0}^{n} h_i^L.$$

For example, see Table 3.4 presenting results for $Poisson(100) - \mathcal{LN}(0,2)$ compound distribution calculated using central, forward and backward differences with step $\delta =$ USD 1. The use of the forward difference f_n^U gives the upper bound for the compound distribution and the use of f_n^L gives the lower bound. Thus the lower

Table 3.4 Example of Panjer recursion calculating the $Poisson(100) - \mathcal{LN}(0, 2)$ compound distributions using central, forward and backward difference discretisation with the step $\delta = 1$

n	H_n^L	H_n	H_n^U
0	3.72008×10^{-44}	2.50419×10^{-28}	1.92875×10^{-22}
1	1.89724×10^{-42}	5.65628×10^{-27}	2.80718×10^{-21}
⋮	⋮	⋮	⋮
5811	0.998953196	0.998983158	0.998999719
5812	0.998953669	0.998983612	0.999000163
⋮	⋮	⋮	⋮
5848	0.9989705	0.998999773	0.999015958
5849	0.998970962	0.999000217	0.999016392
⋮	⋮	⋮	⋮
5913	0.998999942	0.999028056	0.999043605
5914	0.999000385	0.999028482	0.999044022

and upper bounds for a quantile are obtained with f_n^U and f_n^L respectively. In the case of Table 3.4 example, the quantile bound interval is [USD 5811, USD 5914] with the estimate from the central difference USD 5849.

3.3.2 Computational Issues

Underflow[2] in computations of (3.38) will occur for large frequencies during the initialisation of the recursion. This can easily be seen for the case of $Poisson(\lambda)$ and $f_0 = 0$ when $h_0 = \exp(-\lambda)$, that is, the underflow will occur for $\lambda \gtrsim 700$ on a 32bit computer with double precision calculations. Re-scaling h_0 by large factor γ to calculate the recursion (and de-scaling the result) does not really help because overflow will occur for $\gamma h(n)$.

The following identity helps to overcome this problem in the case of Poisson frequency:

$$H^{(m)*}(z; \lambda/m) = H(z; \lambda). \tag{3.41}$$

That is, calculate the compound distribution $H(z; \lambda/m)$ for some large m to avoid underflow. Then preform m convolutions for the obtained distribution directly or via FFT; see Panjer and Willmot [179]. Similar identity is available for negative binomial, $NegBin(r, p)$:

$$H^{(m)*}(z; r/m) = H(z; r). \tag{3.42}$$

[2] Underflow/overflow are the cases when the computer calculations produce a number outside the range of representable numbers leading 0 or $\pm\infty$ outputs respectively.

In the case of binomial, $Bin(M, p)$:

$$H^{(m)*}(z; m_1) * H(z; m_2) = H(z; M), \tag{3.43}$$

where $m_1 = \lfloor M/m \rfloor$ and $m_2 = M - m_1 m$.

To make it more efficient, one can choose $m = 2^k$ so that instead of m convolutions of $H(\cdot)$ only k convolutions are required $H^{(2)*}, H^{(4)*}, \ldots H^{(2^k)*}$, where each term is the convolution of the previous one with itself.

3.3.3 Panjer Extensions

The Panjer recursion formula (3.38) can be extended to a class of frequency distributions $(a, b, 1)$.

Definition 3.1 (Panjer class $(a, b, 1)$**)** The distribution is said to be in $(a, b, 1)$ Panjer class if it satisfies

$$p_n = \left(a + \frac{b}{n}\right) p_{n-1}, \quad \text{for} \quad n \geq 2 \quad \text{and} \quad a, b \in \mathbb{R}. \tag{3.44}$$

Theorem 3.2 (Extended Panjer recursion) *For the frequency distributions in a class* $(a, b, 1)$*:*

$$h_n = \frac{(p_1 - (a+b)p_0) f_n + \sum_{j=1}^{n} (a + bj/n) f_j h_{n-j}}{1 - a f_0}, \quad n \geq 1,$$

$$h_0 = \sum_{k=0}^{\infty} (f_0)^k p_k. \tag{3.45}$$

The distributions of $(a, b, 0)$ class are special cases of $(a, b, 1)$ class. There are two types of frequency distributions in $(a, b, 1)$ class:

- zero-truncated distributions, where $p_0 = 0$: i.e. zero truncated Poisson, zero truncated binomial and zero-truncated negative binomial.
- zero-modified distributions, where $p_0 > 0$: the distributions of $(a, b, 0)$ with modified probability of zero. It can be viewed as a mixture of $(a, b, 0)$ distribution and degenerate distribution concentrated at zero.

Finally, we would like to mention a generalisation of Panjer recursion for the (a, b, l) class

$$p_n = \left(a + \frac{b}{n}\right) p_{n-1}, \quad \text{for} \quad n \geq l + 1. \tag{3.46}$$

For initial values $p_0 = \cdots = p_{l-1} = 0$, and in the case of $f_0 = 0$, it leads to the recursion

$$h_n = p_l f_n^{(l)*} + \sum_{j=1}^{n} (a + bj/n) f_j h_{n-j}, \quad n \geq l.$$

The distribution in this class is, for example, $l-1$ truncated Poisson. For an overview of high order Panjer recursions, see Hess, Liewald and Schmidt [119]. Other types of recursions

$$p_n = \sum_{j=1}^{k} (a_j + b_j/n) p_{n-1}, \quad n \geq 1, \tag{3.47}$$

are discussed in Sundt [226]. Application of the standard Panjer recursion in the case of the generalised frequency distributions such as the extended negative binomial, can lead to numerical instabilities. Generalisation of the Panjer recursion that leads to numerically stable algorithms for these cases is presented in Gerhold, Schmock and Warnung [102]. Discussion on multivariate version of Panjer recursion can be found in Sundt [227] and bivariate cases are discussed in Vernic [236] and Hesselager [120].

3.3.4 Panjer Recursion for Continuous Severity

The Panjer recursion is developed for the case of discrete severities. The analogue of Panjer recursion for the case of continuous severities is given by the following integral equation.

Theorem 3.3 (Panjer recursion for continuous severities) *For frequency distributions in $(a, b, 1)$ class and continuous severity distributions on positive real line:*

$$h(z) = p_1 f(z) + \int_0^x (a + by/z) f(y) h(z-y) dy. \tag{3.48}$$

The proof is presented in Panjer and Willmot ([179], Theorem 6.14.1 and 6.16.1). Note that the above integral equation holds for $(a, b, 0)$ class because it is a special case of $(a, b, 1)$. The integral equation (3.48) is a Volterra integral equation of the second type. There are different methods to solve it described in Panjer and Willmot [179]. A method of solving this equation using hybrid MCMC (minimum variance importance sampling via reversible jump MCMC) is presented in Peters, Johansen and Doucet [185].

3.4 Fast Fourier Transform

The FFT is another efficient method to calculate compound distributions via the inversion of the characteristic function. The method has been known for many decades and originates from the signal processing field. The existence of the algorithm became generally known in the mid-1960s, but it was independently discovered by many researchers much earlier. One of the early books on FFT is

Brigham [37]. A detailed explanation of the method in application to aggregate loss distribution can be found in Robertson [203]. In our experience, operational risk practitioners in banking regard the method as difficult and rarely use it in practice. In fact, it is a very simple algorithm to implement, although to make it really efficient, especially for heavy-tailed distribution, some improvements are required. Below we describe the essential steps and theory required for successful implementation of the FFT for operational risk.

As with Panjer recursion case, FFT works with discrete severity and based on the discrete Fourier transformation defined as follows:

Definition 3.2 (Discrete Fourier transformation) For a sequence $f_0, f_1, \ldots, f_{M-1}$, the discrete Fourier transformation (DFT) is defined as

$$\phi_k = \sum_{m=0}^{M-1} f_m \exp\left(\frac{2\pi i}{M} mk\right), \quad k = 0, 1, \ldots, M-1 \tag{3.49}$$

and the original sequence f_k can be recovered from ϕ_k by the inverse transformation

$$f_k = \frac{1}{M} \sum_{m=0}^{M-1} \phi_m \exp\left(-\frac{2\pi i}{M} mk\right), \quad k = 0, 1, \ldots, M-1. \tag{3.50}$$

Here, $i = \sqrt{-1}$ is a unit imaginary number and M is some truncation point. It is easy to see that to calculate M points of ϕ_m, the number of operations is of the order of M^2, i.e. $O(M^2)$. If M is a power of 2, then DFT can be efficiently calculated via FFT algorithms with the number of computations $O(M \log_2 M)$. This is due to the property that DFT of length M can be represented as the sum of DFT over even points ϕ_k^e and DFT over odd points ϕ_k^o:

$$\begin{aligned}
\phi_k &= \phi_k^e + \exp\left(\frac{2\pi i}{M} k\right) \phi_k^o;\\
\phi_k^e &= \sum_{m=0}^{M/2-1} f_{2m} \exp\left(\frac{2\pi i}{M} mk\right);\\
\phi_k^o &= \sum_{m=0}^{M/2-1} f_{2m+1} \exp\left(\frac{2\pi i}{M} mk\right).
\end{aligned}$$

Subsequently, each of these two DFTs can be calculated as a sum of two DFTs of length $M/4$. For example, ϕ_k^e is calculated as a sum of ϕ_k^{ee} and ϕ_k^{eo}. This procedure is continued until the transforms of the length 1. The latter is simply identity operation. Thus every obtained pattern of odd and even DFTs will be f_m for some m:

$$\phi_k^{eo \cdots oe} = f_m.$$

The bit reversal procedure can be used to find m that corresponds to a specific pattern. That is, set $e = 0$ and $o = 1$, then the reverse pattern of e's and o's is the value of m in binary. Thus the logical steps of FFT, where M is integer power of 2, are as follows:

Algorithm 3.3 (Simple FFT)

1. Sort the data in a bit-reversed order. The obtained points are simply one-point transforms.
2. Combine the neighbour points into non-overlapping pairs to get two-point transforms. Then combine two-point transforms into 4-point transforms and continue subsequently until the final M point transform is obtained. Thus there are $\log_2 M$ iterations and each iteration involves of the order of M operations.

The basic FFT algorithm is very simple and its code is short; see, for example, C code provided in Press, Teukolsky, Vetterling and Flannery ([195], chapter 12). The inverse FFT transformation can be calculated in the same way as FFT (the only differences are sign change and division by M, see (3.49) and (3.50)).

3.4.1 Compound Distribution via FFT

Calculation of the compound distribution via FFT can be done using the following logical steps.

Algorithm 3.4 (Compound Distribution via FFT)

1. Discretise severity to obtain
$$f_0, f_1, \ldots, f_{M-1},$$
where $M = 2^r$ with integer r, and M is the truncation point in the aggregate distribution.
2. Using FFT, calculate the characteristic function of the severity
$$\varphi_0, \ldots, \varphi_{M-1}.$$
3. Calculate the characteristic function of the compound distribution using (3.7), i.e.
$$\chi_m = \psi(\varphi_m), \quad m = 0, 1, \ldots, M-1.$$
4. Perform inverse FFT (which is the same as FFT except the change of sign under the exponent and factor $1/M$) applied to $\chi_0, \ldots, \chi_{M-1}$ to obtain the compound distribution $h_0, h_1, \ldots, h_{M-1}$.

Remark 3.8 To calculate the compound distribution in the case of the severity distribution $F(x)$ with a finite support (i.e. $0 < a \le x \le b < \infty$) one can set $F(x) = 0$ for x outside the support range when calculating discretised severity $f_0, \ldots, f_{M-1}$

using (3.39). For example, this is the case for distribution of losses exceeding some threshold. Note that we need to set $F(x) = 0$ in the range $x \in [0, a)$ due to the finite probability of zero compound loss.

3.4.2 Aliasing Error and Tilting

If there is no truncation error in the severity discretisation, i.e.

$$\sum_{m=0}^{M-1} f_m = 1,$$

then FFT procedure calculates the compound distribution on $m = 0, 1, \ldots, M$. That is, the mass of compound distribution beyond M is "wrapped" and appears in the range $m = 0, \ldots, M-1$ (the so-called *aliasing error*). This error is larger for heavy-tailed severities. To decrease the error for compound distribution on $0, 1, \ldots, n$, one has to take M much larger than n. If the severity distribution is bounded and M is larger than the bound, then one can put zero values for points above the bound (the so-called padding by zeros). Another way to reduce the error is to apply some transformation to increase the tail decay (the so-called *tilting*). The exponential tilting technique for reducing aliasing error under the context of calculating compound distribution was first investigated by Grubel and Hermesmeier [114]. Many authors suggest the following tilting transformation:

$$\widetilde{f}_j = \exp(-j\theta) f_j, \quad j = 0, 1, \ldots, M-1, \tag{3.51}$$

where $\theta > 0$. This transformation commutes with convolution in a sense that convolution of two functions $f(x)$ and $g(x)$ equals the convolution of the transformed functions $\widetilde{f}(x) = f(x)\exp(-\theta x)$ and $\widetilde{g}(x) = g(x)\exp(-\theta x)$ multiplied by $\exp(\theta x)$, i.e.

$$(f * g)(x) = e^{\theta x}(\widetilde{f} * \widetilde{g})(x). \tag{3.52}$$

This can easily be shown using the definition of convolution. Then calculation of the compound distribution is performed using the transformed severity distribution as follows.

Algorithm 3.5 (Compound distribution via FFT with tilting)

1. Define $f_0, f_1, \ldots, f_{M-1}$ for some large M.
2. Perform tilting, i.e. calculate the transformed function $\widetilde{f}_j = \exp(-j\theta) f_j$, $j = 0, 1, \ldots, M-1$.
3. Apply FFT to a set $\widetilde{f}_0, \ldots, \widetilde{f}_{M-1}$ to obtain $\widetilde{\phi}_0, \ldots, \widetilde{\phi}_{M-1}$.
4. Calculate $\widetilde{\chi}_m = \psi(\widetilde{\phi}_m)$, $m = 0, 1, \ldots, M-1$.
5. Apply the inverse FFT to the set $\widetilde{\chi}_0, \ldots, \widetilde{\chi}_{M-1}$, to obtain $\widetilde{h}_0, \widetilde{h}_1, \ldots, \widetilde{h}_{M-1}$.
6. Untilt by calculating final compound distribution as $h_j = \widetilde{h}_j \exp(\theta j)$.

This tilting procedure is very effective in reducing the aliasing error. The parameter θ should be as large as possible but not producing under- or overflow that will occur for very large θ. It was reported in Embrechts and Frei [81] that the choice $M\theta \approx 20$ works well for standard double precision (8 bytes) calculations. Evaluation of the probability generating function $\psi(\cdot)$ of the frequency distribution may lead to the problem of underflow in the case of large frequencies that can be resolved using methods described in Sect. 3.3.2.

Example 3.3 To demonstrate the effectiveness of the tilting, consider the following calculations:

- FFT with the central difference discretisation, where the tail probability compressed into the last point $f_{M-1} = 1 - F(\delta(M-1) - \delta/2)$. Denote the corresponding quantile estimator as $Q^{(1)}_{0.999}$;
- FFT with the central difference discretisation with the tail probability ignored, i.e. $f_{M-1} = F(\delta(M-1)+\delta/2) - F(\delta(M-1)-\delta/2)$. Denote the corresponding quantile estimator as $Q^{(2)}_{0.999}$;
- FFT with the central difference discretisation utilising tilting $Q^{(tilt)}_{0.999}$. The tilting parameter θ is chosen to be $\theta = 20/M$.

The calculation results presented in Table 3.5 demonstrate the efficiency of the tilting. If FFT is performed without tilting then the truncation level for the severity should exceed the quantile significantly. In this particular case it should exceed by approximately factor of 10 to get the exact result for this discretisation step. The latter is obtained by Panjer recursion that does not require the discretisation beyond the calculated quantile. Thus the FFT and Panjer recursion are approximately the same in terms of computing time required for quantile estimate in this case. However, once the tilting is utilised, the cut off level does not need to exceed the quantile significantly to obtain the exact result – making FFT superior to Panjer recursion. In this example, the computing time for FFT with tilting is 0.17 s in comparison with 5.76 s of Panjer recursion, see Table 3.3. Also, in this case, the treatment of the severity tail by ignoring it or absorbing into the last point f_{M-1} does not make any difference when tilting is applied.

Table 3.5 Example of FFT calculating the 0.999 quantile of the $Poisson(100) - \mathcal{LN}(0, 2)$ compound distribution using central difference discretisation with the step $\delta = 0.5$. The exact Panjer recursion for this discretisation step gives $Q_{0.999} = 5851.5$

r	$L = \delta \times 2^r$	$Q^{(1)}_{0.999}$	$Q^{(2)}_{0.999}$	$Q^{(tilt)}_{0.999}$	time (sec)
14	8192	5117	5665.5	5851.5	0.17
15	16384	5703.5	5834	5851.5	0.36
16	32768	5828	5850	5851.5	0.75
17	65536	5848.5	5851.5	5851.5	1.61
18	131072	5851.5	5851.5	5851.5	3.64
19	262144	5851.5	5851.5	5851.5	7.61

3.5 Direct Numerical Integration

In the case of nonnegative severities, the distribution of the compound loss is

$$H(z) = \frac{2}{\pi} \int_0^\infty \mathrm{Re}[\chi(t)] \frac{\sin(tz)}{t} dt, \quad z \geq 0, \tag{3.53}$$

where $\chi(t)$ is a compound distribution characteristic function calculated via the severity characteristic function $\varphi(t)$ using (3.7), see Lemma 3.1 and formula (3.12). The explicit expression of $\mathrm{Re}[\chi(t)]$ for $Poisson(\lambda)$ is

$$\mathrm{Re}[\chi(t)] = e^{-\lambda} \exp(\lambda \mathrm{Re}[\varphi(t)]) \times \cos(\lambda \mathrm{Im}[\varphi(t)]). \tag{3.54}$$

For the cases of negative binomial and binomial distributions, $\mathrm{Re}[\chi(t)]$ is easily obtained through complex variable functions in the relevant computer language. Hereafter, direct calculation of the distribution function for annual loss Z using (3.53) is referred to as *direct numerical integration* (DNI).

Much work has been done in the last few decades in the general area of inverting characteristic functions numerically. Just to mention a few, see the works by Bohman [33]; Seal [211]; Abate and Whitt [1, 2]; Heckman and Meyers [118]; Shephard [214]; Waller, Turnbull and Hardin [237]; and Den Iseger [74]. These papers address various issues such as singularity at the origin; treatment of long tails in the infinite integration; and choices of quadrature rules covering different objectives with different distributions. Craddock, Heath and Platen [64] gave an extensive survey of numerical techniques for inverting characteristic functions. A tailor-made numerical algorithm to integrate (3.53) was presented in Luo and Shevchenko [147] with a specific requirement on accuracy and efficiency in calculating high quantiles such as 0.999 quantile. The method works well both for a wide range of frequencies from very low to very high ($> 10^5$) and heavy-tailed severities.

Each of the many existing techniques has particular strengths and weaknesses, and no method works equally well for all classes of problems. In an operational risk context, for instance, there are special requirements in computing the 0.999 quantile of the aggregate loss distribution. The accuracy demanded is high and at the same time the numerical inversion could be very time consuming due to rapid oscillations and slow decay in the characteristic function. This is the case, for example, for heavy tailed severities. Also, the characteristic function of compound distributions should be calculated numerically through semi-infinite integrations.

Below we describe the essential steps of the DNI method to calculate the annual loss distribution via (3.53).

3.5.1 Forward and Inverse Integrations

The task of the characteristic function inversion is analytically straightforward, but numerically difficult in terms of achieving high accuracy and computational efficiency *simultaneously*. The computation of compound distribution through the characteristic function involves two steps: computing the characteristic function (Fourier transform of the density function, referred to as the *forward integration*) and inverting it (referred to as the *inverse integration*).

Forward integration. This step requires integration (3.5), that is, calculation of the real and imaginary parts of the characteristic function for a severity distribution:

$$\mathrm{Re}[\varphi(t)] = \int_0^\infty f(x)\cos(tx)dx, \quad \mathrm{Im}[\varphi(t)] = \int_0^\infty f(x)\sin(tx)dx. \tag{3.55}$$

Then, the characteristic function of the compound loss is calculated using (3.7). These tasks are relatively simple because the severity density typically has closed-form expression, and is well-behaved having a single mode.

This step can be done more or less routinely and many existing algorithms, including the ones commonly available in many software packages, can be employed. The oscillatory nature of the integrand only comes from the sin( ) or cos( ) functions. This well-behaved weighted oscillatory integrand can be effectively dealt with by the modified Clenshaw-Curtis integration method; see Clenshaw and Curtis [58] and Piessens, Doncker-Kapenga, Überhuber and Kahaner [192]. In this method the oscillatory part of the integrand is transferred to a weight function, the non-oscillatory part is replaced by its expansion in terms of a finite number of Chebyshev polynomials and the modified Chebyshev moments are calculated. If the oscillation is slow when the argument t of the characteristic function is small, the standard Guass-Legendre and Kronrod quadrature formulae are more effective; see Kronrod [139], Golub and Welsh [110], Szegö [231], and Sect. 3.5.2. In general, double precision accuracy can be routinely achieved for the forward integrations.

Inverse integration. This step requires integration (3.53), which is much more challenging task. Changing variable $x = t \times z$, (3.53) can be rewritten as

$$H(z) = \int_0^\infty G(x,z)\sin(x)dx, \quad G(x,z) = \frac{2}{\pi}\frac{\mathrm{Re}[\chi(x/z)]}{x}, \tag{3.56}$$

where $\chi(t)$ depends on $\mathrm{Re}[\varphi(t)]$ and $\mathrm{Im}[\varphi(t)]$ calculated from the forward semi-infinite integrations (3.55) for any required argument t. The total number of forward integrations required by the inversion is usually quite large. This is because in this case the characteristic function could be highly oscillatory due to high frequency and it may decay very slowly due to heavy tails. There are two oscillatory components in the integrand represented by $\sin(x)$ and another part in $\mathrm{Re}[\chi(x/z)]$; see (3.54). It is convenient to treat $\sin(x)$ as the principal oscillatory factor and the other part as secondary. Typically, given z, $\mathrm{Re}[\chi(x/z)]$ decays fast initially and then approaches zero slowly as x approaches infinity; see Fig. 3.2 for the case of $Poisson(10^5)$-$\mathcal{LN}(0,2)$ compound distribution with the value of z corresponding to $H(z) = 0.999$.

Although the oscillation frequency of Re$[\chi]$ increases with λ, this increase is much slower than a linear increase. In fact, at $\lambda = 10^5$ (see Fig. 3.2) the oscillation frequency of Re$[\chi]$ is still smaller than that of $\sin(x)$. This can be quantified by ω, the relative oscillation frequency of Re$[\chi]$ with respect to $\sin(x)$, defined as

$$\omega(x, z) = \lambda \frac{\partial \mathrm{Im}[\varphi(x/z)]}{\partial x},$$

where $\omega < 1$ indicates that the local oscillation frequency is smaller than that of $\sin(x)$. Figure 3.2 shows a plot of ω as a function of x. It shows that not only ω is less than one in this case, but also that it appears to decay linearly as x increases, justifying treatment of Re$[\chi]$ as the secondary oscillator.

To calculate (3.56), one could apply the same standard general purpose adaptive integration routines as for the forward integration. However, this is typically not efficient because it does not address irregular oscillation specifically and can lead to an excessive number of integrand evaluations. The approach taken in Luo and Shevchenko [147] divides the integration range of (3.56) into intervals with an equal length of π (referred to as π-cycle) and truncates at $2K$ π-cycles:

$$H(z) \approx \sum_{k=0}^{2K-1} H_k, \quad H_k = \int_{k\pi}^{(k+1)\pi} G(x)\sin(x)dx. \tag{3.57}$$

Within each π-cycle, the secondary oscillation could be dominating for some early cycles, thus the π-cycle could in fact contain multiple cycles due to the "secondary" oscillation. Thus a further sub-division is warranted. Sub-dividing interval $(k\pi,\ (k+1)\pi)$ into n_k segments of equal length of $\Delta_k = \pi/n_k$, (3.57) can be written as

$$H_k = \sum_{j=1}^{n_k} H_k^{(j)}, \quad H_k^{(j)} = \int_{a_{k,j}}^{b_{k,j}} G(x)\sin(x)dx, \tag{3.58}$$

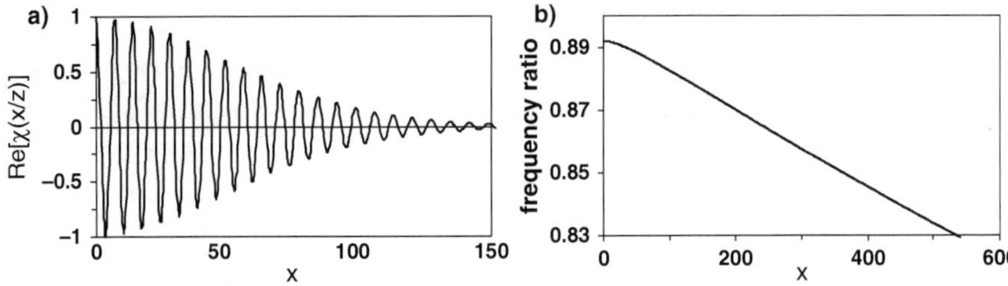

Fig. 3.2 Re$[\chi(x/z)]$ (*left figure*) and frequency ratio $\omega(x, z)$ (*right figure*) for $z = 8.22 \times 10^5 \approx Q_{0.999}$ in the case of $Poisson(10^5) - \mathcal{LN}(0, 2)$

where

$$a_{k,j} = k\pi + (j-1)\Delta_k, \quad b_{k,j} = a_{k,j} + \Delta_k.$$

The above calculation will be most effective if the sub-division is made adaptive for each π-cycle according to the changing behaviour of $G(x)$. Assuming that for the first π-cycle ($k = 0$) we have initial partition n_0, Luo and Shevchenko [147] recommends making n_k adaptive for the subsequent cycles by the following two simple rules:

- Rule 1. Let n_k be proportional to the number of π-cycles of the secondary oscillation – the number of oscillations in $G(x)$ within each principal π-cycle.
- Rule 2. Let n_k be proportional to the magnitude of the maximum gradient of $G(x)$ within each principal π-cycle.

Application of Rule 1 and Rule 2 requires correct counting of secondary cycles and good approximation of the local gradient in $G(x)$. Both can be achieved with a significant number of points at which $G(x)$ is computed within each cycle using, for example, the m-point Gaussian quadrature described in the next section.

Remark 3.9 (Accuracy requirement) Accurate calculation of the quantile as an inverse of the distribution function requires high precision in evaluation of the distribution function. To demonstrate, consider the lognormal distribution $\mathcal{LN}(0,2)$. In this case, the "exact" 0.999 quantile $q_{0.999} = 483.2164\ldots$. However, at $\alpha = 0.99902$, the quantile becomes $q_\alpha = 489.045\ldots$. That is, a mere 0.002% change in the distribution function value causes more than 1% change in the quantile value, which is an amplification of the error by 500 times in percentage terms. In other words, achieving the error for the 0.999 quantile within 1% requires calculation of the distribution function to be accurate to the fifth digit. Formally, the error propagation from the distribution function level to the quantile value can be estimated by the relation between the density, $f(x)$, and its distribution function, $F(x)$: $dF/dx = f(x)$. In the above example, $x = 483.2164\ldots$ and $1/f(x) = \sigma x\sqrt{2\pi}\exp[0.5(\ln x/\sigma)^2] \approx 287023$. That is, in absolute terms, an error in the distribution function estimation will be amplified by 287023 times in the error for the corresponding 0.999 quantile. In the case of a compound distribution, the requirement for accuracy in the distribution function could be even higher, because $1/f(x)$ could be larger at $x = q_{0.999}$. In fact, for compound distribution with high frequency and heavy-tailed severity, it is often observed that distribution function correct to the fifth digit is not accurate enough for an accurate estimation of the 0.999 quantiles.

Remark 3.10 (Error sources) The final result of the inverse integration has three error sources: the discretisation error of the Gauss quadrature; the error from the tail approximation; and the error propagated from the error of the forward integration. These were analysed in Luo and Shevchenko [147]. It was shown that the propagation error is proportional to the forward integration error bound. At the extreme case of $\lambda = 10^6$, a single precision can still be readily achieved if the forward integration has a double precision. For very large λ, the propagation error is likely the largest among the three error sources. Though some formulas for error bounds were derived, these are not very useful in practise because high order derivatives are involved, which is typical for analytical error bounds. An established and satisfactory practice is to use finer grids to estimate the error of the coarse grids.

3.5.2 Gaussian Quadrature for Subdivisions

With a proper sub-division, even a simple trapezoidal rule can be applied to get a good approximation for integration over the sub-division $H_k^{(j)}$ in (3.58). However, higher order numerical quadrature can achieve higher accuracy for the same computing effort or it requires less computing effort for the same accuracy. The m-point Gaussian quadrature makes the computed integral exact for all polynomials of degree $2m-1$ or less. In particular:

$$\int_a^b g(x)dx \approx \frac{\Delta}{2}\sum_{i=1}^{m} w_i g\left((a+b+\zeta_i\Delta)/2\right), \tag{3.59}$$

where $0 < w_i < 1$ and $-1 < \zeta_i < 1$ are the ith weight and the ith abscissa of the Gaussian quadrature respectively, $\Delta = b - a$ and m is the order of the Gaussian quadrature. For completeness, Table 3.6 presents 7-point Gaussian quadrature weights and abscissas.

The efficiency of the Gaussian quadrature is much superior to the trapezoidal rule. For instance, integrating the function $\sin(3x)$ over the interval $(0, \pi)$, the 7-point Gaussian quadrature has a relative error less than 10^{-5}, while the trapezoidal rule requires about 900 function evaluations (grid spacing $\delta x = \pi/900$) to achieve a similar accuracy. The reduction of the number of integrand function evaluations is important for a fast integration of (3.57), because the integrand itself is a time consuming semi-infinite numerical integration. The error of the m-point Gaussian quadrature rule can be accurately estimated if the $2m$ order derivative of the integrand can be computed [132, 223]. In general, it is difficult to estimate the $2m$ order derivative and the actual error may be much less than a bound established by the derivative. As it has already been mentioned, a common practice is to use two

Table 3.6 The weights w_i and abscissas ζ_i of the 7-point Gaussian quadrature

i	ζ_i	w_i
1	−0.949107912342759	0.129484966168870
2	−0.741531185599394	0.279705391489277
3	−0.405845151377397	0.381830050505119
4	0.0	0.417959183673469
5	0.405845151377397	0.381830050505119
6	0.741531185599394	0.279705391489277
7	0.949107912342759	0.129484966168870

numerical evaluations with the grid sizes different by the factor of two and estimate the error as the difference between the two results. Equivalently, different orders of quadrature can be used to estimate error. Often, Guass-Kronrod quadrature is used for this purpose. Table 3.7 gives 15-point Guass-Kronrod quadrature weights and abscissas.

Let δ_m^G denote the error bound for the m-order Gauss quadrature and δ_{2m+1}^{GK} be the error bound for the corresponding Guass-Kronrod quadrature. Brass and Förster [35] proved that

$$\delta_{2m+1}^{GK}/\delta_m^G \leq const \times \sqrt[4]{m}\,(1/3.493)^m \,.$$

Because δ_{2m+1}^{GK} is smaller than δ_m^G by at least an order of magnitude, the difference between Gauss-Kronrod and Gauss quadrature serves as a good estimate for δ_m^G. Adaptive integration functions in many numerical software packages use this estimate to achieve an overall error bound below the user-specified tolerance. For example, the *IMSL* subroutine *QDAG* subdivides a given interval and uses the $(2m+1)$-point Gauss-Kronrod rule to estimate the integral over each subinterval. The error for each subinterval is estimated by comparison with the $m-$point Gauss quadrature rule. The subinterval with the largest estimated error is then bisected and the same procedure is applied to both halves. The bisection process is continued until the error criterion is satisfied, or the subintervals become too small, or the maximum number of subintervals allowed is reached. As it has already been mentioned, this numerical functions can successfully be applied for the forward integration but is not efficient for the inverse integration.

Typically, even a simple 7-point Gaussian quadrature ($m = 7$), which calculates all polynomials of degree 13 or less exactly, can successfully be used to calculate $H_k^{(j)}$ in (3.57, 3.58).

Table 3.7 The weights w_i and abscissas ζ_i of the 15-point Guass-Kronrod quadrature

i	ζ_i	w_i
1	−0.991455371120813	0.022935322010529
2	−0.949107912342759	0.063092092629979
3	−0.864864423359769	0.104790010322250
4	−0.741531185599394	0.140653259715525
5	−0.586087235467691	0.169004726639267
6	−0.405845151377397	0.190350578064785
7	−0.207784955007898	0.204432940075298
8	0.0	0.209482141084728
9	0.207784955007898	0.204432940075298
10	0.405845151377397	0.190350578064785
11	0.586087235467691	0.169004726639267
12	0.741531185599394	0.140653259715525
13	0.864864423359769	0.104790010322250
14	0.949107912342759	0.063092092629979
15	0.991455371120813	0.022935322010529

3.5.3 Tail Integration

The truncation error of using (3.57) is

$$H_T = \int_{2K\pi}^{\infty} G(x)\sin(x)dx. \tag{3.60}$$

For higher accuracy, instead of increasing truncation length at the cost of computing time, one can try to calculate the tail integration H_T approximately or use tilting transform (3.51). Integration of (3.60) by parts, gives

$$\int_{2K\pi}^{\infty} G(x)\sin(x)dx = G(2K\pi) + \sum_{j=1}^{k-1}(-1)^j G^{(2j)}(2K\pi) + (-1)^k \int_{2K\pi}^{\infty} G^{(2k)}(x)\sin(x)dx, \tag{3.61}$$

where $k \geq 1$, $G^{(2j)}(2K\pi)$ is the $2j$-th order derivative of $G(x)$ at the truncation point. Under some conditions, as $K \to \infty$,

$$\int_{2K\pi}^{\infty} G(x)\sin(x)dx \to G(2K\pi) + \sum_{j=1}^{\infty}(-1)^j G^{(2j)}(2K\pi).$$

For example, if we assume that for some $\gamma < 0$, $G^{(m)}(x) = O(x^{\gamma-m})$, $m = 0, 1, 2, \ldots$ as $K \to \infty$, then the series converges to the integral. However, this is not true for some functions, such as $\exp(-x)$. Though, typically in this case the truncation error is not material.

It appears that often (see [147, 149]) the very first term in (3.61) gives a very good approximation

$$H_T = \int_{2K\pi}^{\infty} G(x)\sin(x)dx \approx G(2K\pi) \tag{3.62}$$

for the tail integration or does not have a material impact on the overall integration. This elegant result means that we only need to evaluate the integrand at one single point $x = 2\pi K$ for the entire tail integration. One can consider this as an assumption that $G(x)$ is well approximated by a function piece-wise linear within each π cycle. The approximation (3.62) can be improved by including further terms if derivatives are easy to calculate, e.g. $H_T \approx G(2K\pi) - G^{(2)}(2K\pi)$.

Thus the total integral approximation (3.57) can be improved by including tail correction giving

$$H(z) \approx \sum_{k=0}^{2K-1} H_k + G(2N\pi). \tag{3.63}$$

Remark 3.11 If the oscillating factor is $\cos(x)$ instead of $\sin(x)$, one can still derive a one-point formula similar to (3.61) by starting the tail integration at $(2K - 1/2)\pi$ instead of $2K\pi$. In this case, the tail integration is

$$\int_{(2K-1/2)\pi}^{\infty} G(x)\cos(x)dx \approx G\left((2K - 1/2)\pi\right). \tag{3.64}$$

Also, the tail integration approximation can be applied to the left tail (integrating from $-\infty$ to $-2K\pi$) as well, if such integration is required.

Remark 3.12 Of course there are more elaborate methods to treat the truncation error which are superior to a simple approximation (3.62) in terms of better accuracy and broader applicability, such as some of the extrapolation methods proposed in Wynn [241], Sidi [219, 220].

Example 3.4 As an example of effectiveness of the above tail integration approximation consider the integrals

$$I_E = \int_0^{\infty} G(x)\sin(x)dx, \quad \tilde{I}(2K\pi) = \int_0^{2\pi K} G(x)\sin(x)dx, \tag{3.65}$$

where $G(x) = 1/\sqrt{x}$. The exact tail integration can be computed from $I_T(2K\pi) = I_E - \tilde{I}(2K\pi)$. We compare $\tilde{I}(2K\pi) + G(2K\pi)$ with $\tilde{I}(2K\pi)$ and compare both of them with the exact semi-infinite integration I_E. The error of using (3.62) is $\varepsilon_T = I_E - [\tilde{I}(2K\pi) + G(2K\pi)]$. Here, we have a closed form for the total integral $I_E = \sqrt{\pi/2}$ and $\tilde{I}$ was accurately computed by an adaptive integration function from *IMSL* software package.

Figure 3.3 compares the tail integration $I_T(2K\pi)$ with a one-point value $G(2K\pi)$. One can see that, one-point approximation does an extremely good job. Even at the shortest truncation length of just 2π (i.e. $K = 1$), one-point approximation is very close to the exact semi-infinite tail integration. The relative error $\delta_T/I_E(2K\pi)$ is about 1% at $K = 1$ and it is about 0.002% at $K = 10$. Apparently, if the extra correction term $G^{(2)}(2K\pi)$ is included, the error δ_T reduces further by an order of magnitude at $K = 1$ and by several orders of magnitude at $K = 10$.

Figure 3.3 shows $\tilde{I}(2K\pi)$ and $\tilde{I} + G(2K\pi)$, along with the correct value of the full integration $I_E = \sqrt{\pi/2}$. The contrast between results with and without the one-point tail approximation is striking. At the shortest truncation length of 2π $(K = 1)$, the relative error due to truncation for the truncated integration $(I_E - \tilde{I}(2K\pi))/I_E$ is more than 30%, but with the tail approximation added, the relative error $(I_E - \tilde{I}(2K\pi) - G(2K\pi))/I_E$ reduces to 0.5%. At 100π, the largest truncation length

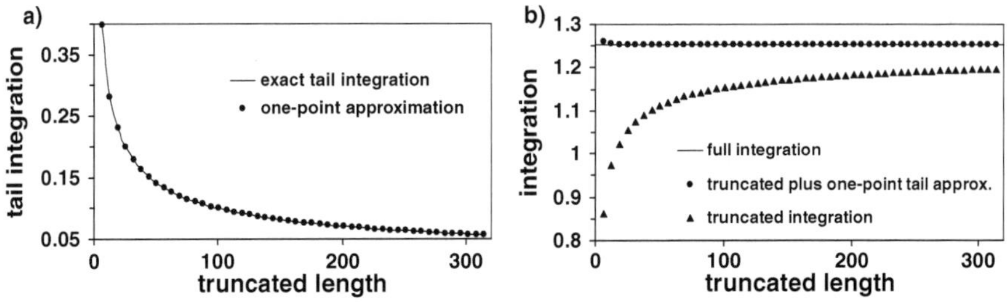

Fig. 3.3 $G(x) = 1/\sqrt{x}$. *Left figure* **(a)** – comparison between exact tail integration $\int_{2\pi K}^{\infty} G(x)\sin(x)dx$ and a simple one-point approximation (3.62), $G(2K\pi)$. *Right figure* **(b)** – comparison between truncated integration $\widetilde{I}(2K\pi) = \int_0^{2\pi K} G(x)\sin(x)dx$ and the truncated integration plus the one-point approximation of tail integration, $\widetilde{I}(2K\pi) + G(2K\pi)$. The *solid line* represents the exact value of the full integration without truncation error, $\widetilde{I}(\infty) = \sqrt{\pi/2}$. The truncated length is $l_T = 2K\pi$ with $2 \leq K \leq 50$

shown in Fig. 3.3, the relative error due to truncation is still more than 4%. After one-point correction is added, the relative error reduces to less than 0.5×10^{-6}.

Another way to look at these comparisons, which is relevant for integrating heavy-tailed functions, is to consider the required truncation length for the truncated integration to achieve the same accuracy as the one with one-point correction. For the truncated integration $\widetilde{I}(2K\pi)$ to achieve the same accuracy of $\widetilde{I}(2\pi) + G(2\pi)$ (i.e. integration truncated at one-cycle plus the "magic point"), the integration length should be extended to 7700π. For $\widetilde{I}(2K\pi)$ to achieve the same accuracy of $\widetilde{I}(100\pi) + G(100\pi)$, the integration length has to be extended to more than $10^{12}\pi$! On the other hand, if we add the tail approximation $G(7700\pi)$ to $\widetilde{I}(7700\pi)$, the relative error reduces from 0.5% to less than 10^{-11}! This error reduction requires virtually no extra computing, since calculation of $G(7700\pi) = 1/\sqrt{7700\pi}$ is trivial.

Example 3.5 Table 3.8 shows the convergence of DNI results (seven digits), for truncation lengths $2 \leq K \leq 80$ in the cases of tail correction included and ignored. One can see a material improvement from the tail correction. Also, as the truncation length increases, both estimators with the tail correction and without converge.

Table 3.8 Convergence in DNI estimates of $H(z = 5{,}853.1)$ for $Poisson(100)$-$\mathcal{LN}(0, 2)$ in the case of $n_0 = 1$ and different truncation length K. $\widehat{H}_{tail}$ is the estimate with the tail correction and $\widehat{H}$ is the estimate without the tail correction

K	$\widehat{H}$	$\widehat{H}_{tail}$	$Time(s)$
2	0.9938318	0.9999174	0.0625
3	1.0093983	0.9993260	0.094
4	1.0110203	0.9991075	0.125
5	1.0080086	0.9990135	0.141
10	0.9980471	0.9989910	0.297
20	0.9990605	0.9990002	0.578
40	0.9989996	0.9990000	1.109
80	0.9990000	0.9990000	2.156

In this particular case we calculate compound distribution $Poisson(100)$-$\mathcal{LN}(0, 2)$ at the level $z = 5853.1$. The latter is the value that corresponds to the 0.999 quantile (within 1st decimal place) of this distribution as has already been calculated by Panjer recursion; see Table 3.3. Of course, to calculate the quantile at the 0.999 level using DNI, a search algorithm such as bisection should be used that will require evaluation of distribution function many times (of the order of 10) increasing computing time. Comparing this with Tables 3.3 and 3.5, one can see that for this case DNI is faster than Panjer recursion while slower than FFT (with tilting) by a factor of 10.

3.6 Comparison of Numerical Methods

For comparison purposes, Tables 3.9 and 3.10 present results for the 0.999 quantile of compound distributions $Poisson(\lambda)$-$\mathcal{LN}(0, 2)$ and $Poisson(\lambda)$-$GPD(1, 1)$ (with $\lambda = 0.1, 10, 10^3$), calculated by the DNI, FFT, Panjer and MC methods. Note that, with the shape parameter $\xi = 1$, $GPD(\xi, \beta)$ has infinite mean and all higher moments. For DNI, FFT and Panjer recursion methods, the results, accurate up to 5 significant digits, were obtained as follows:

- For DNI algorithm we start with a relatively coarse grid ($n_0 = 1$) and short truncation length $K = 25$, and keep halving the grid size and doubling the truncation length until the difference in the 0.999 quantile is within required accuracy. The DNI algorithm computes distribution function, $H(z)$, for any given level z by (3.53), one point at a time. Thus with DNI we have to resort to an iterative procedure to inverse (3.53). This requires evaluating (3.53) many times depending on the search algorithm employed and the initial guess. Here, a standard bisection

Table 3.9 The estimates of the 0.999 quantile, $Q_{0.999}$, for $Poisson(\lambda)$-$\mathcal{LN}(0, 2)$, calculated using DNI, FFT, Panjer recursion and MC methods. Standard errors of MC estimates are given in brackets next to the estimator

	λ	0.1	10	1,000
DNI	$Q_{0.999}$	105.36	1, 779.1	21, 149
	time	15.6 s	6 s	25 s
	$K\backslash n_0$	50\2	25\1	25\1
MC	$Q_{0.999}$	105.45(0.26)	1, 777(9)	21, 094(185)
	time	3 min	3.9 min	11.7 min
	N_{MC}	10^8	10^7	10^6
Panjer	$Q_{0.999}$	105.36	1, 779.1	21, 149
	time	7.6 s	8.5 s	3.6 h
	h	2^{-7}	2^{-3}	2^{-4}
FFT	$Q_{0.999}$	105.36	1, 779.1	21, 149
	time	0.17 s	0.19 s	7.9 s
	h	2^{-7}	2^{-3}	2^{-4}
	M	2^{14}	2^{14}	2^{19}

Table 3.10 The estimates of the 0.999 quantile, $Q_{0.999}$, for $Poisson(\lambda)$-$GPD(1,1)$, calculated using DNI, FFT, Panjer recursion and MC methods. Standard errors of MC estimates are given in brackets next to the estimator

	λ	0.1	10	1,000
DNI	$Q_{0.999}$	99.352	10, 081	1.0128×10^6
	time	21 s	29 s	52 s
	$K\backslash n_0$	100\2	100\2	100\1
MC	$Q_{0.999}$	99.9(0.3)	10, 167(89)	$1.0089(0.026) \times 10^6$
	time	3.1 min	3.6 min	7.8 min
	N_{MC}	10^8	10^7	10^6
Panjer	$Q_{0.999}$	99.352	10, 081	1.0128×10^6
	time	6.9 s	4.4 s	15 h
	h	2^{-7}	1	1
FFT	$Q_{0.999}$	99.352	10, 081	1.0128×10^6
	time	0.13 s	0.13 s	28 s
	h	2^{-7}	1	1
	M	2^{14}	2^{14}	2^{21}

algorithm is employed. Other methods (MC, Panjer recursion and FFT) have the advantage that they obtain the whole distribution in a single run.
- For Panjer recursion, starting with a large step (e.g. $\delta = 8$) the step δ is successively reduced until the change in the result is smaller than the required accuracy.
- For FFT with tilting, the same step δ is used as the one in the Panjer recursion. If we would not know the Panjer recursion results, then we would successively reduce the step δ (starting with some large step) until the change in the result is smaller than the required accuracy. The truncation length $M = 2^r$ has to be large enough so that $\delta M > \widehat{Q}_q$ is satisfied. We use the smallest possible integer r that allows to identify the quantile, typically such that $\delta M \approx 2\widehat{Q}_q$. Here, $\widehat{Q}_q$ is the quantile to be computed, which is not known a priori and some extra iteration is typically required. Also, the tilting parameter is set to $\theta = 20/M$.
- For the MC estimates, the number of simulations, N_{MC} (denoted by K in Sect. 3.2), ranges from 10^6 to 10^8, so that calculations are accomplished within $\approx$ 10 min. The error of the MC estimate is approximately proportional to $1/\sqrt{N_{MC}}$ and the calculation time is approximately proportional to N_{MC}. Thus the obtained results allow to judge how many simulations (time) is required to achieve a specific accuracy.

The agreement between FFT, Panjer recursion and DNI estimates is perfect. Also, the difference between these results and corresponding MC estimates is always within the two MC standard errors. However, the CPU time is very different across the methods:

- The quoted CPU time for the MC results is of the order 10 min. However, it is clear from the standard error results (recalling that the error is proportional to $1/\sqrt{N_{MC}}$) that the CPU time, required to get the results accurate up to five significant digits, would be of the order of several days. Thus MC is the slowest method.

- Typically, the CPU time for both Panjer recursion and FFT increase as λ increases, while CPU time for DNI does not change significantly.
- FFT is the fastest method, though at very high frequency $\lambda = 10^3$, DNI performance is of a similar order. As reported in Luo and Shevchenko [147], DNI becomes faster than FFT for higher frequencies $\lambda > 10^3$.
- Panjer recursion is always slower than FFT. It is faster than DNI for small frequencies and much slower for high frequencies.

Finally note that, the FFT, Panjer recursion and DNI results were obtained by successive reduction of grid size (starting with a coarse grid) until the required accuracy is achieved. The quoted CPU time is for the last iteration in this procedure. Thus the results for CPU time should be treated as indicative only.

For comparison of FFT and Panjer, also see Embrechts and Frei [81], and Bühlmann [43].

3.7 Closed-Form Approximation

There are several well-known approximations for the compound loss distribution. These can be used with different success depending on the quantity to be calculated and distribution types. Even if the accuracy is not good, these approximations are certainly useful from the methodological point of view in helping to understand the model properties. Also, the quantile estimate derived from these approximations can successfully be used to set a cut-off level for FFT algorithms that will subsequently determine the quantile more precisely.

3.7.1 Normal and Translated Gamma Approximations

Many parametric distributions can be used as an approximation for a compound loss distribution by moment matching. This is because the moments of the compound loss can be calculated in closed form. In particular, the first four moments are given in Proposition 3.1. Of course these can only be used if the required moments exist which is not the case for some heavy-tailed risks with infinite moments. Below we mention normal and translated gamma approximations.

Normal approximation. As the severities $X_1, X_2, \ldots$ are independent and identically distributed, at very high frequencies the central limit theory is expected to provide a good approximation to the distribution of the annual loss Z (if the second moment of severities is finite). Then the compound distribution is approximated by the normal distribution with the mean and variance given in Proposition 3.1, that is,

$$H(z) \approx \mathcal{N}(\mathrm{E}[Z], \sqrt{\mathrm{Var}[Z]}). \tag{3.66}$$

This is an asymptotic result and a priori we do not know how well it will perform for specific distribution types and distribution parameter values. Also, it cannot be used for the cases where variance or mean are infinite.

Example 3.6 If N is distributed from $Poisson(\lambda)$ and $X_1, \ldots, X_N$ are independent random variables from $\mathcal{LN}(\mu, \sigma)$, then

$$\mathrm{E}[Z] = \lambda \exp(\mu + 0.5\sigma^2), \quad \mathrm{Var}[Z] = \lambda \exp(2\mu + 2\sigma^2). \tag{3.67}$$

Translated gamma approximation. From (3.21), the skewness of the compound distribution, in the case of Poisson distributed frequencies, is

$$\frac{\mathrm{E}[(Z - \mathrm{E}[Z])^3]}{(\mathrm{Var}[Z])^{3/2}} = \frac{\lambda \mathrm{E}[X^3]}{\left(\lambda \mathrm{E}[X^2]\right)^{3/2}} > 0, \tag{3.68}$$

that approaches zero as λ increases but finite positive for finite $\lambda > 0$. To improve the normal approximation (3.66), the compound loss can be approximated by the shifted gamma distribution which has a positive skewness, that is, Z is approximated as $Y + a$ where a is a shift and Y is a random variable from $Gamma(\alpha, \beta)$. The three parameters are estimated by matching the mean, variance and skewness of the approximate distribution and the correct one:

$$a + \alpha\beta = \mathrm{E}[Z]; \quad \alpha\beta^2 = \mathrm{Var}[Z]; \quad \frac{2}{\sqrt{\alpha}} = \mathrm{E}[(Z - \mathrm{E}[Z])^3] / (\mathrm{Var}[Z])^{3/2}. \tag{3.69}$$

This approximation requires the existence of the first three moments and thus cannot be used if the third moment does not exist.

Example 3.7 If frequencies are Poisson distributed, $N \sim Poisson(\lambda)$, then

$$a + \alpha\beta = \lambda \mathrm{E}[X]; \quad \alpha\beta^2 = \lambda \mathrm{E}[X^2]; \quad \frac{2}{\sqrt{\alpha}} = \lambda \mathrm{E}[X^3] / \left(\lambda \mathrm{E}[X^2]\right)^{3/2}. \tag{3.70}$$

3.7.2 VaR Closed-Form Approximation

If severities $X_1, \ldots, X_N$ are independent and identically distributed from the subexponential (heavy tail) distribution $F(x)$, and frequency distribution satisfies

$$\sum_{n=0}^{\infty} (1 + \epsilon)^n \Pr[N = n] < \infty$$

for some $\epsilon > 0$, then the tail of the compound distribution $H(z)$, of the compound loss $Z = X_1 + \cdots + X_N$, is related to the severity tail as

$$1 - H(z) \to \mathrm{E}[N](1 - F(z)), \quad \text{as} \quad z \to \infty; \tag{3.71}$$

see Theorem 1.3.9 in Embrechts, Klüppelberg and Mikosch [83][3]. This will be discussed more in Sect. 6.7. The validity of this asymptotic result was demonstrated for the cases when N is distributed from Poisson, binomial or negative binomial. This approximation can be used to calculate the quantiles of the annual loss as

$$\mathrm{VaR}_\alpha[Z] \to F^{-1}\left(1 - \frac{1-\alpha}{\mathrm{E}[N]}\right), \quad \text{as} \quad \alpha \to 1. \tag{3.72}$$

For application in the operational risk context, see Böcker and Klüppelberg [29]. Under the assumption that the severity has a finite mean, Böcker and Sprittulla [32] derived a correction reducing the approximation error of (3.72).

Example 3.8 Consider a heavy-tailed $Poisson(\lambda)$-$GPD(\xi, \beta)$ compound distribution. In this case, (3.72) gives

$$\mathrm{VaR}_\alpha[Z] \to \frac{\beta}{\xi}\left(\frac{\lambda}{1-\alpha}\right)^{\xi}, \quad \text{as} \quad \alpha \to 1. \tag{3.73}$$

This implies a simple scaling, $\mathrm{VaR}_\alpha[Z] \propto \lambda^{\xi}$, with respect to the event intensity λ for large α.

Example 3.9 To demonstrate the accuracy of the above approximations, consider compound distribution $Poisson(\lambda = 100)$-$\mathcal{LN}(\mu = 0, \sigma = 2)$ with relatively heavy tail severity. Calculating moments of the lognormal distribution $\mathrm{E}[X^m]$ using (3.22) and substituting into (3.21) gives

$$\mathrm{E}[Z] \approx 738.9056, \quad \mathrm{Var}[Z] \approx 298095.7987,$$
$$\mathrm{E}[(Z - \mathrm{E}[Z])^3]/(\mathrm{Var}[Z])^{3/2} \approx 40.3428.$$

Approximating the compound distribution by the normal distribution with these mean and variance gives normal approximation. Approximating the compound distribution by the translated gamma distribution (3.69) with these mean, variance and skewness gives

$$\alpha \approx 0.002457, \quad \beta \approx 11013.2329, \quad a \approx 711.8385.$$

Figure 3.4a shows the normal and translated gamma approximations for the tail of the compound distribution. These are compared with the asymptotic result for heavy tail distributions (3.71) and "*exact*" values obtained by FFT. It is easy to see

[3] Note that often, in the relevant literature, notation "$\sim$" is used to indicate that the ratio of the left- and righthand sides converge to 1; here we use "$\to$" to avoid confusion with notation used to indicate that a random variable is distributed from a distribution.

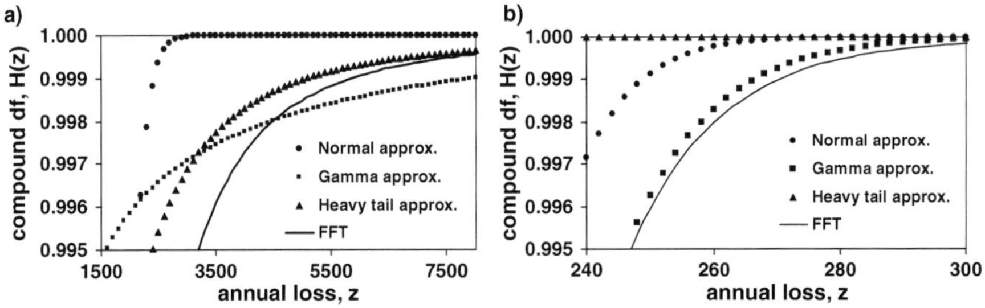

Fig. 3.4 Different approximations for the tail of the $Poisson(100) - \mathcal{LN}(0, \sigma)$ distribution for **(a)** $\sigma = 2$; and **(b)** less heavier tail $\sigma = 1$. See Example 3.9 for details

that the heavy tail asymptotic approximation (3.71) converges to the *exact* result for large quantile level $\alpha \to 1$, while the normal and gamma approximations perform badly. The results for the case of not so heavy tail, when the severity distribution is $\mathcal{LN}(0, 1)$, are shown in Fig. 3.4b. Here, the gamma approximation outperforms normal approximation and heavy tail approximation is very bad. The accuracy of the heavy tail approximation (3.71) improves for more heavy-tailed distributions, such as GPD with infinite variance or even infinite mean.

Problems[4]

3.1 ($\star$) Suppose that a non-negative random variable Z is from a distribution $H(z)$ whose characteristic function is $\chi(t)$. Prove that the expected exceedance over the level L is

$$\mathrm{E}[Z|Z \geq L] = \frac{1}{1 - H(L)} \left[\mathrm{E}[Z] - H(L)L + \frac{2L}{\pi} \int_0^\infty \mathrm{Re}[\chi(x/L)] \frac{1 - \cos x}{x^2} dx \right],$$

which is the expected shortfall if $L = \mathrm{VaR}_\alpha[Z]$. Assume that $H(z)$ is continuous for $z \geq L$.

3.2 ($\star$) Simulate 10^5 independent samples from $\mathcal{LN}(0, 3)$. Estimate the 0.99 and 0.999 quantiles and their numerical errors. For the latter, use the conservative confidence interval (3.29). Calculate the conservative interval bounds using exact formula (3.30) and compare with the normal approximation (3.31). Repeat estimation using 4×10^5 simulations.

3.3 ($\star$) Using simulated samples of Z from Problem 3.2, estimate the expected shortfalls above the 0.99 and 0.999 quantiles.

3.4 ($\star$) Simulate 10^5 independent samples of $Z = X_1 + \ldots + X_N$, where $X_i \sim \mathcal{LN}(0, 1.5)$ and $N \sim Poisson(5)$, X_i and N are independent. Estimate the 0.99 and

[4] Problem difficulty is indicated by asterisks: ($\star$) – low; ($\star\star$) – medium, ($\star\star\star$) – high.

0.999 quantiles of Z. Estimate numerical errors of the estimated quantile using the conservative confidence interval (3.29). Repeat estimation using 4×10^5 simulations.

3.5 ($\star\star$) Estimate the 0.99 and 0.999 quantiles of $Z = X_1 + \ldots + X_N$, where $X_i \sim \mathcal{LN}(0, 1.5)$ and $N \sim Poisson(5)$ (i.e. the same as in Problem 3.4), using Panjer recursion. Compare with the Monte Carlo results for Problem 3.4.

3.6 ($\star\star\star$) Using FFT, estimate the 0.99 and 0.999 quantiles of $Z = X_1 + \ldots + X_N$, where $X_i \sim \mathcal{LN}(0, 1.5)$ and $N \sim Poisson(5)$ (i.e. the same as in Problem 3.4). Compare with the Monte Carlo and Panjer recursion results from Problems 3.4 and 3.5. Obtain estimates using FFT with and without tilting.

3.7 ($\star\star\star$) Using Panjer recursion, estimate the 0.99 and 0.999 quantiles of a random variable $Z = X_1 + \ldots + X_N$, where

- $X_1, \ldots, X_N$ are independent random variables from $\mathcal{LN}(0, 1.5)$, and independent of N.
- N is a random variable from a zero truncated Poisson distribution

$$\Pr[N = 0] = 0 \quad \text{and} \quad \Pr[N = k] = \frac{p_k}{1 - p_0}, \quad k = 1, 2, \ldots,$$

  where p_k is $Poisson(5)$.

Estimate numerical error of the quantile estimates.

3.8 ($\star\star\star$) Using Panjer recursion, estimate the 0.99 and 0.999 quantiles of a random variable $Z = X_1 + \ldots + X_N$, where

- $X_1, \ldots, X_N$ are independent random variables from $\mathcal{LN}(0, 1.5)$, and independent of N.
- N is a random variable from a zero modified Poisson distribution

$$\Pr[N = 0] = q \quad \text{and} \quad \Pr[N = k] = (1 - q)\frac{p_k}{1 - p_0}, \quad k = 1, 2, \ldots,$$

  where p_k is $Poisson(5)$ and $q = 0.1$.

Estimate numerical error of the quantile estimates.

3.9 ($\star\star\star$) Prove that the first four moments of the compound random variable $Z = X_1 + \cdots + X_N$ are given by the Proposition 3.1. Here, we assume that $X_1, \ldots, X_N$ are independent and identically distributed, and independent from a random frequency N. Hint: use the formula (3.13) that calculates moments via the characteristic function (3.7), severity characteristic function (3.5) and probability generating function (3.6).

3.10 ($\star\star$) Using the normal and translated gamma approximations, estimate the 0.99 and 0.999 quantiles of $Z = X_1 + \ldots + X_N$, where $X_i \sim \mathcal{LN}(0, 1.5)$ and $N \sim Poisson(5)$ (i.e. the same as in Problem 3.4). Compare with the Monte Carlo, Panjer recursion and FFT results from Problems 3.4, 3.5 and 3.6.

Chapter 4
Bayesian Approach for LDA

Essentially, all models are wrong but some of them useful.
George Box

Abstract To meet the Basel II regulatory requirements for the Advanced Measurement Approaches, a bank's internal model must include the use of internal data, relevant external data, scenario analysis and factors reflecting the business environment and internal control systems. Bayesian inference is a statistical technique well suited for combining different data sources. This chapter presents examples of the Bayesian inference and closely related credibility theory methods for quantifying operational risk.

4.1 Introduction

Basel II AMA includes the following requirement[1] ([17], p. 152):

> Any operational risk measurement system must have certain key features to meet the supervisory soundness standard set out in this section. These elements must include the use of internal data, relevant external data, scenario analysis and factors reflecting the business environment and internal control systems.

Combining these different data sources for model estimation is certainly one of the main challenges in operational risk. This was emphasised in interviews with industry executives in September 2006 (Davis [70]):

> [...] Another big challenge for us is how to mix the internal data with external data; this is something that is still a big problem because I don't think anybody has a solution for that at the moment. [...] What can we do when we don't have enough data [...] How do I use a small amount of data when I can have external data with scenario generation? [...] I think it is one of the big challenges for operational risk managers at the moment.

[1] The original text is available free of charge on the BIS website www.BIS.org/bcbs/publ.htm

P. Shevchenko, *Modelling Operational Risk Using Bayesian Inference*,
DOI 10.1007/978-3-642-15923-7_4,

Under the Loss Distribution Approach (LDA), banks should quantify distributions for frequency and severity of operational losses for each risk cell (business line/event type) over a 1-year time horizon. The commonly used LDA model for the annual loss is a compound loss (2.1). In this chapter, we consider a single risk cell (business line/event type). As a reminder, the annual loss in a risk cell under the LDA model is

$$Z_t = \sum_{i=1}^{N_t} X_i(t). \tag{4.1}$$

Here: $t = 1, 2, \ldots, T, T + 1$ is discrete time in annual units and $T + 1$ refers to the next year; N_t is the annual number of events (frequency) modelled as a random variable from some discrete distribution (typically Poisson); and $X_i(t)$ are the severities of the events modelled as random variables from a continuous distribution. The case of many risks cells with dependence will be considered in Chap. 7.

Several studies, for example, Moscadelli [166] and Dutta and Perry [77], analysed operational risk data collected over many banks by Basel II business line and event type. While analyses of collective data may provide a picture for the whole banking industry, estimation of frequency and severity distributions of operational risks for each risk cell is a challenging task for a single bank, especially for low-frequency/high-severity losses. The banks internal data (usually truncated below approximately USD 20,000) are available typically over several years and contain few (or no) low-frequency/high-severity losses. The external data (losses experienced by other banks) are available through third party databases, but these are difficult to adapt directly to internal processes due to different volumes, thresholds, etc. Moreover, the data have a survival bias as typically the data of all collapsed companies are not available. It is difficult to estimate distributions using these data only.

It is also clear that estimation based on historical losses is backward looking and has limited ability to predict the future due to a constantly changing banking environment. For example, assume that a new policy was introduced in the bank, aiming to decrease the operational risk losses. Then it cannot be captured in the model based on the loss data only. As another example, assume that the annual intensity of risk events is 1/100. A bank started to collect data 2 years ago and by chance this risk event occurred within this period. Formally, applying the loss data approach, the intensity of this risk might be estimated as 1/2. This is clearly overestimated, yet it is important to take this event into account.

It is very important to have scenario analysis/expert judgments incorporated into the model. These judgments may provide valuable information for forecasting and decision making, especially for risk cells lacking internal loss data. In fact, it is mandatory to include scenario analysis into the model to meet the regulatory requirements. In the past, quantification of operational risk was based on such expert judgments only. Scenario analysis is a process undertaken by banks to identify risks, to analyse past events experienced both internally and by other banks (including near miss losses), and to consider current and planned controls. Usually, it involves workshops and templates to identify weaknesses, strengths and other factors. As a result some rough quantitative assessment of risk frequency and severity distributions is obtained from expert opinions. By itself, scenario analysis is very subjective and should be combined (supported) by the actual loss data analysis.

Bayesian inference is a statistical technique that can be used to incorporate expert opinions into data analysis and to combine different data sources. It is also a convenient method to account for parameter uncertainty, which is critical for low-frequency/high-severity operational risks with small datasets. The main concept and notation of this method have already been introduced in Sect. 2.9. This chapter focuses on description of this technique within the context of operational risk and provides several examples of its application for operational risk quantification. There is a broad literature covering Bayesian inference and its applications for the insurance industry as well as other areas. For a good introduction to the Bayesian inference method, see Berger [27] and Robert [200]; for the closely related methods of credibility theory, see Bühlmann and Gisler [44]. The method allows for structural modelling where expert opinions are incorporated into the analysis via specifying distributions (so-called prior distributions) for model parameters. These are updated by the data as they become available. Given new information (for example, new policy control is introduced), the expert may reassess the prior distributions to incorporate this information into a model.

The Bayesian inference methods, in the context of operational risk, have been briefly mentioned in the early literature. Books such as King ([134], chapter 12), Cruz ([65], chapter 10) and Panjer ([181], section 10.5) have short sections on a basic concept of Bayesian method. Bayesian method was implicitly used to estimate operational risk frequency in the working paper of Frachot and Roncalli [96]. However, the Bayesian methods have not really merged into operational risk literature as a recurrent research tool until approximately 2006. One of the first publications to present detailed and illustrative examples of the Bayesian inference methodology for estimation of the operational risk frequency and severity, and subsequent estimation of the capital, was a paper by Shevchenko and Wüthrich [218]. Then, an example of a "toy" model for operational risk, based on the closely related credibility theory, was presented in Bühlmann, Shevchenko and Wüthrich [45]. The Bayesian methodology was extended to combine three data sources (expert opinion, internal and external data) in Lambrigger, Shevchenko and Wüthrich [141]; and developed further in Peters, Shevchenko and Wüthrich [187] for a multivariate case with dependence between risks. Currently, the use of Bayesian methods for modelling operational risk is an active research line. This can be seen over the last few years, for example, in *The Journal of Operational Risk* available at www.journalofoperationalrisk.com. Also, there are several publications on the use of Bayesian belief networks for operational risk which are not discussed in this book; for references see Sect. 1.5.

This chapter demonstrates how to combine two data sources (either expert opinion and internal data or external data and internal data), which is a standard Bayesian method. Then, the methodology is extended to the model combining three data sources (internal data, external data and expert opinions) simultaneously. Next, methods of credibility theory are presented: these should be very useful if the data are so limited that no reliable quantification of prior distribution can be made. Finally, estimation of the annual loss distribution is described.

4.2 Combining Different Data Sources

Combining different sources of information is critical for estimation of operational risks, especially for low-frequency/high-severity risks. These data sources are multiple expert opinions, internal data, external data, and factors of business environment and control systems. Conceptually, the following ways have been proposed to process different data sources of information; see for example Berger ([27], sections 4.11 and 4.12):

- numerous ad-hoc procedures;
- Bayesian methods; and
- general non-probabilistic methods such as Dempster-Shafer theory.

Some of the ad-hoc procedures will be presented shortly and Bayesian methods are the main focus of this chapter. Dempster-Shafer theory is based on the so-called belief functions and *Dempster's rule* for combining evidence; see Dempster [73] and Shafer [212]. It is often referred to as a generalisation of Bayesian method. Closely related ideas of "probability-boxes" (referred to as *"p-boxes"*) attempt to model uncertainty by constructing the bounds on cumulative distribution functions. For a good summary on the methods for obtaining Dempster-Shafer structures and "p-boxes", and aggregation methods handling a conflict between the objects from different sources, see Ferson et al. [94]. Some writers consider this approach as unnecessary elaboration that can be handled within the Bayesian paradigm through Baysian robustness (section 4.7 in Berger [27]). These methods are attractive for operational risk (though they have not appeared in the operational risk literature yet) but will not be considered in this book and the reader is referred to the literature mentioned above.

4.2.1 Ad-hoc Combining

Often in practice, accounting for factors reflecting the business environment and internal control systems is achieved via scaling of data. Then ad-hoc procedures are used to combine internal data, external data and expert opinions. For example:

- Fit the severity distribution to the combined samples of internal and external data and fit the frequency distribution using internal data only.
- Estimate the Poisson annual intensity for the frequency distribution as $w\lambda_{int} + (1-w)\lambda_{ext}$, where the intensities λ_{ext} and λ_{int} are implied by the external and internal data respectively, using expert specified weight w.

- Estimate the severity distribution as a mixture

$$w_1 F_{SA}(x) + w_2 F_I(x) + (1 - w_1 - w_2) F_E(x),$$

where $F_{SA}(x)$, $F_I(x)$ and $F_E(x)$ are the distributions identified by scenario analysis, internal data and external data respectively, using expert specified weights w_1 and w_2.
- Apply the *minimum variance principle*, where the combined estimator is a linear combination of the individual estimators obtained from internal data, external data and expert opinion separately with the weights chosen to minimise the variance of the combined estimator.

Probably the easiest to use and most flexible procedure is the minimum variance principle. The rationale behind the principle is as follows. Consider two unbiased independent estimators $\widehat{\Theta}^{(1)}$ and $\widehat{\Theta}^{(2)}$ for parameter θ, i.e. $\mathrm{E}[\widehat{\Theta}^{(k)}] = \theta$ and $\mathrm{Var}[\widehat{\Theta}^{(k)}] = \sigma_k^2$, $k = 1, 2$. Then the combined unbiased linear estimator and its variance are

$$\widehat{\Theta}_{tot} = w_1 \widehat{\Theta}^{(1)} + w_2 \widehat{\Theta}^{(2)}, \quad w_1 + w_2 = 1 \tag{4.2}$$

$$\mathrm{Var}[\widehat{\Theta}_{tot}] = w_1^2 \sigma_1^2 + (1 - w_1)^2 \sigma_2^2. \tag{4.3}$$

It is easy to find the weights minimising $\mathrm{Var}[\widehat{\Theta}_{tot}]$:

$$w_1 = \frac{\sigma_2^2}{\sigma_1^2 + \sigma_2^2} \text{ and } w_2 = \frac{\sigma_1^2}{\sigma_1^2 + \sigma_2^2}.$$

The weights behave as expected in practice. In particular, $w_1 \to 1$ if $\sigma_1^2/\sigma_2^2 \to 0$ (σ_1^2/σ_2^2 is the uncertainty of the estimator $\widehat{\Theta}^{(1)}$ over the uncertainty of $\widehat{\Theta}^{(2)}$) and $w_1 \to 0$ if $\sigma_2^2/\sigma_1^2 \to 0$. This method can easily be extended to combine three or more estimators using the following theorem.

Theorem 4.1 (Minimum variance estimator) *Assume that we have* $\widehat{\Theta}^{(i)}$, $i = 1, 2, \ldots, K$ *unbiased and independent estimators of* θ *with variances* $\sigma_i^2 = \mathrm{Var}[\Theta^{(i)}]$. *Then the linear estimator*

$$\widehat{\Theta}_{tot} = w_1 \widehat{\Theta}^{(1)} + \cdots + w_K \widehat{\Theta}^{(K)},$$

is unbiased and has a minimum variance if

$$w_i = \frac{1/\sigma_i^2}{\sum_{k=1}^{K} (1/\sigma_k^2)}.$$

In this case, $w_1 + \cdots + w_K = 1$ *and*

$$\mathrm{Var}[\widehat{\Theta}_{tot}] = \left(\sum_{k=1}^{K} \frac{1}{\sigma_k^2} \right)^{-1}.$$

Proof See e.g. Lemma 3.4 in Wüthrich and Merz [240]. □

Heuristically, this can be applied to almost any quantity, including a distribution parameter or distribution characteristic such as mean, variance or quantile. The assumption that the estimators are unbiased estimators for θ is probably reasonable when combining estimators from different experts (or from expert and internal data). However, it is certainly questionable if applied to combine estimators from the external and internal data. The following sections focus on the Bayesian inference method that can be used to combine these data sources in a consistent statistical framework.

4.2.2 Example of Scenario Analysis

Expert opinions on potential losses and corresponding probabilities are often expressed using the following approaches:

- opinion on the distribution parameter;
- opinions on the number of losses with the amount to be within some ranges;
- separate opinions on the frequency of the losses and quantiles of the severity;
- opinion on how often the loss exceeding some level may occur.

Expert elicitation is certainly one of the challenges in operational risk because many managers and employees may not have a sound knowledge of statistics and probability theory. This may lead to misleading and misunderstanding. It is important that questions answered by experts are simple and well understood by respondents. There are psychological aspects involved. There is a vast literature on expert elicitation published by statisticians, especially in areas such as security and ecology. For a good review, see O'Hagan et al. [178].

However, published studies on the use of expert elicitation for operational risk LDA are scarce. Among the few are Frachot, Moudoulaud and Roncalli [95]; Alderweireld, Garcia and Léonard [6]; Steinhoff and Baule [222]; and Peters and Hübner [191]. These studies suggest that questions on *"how often the loss exceeding some level may occur"* are well understood by operational risk experts. Here, experts express the opinion that a loss of amount L or higher is expected to occur every d years. If there are M experts then we have M opinions $(L_1, d_1), \ldots, (L_M, d_M)$. These opinions can be used to fit assumed frequency and severity distributions. For example, assume that the frequency is modelled by $Poisson(\lambda)$ and severity is modelled by distribution $F(x|\boldsymbol{\theta})$. Then, the number of losses exceeding level L_i is distributed from $Poisson(\lambda(1-F(L_i|\boldsymbol{\theta})))$. That is, the expected number of losses exceeding L_i per year is $\widetilde{\lambda} = \lambda(1 - F(L_i|\boldsymbol{\theta}))$. This is typically interpreted that the loss exceeding L_i occurs (on average) every $1/\widetilde{\lambda}$ years or the expected duration between losses exceeding L_i is $1/\widetilde{\lambda}$. Then the parameters $(\lambda, \boldsymbol{\theta})$ can be estimated as

$$(\widehat{\lambda}, \widehat{\boldsymbol{\theta}}) = \arg\min_{\lambda,\boldsymbol{\theta}} \sum_{j=1}^{M} w_j \left(d_j - \frac{1}{\lambda\left(1 - F(L_j|\boldsymbol{\theta})\right)} \right)^2, \tag{4.4}$$

where w_j is the weight associated with the j-th opinion. The above-mentioned literature suggests to use a weight w_j equal to the inverse of the variance estimate of the duration between events exceeding L_j, i.e. $w_j = 1/d_j$. If the severity is assumed to be from a two-parameter distribution, then one can fit all three model parameters (frequency and severity) using three or more opinions. However, the above method does not allow for estimation of parameter uncertainty (prior distribution) if a Bayesian approach is undertaken. For the latter, it is important that experts specify not just the expected duration d_j, but also the uncertainty of their estimates. This will be discussed more in Sect. 4.3.1.

4.3 Bayesian Method to Combine Two Data Sources

The Bayesian inference method can be used to combine different data sources in a consistent statistical framework. The main concept of the Bayesian approach has already been introduced in Sect. 2.9. Now we consider the approach in detail.

Consider a random vector of data $\mathbf{X} = (X_1, X_2, \ldots, X_n)'$ whose joint density, for a given vector of parameters $\boldsymbol{\Theta} = (\Theta_1, \Theta_2, \ldots, \Theta_K)'$, is $h(\mathbf{x}|\boldsymbol{\theta})$. In the Bayesian approach, both observations and parameters are considered to be random. Then the joint density is

$$h(\mathbf{x}, \boldsymbol{\theta}) = h(\mathbf{x}|\boldsymbol{\theta})\pi(\boldsymbol{\theta}) = \pi(\boldsymbol{\theta}|\mathbf{x})h(\mathbf{x}), \tag{4.5}$$

where:

- $\pi(\boldsymbol{\theta})$ is the probability density of the parameters, a so-called prior density function. Typically, $\pi(\boldsymbol{\theta})$ depends on a set of further parameters that are called hyper-parameters, omitted here for simplicity of notation;
- $\pi(\boldsymbol{\theta}|\mathbf{x})$ is the density of parameters given data $\mathbf{X}$, a so-called posterior density;
- $h(\mathbf{x}, \boldsymbol{\theta})$ is the joint density of observed data and parameters;
- $h(\mathbf{x}|\boldsymbol{\theta})$ is the density of observations for given parameters. This is the same as a likelihood function if considered as a function of $\boldsymbol{\theta}$, i.e. $\ell_{\mathbf{x}}(\boldsymbol{\theta}) = h(\mathbf{x}|\boldsymbol{\theta})$;
- $h(\mathbf{x})$ is a marginal density of $\mathbf{X}$ that can be written as

$$h(\mathbf{x}) = \int h(\mathbf{x}|\boldsymbol{\theta})\pi(\boldsymbol{\theta})d\boldsymbol{\theta}. \tag{4.6}$$

For simplicity of notation, we consider continuous $\pi(\boldsymbol{\theta})$ only. If $\pi(\boldsymbol{\theta})$ is a discrete probability function, then the integration in the above expression should be replaced by a corresponding summation; see Definition 2.9.

Predictive distribution. The objective (in the context of operational risk) is to estimate the predictive distribution (frequency and severity) of a future observation X_{n+1} conditional on all available information $\mathbf{X} = (X_1, X_2, \ldots, X_n)$. Assume that conditionally, given $\boldsymbol{\Theta}$, X_{n+1} and $\mathbf{X}$ are independent, and X_{n+1} has a density $f(x_{n+1}|\boldsymbol{\theta})$. It is even common to assume that $X_1, X_2, \ldots, X_n, X_{n+1}$ are all conditionally independent (given $\boldsymbol{\Theta}$) and identically distributed. Then the conditional density of X_{n+1}, given data $\mathbf{X} = \mathbf{x}$, is

$$f(x_{n+1}|\mathbf{x}) = \int f(x_{n+1}|\boldsymbol{\theta})\pi(\boldsymbol{\theta}|\mathbf{x})d\boldsymbol{\theta}. \tag{4.7}$$

If X_{n+1} and $\mathbf{X}$ are not independent, then the predictive density should be written as

$$f(x_{n+1}|\mathbf{x}) = \int f(x_{n+1}|\boldsymbol{\theta}, \mathbf{x})\pi(\boldsymbol{\theta}|\mathbf{x})d\boldsymbol{\theta}. \tag{4.8}$$

Posterior distribution. Bayes's theorem (see Theorem 2.3) says that the posterior density can be calculated from (4.5) as

$$\pi(\boldsymbol{\theta}|\mathbf{x}) = h(\mathbf{x}|\boldsymbol{\theta})\pi(\boldsymbol{\theta})/h(\mathbf{x}). \tag{4.9}$$

Here, $h(\mathbf{x})$ plays the role of a normalisation constant. Thus the posterior distribution can be viewed as a product of a prior knowledge with a likelihood function for observed data.

In the context of operational risk, one can follow the following three logical steps:

- The prior $\pi(\boldsymbol{\theta})$ should be estimated by scenario analysis (expert opinions with reference to external data).
- Then the prior should be weighted with the observed data using formula (4.9) to get the posterior $\pi(\boldsymbol{\theta}|\mathbf{x})$.
- Formula (4.7) is then used to calculate the predictive density of X_{n+1} given the data $\mathbf{X}$.

Remark 4.1 Of course, the posterior density can be used to find parameter point estimators. Typically, these are the mean, mode or median of the posterior; see Sect. 2.9.3. The use of the posterior mean as the point parameter estimator is optimal in a sense that the mean square error of prediction is minimised. For more on this topic, see Sect. 2.10 or Bühlmann and Gisler ([44], section 2.3). However, in the case of operational risk, it is more appealing to use the whole posterior to calculate the predictive distribution (4.7).

The iterative update procedure for priors. If the data $X_1, X_2, \ldots, X_n$ are conditionally (given $\boldsymbol{\Theta} = \boldsymbol{\theta}$) independent and X_k is distributed with a density $f_k(\cdot|\boldsymbol{\theta})$, then the joint density of the data for given $\boldsymbol{\theta}$ can be written as $h(\mathbf{x}|\boldsymbol{\theta}) = \prod_{i=1}^{n} f_i(x_i|\boldsymbol{\theta})$. Denote the posterior density calculated after k observations as $\pi_k(\boldsymbol{\theta}|x_1,\ldots,x_k)$, then using (4.9), observe that

$$\pi_k(\boldsymbol{\theta}|x_1,\ldots,x_k) \propto \pi(\boldsymbol{\theta})\prod_{i=1}^{k} f_i(x_i|\boldsymbol{\theta}) \\ \propto \pi_{k-1}(\boldsymbol{\theta}|x_1,\ldots,x_{k-1})f_k(x_k|\boldsymbol{\theta}). \quad (4.10)$$

It is easy to see from (4.10), that the updating procedure which calculates the posteriors from priors can be done iteratively. Only the posterior distribution calculated after $k-1$ observations and the k-th observation are needed to calculate the posterior distribution after k observations. Thus the loss history over many years is not required, making the model easier to understand and manage, and allowing experts to adjust the priors at every step. Formally, the posterior distribution calculated after $k-1$ observations can be treated as a prior distribution for the k-th observation. In practice, initially, we start with the prior $\pi(\boldsymbol{\theta})$ identified by expert opinions and external data only. Then, the posterior $\pi(\boldsymbol{\theta}|\mathbf{x})$ is calculated, using (4.9), when actual data are observed. If there is a reason (for example, the new control policy introduced in a bank), then this posterior distribution can be adjusted by an expert and treated as the prior distribution for subsequent observations. Examples will be presented in the following sections.

Conjugate prior distributions. So-called conjugate distributions (see Definition 2.22) are very useful in practice when Bayesian inference is applied. Below we present conjugate pairs (Poisson-gamma, lognormal-normal, Pareto-gamma) that are good illustrative examples for modelling frequencies and severities in operational risk. Several other pairs (binomial-beta, gamma-gamma, exponential-gamma) can be found, for example, in Bühlmann and Gisler [44]. In all these cases, the prior and posterior distributions have the same type and the posterior distribution parameters are easily calculated using the prior distribution parameters and observations (or recursively using (4.10)).

4.3.1 Estimating Prior: Pure Bayesian Approach

In general, the structural parameters of the prior distributions can be estimated subjectively using expert opinions (*pure Bayesian approach*) or using data (*empirical Bayesian approach*). In a pure Bayesian approach, the prior distribution is specified subjectively (that is, in the context of operational risk, using expert opinions). Berger [27] lists several methods.

- *Histogram approach*: split the space of the parameter $\boldsymbol{\theta}$ into intervals and specify the subjective probability for each interval. From this, the smooth density of the prior distribution can be determined.
- *Relative Likelihood Approach*: compare the intuitive likelihoods of the different values of $\boldsymbol{\theta}$. Again, the smooth density of prior distribution can be determined. It is difficult to apply this method in the case of unbounded parameters.

- *CDF determinations*: subjectively construct the distribution function for the prior and sketch a smooth curve.
- *Matching a Given Functional Form*: find the prior distribution parameters assuming some functional form for the prior distribution to match prior beliefs (on the moments, quantiles, etc) as close as possible.

In this chapter, the method of matching a given functional form will be used often. The use of a particular method is determined by a specific problem and expert experience. Usually, if the expected values for the quantiles (or mean) and their uncertainties are estimated by the expert then it is possible to fit the priors.

Often, expert opinions are specified for some quantities such as quantiles or other risk characteristics rather than for the parameters directly. In this case it might be better to assume some priors for these quantities that will imply a prior for the parameters. In general, given model parameters $\boldsymbol{\theta} = (\theta_1, \ldots, \theta_n)$, assume that there are risk characteristics $d_i = g_i(\boldsymbol{\theta})$, $i = 1, 2, \ldots, n$ that are well understood by experts. These could be some quantiles, expected values, expected durations between losses exceeding high thresholds, etc. Now, if experts specify the joint prior $\pi(d_1, \ldots, d_n)$, then using transformation method the prior for $\theta_1, \ldots, \theta_n$ is

$$\pi(\boldsymbol{\theta}) = \pi(g_1(\boldsymbol{\theta}), \ldots, g_n(\boldsymbol{\theta})) \left| \frac{\partial\,(g_1(\boldsymbol{\theta}), \ldots, g_n(\boldsymbol{\theta}))}{\partial\,(\theta_1, \ldots, \theta_n)} \right|, \tag{4.11}$$

where $|\partial\,(g_1(\boldsymbol{\theta}), \ldots, g_n(\boldsymbol{\theta}))/\partial\,(\theta_1, \ldots, \theta_n)|$ is the Jacobian determinant of the transformation. Essentially, the main difficulty in specifying a joint prior is due to a possible dependence between the parameters. It is convenient to choose the characteristics (for specification of the prior) such that independence can be assumed. For example, if the prior for the quantiles $q_1, \ldots, q_n$ (corresponding to probability levels $p_1 < p_2 < \cdots < p_n$) is to be specified, then to account for the ordering it might be better to consider the differences

$$d_1 = q_1, d_2 = q_2 - q_1, \ldots, d_n = q_n - q_{n-1}.$$

Then, it is reasonable to assume independence between these differences and impose constraints $d_i > 0$, $i = 2, \ldots, n$. If experts specify the marginal priors $\pi(d_1), \pi(d_2), \ldots, \pi(d_n)$ (e.g. gamma priors) then the full joint prior is

$$\pi(d_1, \ldots, d_n) = \pi(d_1) \times \pi(d_2) \times \cdots \times \pi(d_n)$$

and the prior for parameters $\boldsymbol{\theta}$ is calculated by transformation using (4.11). To specify the i-th prior $\pi(d_i)$, an expert may use the approaches listed above. For example, if $\pi(d_i)$ is $Gamma(\alpha_i, \beta_i)$, then the expert may provide the mean and variational coefficient for $\pi(d_i)$ (or median and 0.95 quantile) that should be enough to determine α_i and β_i.

A very appealing example demonstrating the importance of subjective prior information was given in Savage [208]:

1. A lady, who adds milk to her tea, claims to be able to tell whether the tea or milk was poured into the cup first. In all ten trials, her answer is correct.
2. A music expert claims to be able to distinguish a page of Haydn score from a page of Mozart score. In all ten trials, he makes a correct determination.
3. A drunken friend says that he can predict the outcome of a coin flip. In all ten trials, his prediction is correct.

In all three cases, the unknown parameter to identify is the probability of the correct answer. Classical statistical approach (based on hypothesis testing) ignoring our prior opinion would give a very strong evidence that all these claims are correct. We would not doubt this result for situation 2. However, for situation 3, our prior opinion is that this prediction is impossible and we would tend to ignore the empirical evidence. Different people may give a different prior opinion for situation 1. Anyway, in all these cases, prior information is certainly valuable.

4.3.2 Estimating Prior: Empirical Bayesian Approach

Under empirical Bayesian approach, the parameter $\boldsymbol{\theta}$ is treated as a random sample from the prior distribution. Then using collective data of *similar* risks, the parameters of the prior are estimated using a marginal distribution of observations. Depending on the model setup, the data can be collective industry data, collective data in the bank, etc.

To explain, consider K similar risks where each risk has own risk profile $\boldsymbol{\Theta}^{(i)}$, $i = 1, \ldots, K$; see Fig. 4.1. Given $\boldsymbol{\Theta}^{(i)} = \boldsymbol{\theta}^{(i)}$, the risk data $X_1^{(i)}, X_2^{(i)}, \ldots$ are generated from the distribution $F(x|\boldsymbol{\theta}^{(i)})$. The risks are different having different risk profiles $\boldsymbol{\theta}^{(i)}$, but what they have in common is that $\boldsymbol{\Theta}^{(1)}, \ldots, \boldsymbol{\Theta}^{(K)}$ are distributed from the same density $\pi(\boldsymbol{\theta})$. Then, one can find the unconditional distribution of the data $\mathbf{X}$ and fit the prior distribution using all data (across all similar risks). This could be done, for example, by the maximum likelihood method or the method of moments or even empirically. This approach will be discussed in detail in Sect. 4.4.

4.3.3 Poisson Frequency

Consider the annual number of events for a risk in one bank in year t modelled as a random variable from the Poisson distribution $Poisson(\lambda)$. The intensity parameter λ is not known and the Bayesian approach models it as a random variable Λ. Then the following model for years $t = 1, 2, \ldots, T, T+1$ (where $T+1$ corresponds to the next year) can be considered.

Model Assumptions 4.1

- *Suppose that, given $\Lambda = \lambda$, the data $N_1, \ldots, N_{T+1}$ are independent random variables from the Poisson distribution, $Poisson(\lambda)$:*

$$\Pr[N_t = n|\lambda] = e^{-\lambda}\frac{\lambda^n}{n!}, \quad \lambda \geq 0. \tag{4.12}$$

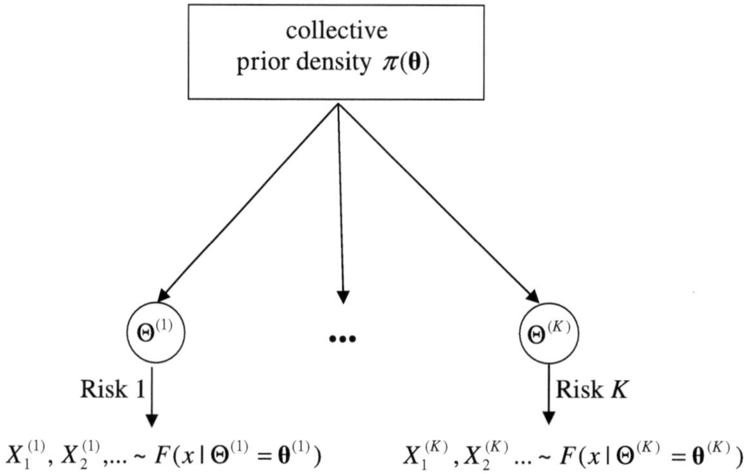

Fig. 4.1 Interpretation of the prior density $\pi(\boldsymbol{\theta})$ using empirical Bayes approach. $\boldsymbol{\Theta}^{(i)}$ is the risk profile of the i-th risk. Given $\boldsymbol{\Theta}^{(i)} = \boldsymbol{\theta}^{(i)}$, the risk data $X_1^{(i)}, X_2^{(i)}, \ldots$ are generated from the distribution $F(x|\boldsymbol{\theta}^{(i)})$. The risks are different having different risk profiles $\boldsymbol{\theta}^{(i)}$, but what they have in common is that $\boldsymbol{\Theta}^{(1)}, \ldots, \boldsymbol{\Theta}^{(K)}$ are distributed from the same density $\pi(\boldsymbol{\theta})$

- *The prior distribution for Λ is a gamma distribution, $Gamma(\alpha, \beta)$, with a density*

$$\pi(\lambda) = \frac{(\lambda/\beta)^{\alpha-1}}{\Gamma(\alpha)\beta} \exp(-\lambda/\beta), \quad \lambda > 0, \alpha > 0, \beta > 0. \tag{4.13}$$

That is, λ plays the role of $\boldsymbol{\theta}$ and $\mathbf{N} = (N_1, \ldots, N_T)'$ the role of $\mathbf{X}$ in (4.9).

Posterior. Given $\Lambda = \lambda$, under the Model Assumptions 4.1, $N_1, \ldots, N_T$ are independent and their joint density, at $\mathbf{N} = \mathbf{n}$, is given by

$$h(\mathbf{n}|\lambda) = \prod_{i=1}^{T} e^{-\lambda} \frac{\lambda^{n_i}}{n_i!}. \tag{4.14}$$

Thus, using formula (4.9), the posterior density is

$$\pi(\lambda|\mathbf{n}) \propto \frac{(\lambda/\beta)^{\alpha-1}}{\Gamma(\alpha)\beta} \exp(-\lambda/\beta) \prod_{i=1}^{T} e^{-\lambda} \frac{\lambda^{n_i}}{n_i!} \propto \lambda^{\alpha_T - 1} \exp(-\lambda/\beta_T), \tag{4.15}$$

which is $Gamma(\alpha_T, \beta_T)$, i.e. the same as the prior distribution with updated parameters α_T and β_T given by:

$$\alpha \to \alpha_T = \alpha + \sum_{i=1}^{T} n_i, \quad \beta \to \beta_T = \frac{\beta}{1 + \beta \times T}. \tag{4.16}$$

Improper constant prior. It is easy to see that, if the prior is constant (improper prior), i.e. $\pi(\lambda|\mathbf{n}) \propto h(\mathbf{n}|\lambda)$, then the posterior is $Gamma(\alpha_T, \beta_T)$ with

$$\alpha_T = 1 + \sum_{i=1}^{T} n_i, \quad \beta_T = \frac{1}{T}. \tag{4.17}$$

In this case, the mode of the posterior $\pi(\lambda|\mathbf{n})$ is

$$\widehat{\lambda}_T^{\text{MAP}} = (\alpha_T - 1)\beta_T = \frac{1}{T}\sum_{i=1}^{T} n_i, \tag{4.18}$$

which is the same as the maximum likelihood estimate (MLE) $\widehat{\lambda}_T^{\text{MLE}}$ of λ.

Predictive distribution. Given data, the full predictive distribution for N_{T+1} is negative binomial, $NegBin(\alpha_T, 1/(1 + \beta_T))$:

$$\begin{aligned}
\Pr[N_{T+1} = m|\mathbf{N} = \mathbf{n}] &= \int f(m|\lambda)\pi(\lambda|\mathbf{n})d\lambda \\
&= \int e^{-\lambda}\frac{\lambda^m}{m!}\frac{\lambda^{\alpha_T - 1}}{(\beta_T)^{\alpha_T}\Gamma(\alpha_T)}e^{-\lambda/\beta_T}d\lambda \\
&= \frac{(\beta_T)^{-\alpha_T}}{\Gamma(\alpha_T)m!}\int e^{-(1+1/\beta_T)\lambda}\lambda^{\alpha_T + m - 1}d\lambda \\
&= \frac{\Gamma(\alpha_T + m)}{\Gamma(\alpha_T)m!}\left(\frac{1}{1 + \beta_T}\right)^{\alpha_T}\left(\frac{\beta_T}{1 + \beta_T}\right)^m.
\end{aligned} \tag{4.19}$$

It is assumed that given $\Lambda = \lambda$, N_{T+1} and $\mathbf{N}$ are independent. The expected number of events over the next year, given past observations, $\mathrm{E}[N_{T+1}|\mathbf{N}]$, i.e. mean of $NegBin(\alpha_T, 1/(1 + \beta_T))$ (which is also a mean of the posterior distribution in this case), allows for a good interpretation as follows:

$$\begin{aligned}
\mathrm{E}[N_{T+1}|\mathbf{N} = \mathbf{n}] = \mathrm{E}[\lambda|\mathbf{N} = \mathbf{n}] = \alpha_T\beta_T &= \beta\frac{\alpha + \sum_{i=1}^{T} n_i}{1 + \beta \times T} \\
&= w_T\widehat{\lambda}_T^{\text{MLE}} + (1 - w_T)\lambda_0.
\end{aligned} \tag{4.20}$$

Here:

- $\widehat{\lambda}_T^{\text{MLE}} = \frac{1}{T}\sum_{i=1}^{T} n_i$ is the estimate of λ using the observed counts only;
- $\lambda_0 = \alpha\beta$ is the estimate of λ using a prior distribution only (e.g. specified by expert); and
- $w_T = \frac{T\beta}{T\beta+1}$ is the credibility weight in [0,1) used to combine λ_0 and $\widehat{\lambda}_T^{\text{MLE}}$.

Remark 4.2

- As the number of observed years T increases, the credibility weight w_T increases and vice versa. That is, the more observations we have, the greater credibility weight we assign to the estimator based on the observed counts, while the lesser credibility weight is attached to the expert opinion estimate. Also, the larger the volatility of the expert opinion (larger β), the greater credibility weight is assigned to observations.
- Recursive calculation of the posterior distribution is very simple. That is, consider observed annual counts $n_1, n_2, \ldots, n_k, \ldots$, where n_k is the number of events in the k-th year. Assume that the prior $Gamma(\alpha, \beta)$ is specified initially, then the posterior $\pi(\lambda|n_1, \ldots, n_k)$ after the k-th year is a gamma distribution, $Gamma(\alpha_k, \beta_k)$, with $\alpha_k = \alpha + \sum_{i=1}^{k} n_i$ and $\beta_k = \beta/(1+\beta \times k)$. Observe that,

$$\alpha_k = \alpha_{k-1} + n_k, \quad \beta_k = \frac{\beta_{k-1}}{1+\beta_{k-1}}. \tag{4.21}$$

This leads to a very efficient recursive scheme, where the calculation of posterior distribution parameters is based on the most recent observation and parameters of posterior distribution calculated just before this observation.

Estimating prior. Suppose that the annual frequency of the operational risk losses N is modelled by the Poisson distribution, $Poisson(\Lambda = \lambda)$, and the prior density $\pi(\lambda)$ for Λ is $Gamma(\alpha, \beta)$. Then, $\mathrm{E}[N|\Lambda] = \Lambda$ and $\mathrm{E}[\Lambda] = \alpha \times \beta$. The expert may estimate the expected number of events but cannot be certain in the estimate. One could say that the expert's "best" estimate for the expected number of events corresponds to $\mathrm{E}[\mathrm{E}[N|\Lambda]] = \mathrm{E}[\Lambda]$. If the expert specifies $\mathrm{E}[\Lambda]$ and an uncertainty that the "true" λ for next year is within the interval $[a,b]$ with a probability $\Pr[a \le \Lambda \le b] = p$ (it may be convenient to set $p = 2/3$), then the equations

$$\begin{aligned} &\mathrm{E}[\Lambda] = \alpha \times \beta, \\ &\Pr[a \le \Lambda \le b] = p = \int_a^b \pi(\lambda|\alpha, \beta) d\lambda = F_{\alpha,\beta}^{(G)}(b) - F_{\alpha,\beta}^{(G)}(a) \end{aligned} \tag{4.22}$$

can be solved numerically to estimate the structural parameters α and β. Here, $F_{\alpha,\beta}^{(G)}(\cdot)$ is the gamma distribution, $Gamma(\alpha, \beta)$, i.e.

$$F_{\alpha,\beta}^{(G)}[y] = \int_0^y \frac{x^{\alpha-1}}{\Gamma(\alpha)\beta^{\alpha}} \exp\left(-\frac{x}{\beta}\right) dx.$$

In the insurance industry, the uncertainty for the "true" λ is often measured in terms of the coefficient of variation, $\mathrm{Vco}[\Lambda] = \sqrt{\mathrm{Var}[\Lambda]}/\mathrm{E}[\Lambda]$. Given the expert estimates for $\mathrm{E}[\Lambda] = \alpha\beta$ and $\mathrm{Vco}[\Lambda] = 1/\sqrt{\alpha}$, the structural parameters α and β are easily estimated.

Example 4.1 If the expert specifies $\mathrm{E}[\Lambda] = 0.5$ and $\Pr[0.25 \leq \Lambda \leq 0.75] = 2/3$, then we can fit a prior distribution $Gamma(\alpha \approx 3.407,\ \beta \approx 0.147)$ by solving (4.22). Assume now that the bank experienced no losses over the first year (after the prior distribution was estimated). Then, using formulas (4.21), the posterior distribution parameters are $\widehat{\alpha}_1 \approx 3.407 + 0 = 3.407$, $\widehat{\beta}_1 \approx 0.147/(1+0.147) \approx 0.128$ and the estimated arrival rate using the posterior distribution is $\widehat{\lambda}_1 = \widehat{\alpha}_1 \times \widehat{\beta}_1 \approx 0.436$. If during the next year no losses are observed again, then the posterior distribution parameters are $\widehat{\alpha}_2 = \widehat{\alpha}_1 + 0 \approx 3.407$, $\widehat{\beta}_2 = \widehat{\beta}_1/(1+\widehat{\beta}_1) \approx 0.113$ and $\widehat{\lambda}_2 = \widehat{\alpha}_2 \times \widehat{\beta}_2 \approx 0.385$. Subsequent observations will update the arrival rate estimator correspondingly using formulas (4.21). Thus, starting from the expert specified prior, observations regularly update (refine) the posterior distribution. The expert might reassess the posterior distribution at any point in time (the posterior distribution can be treated as a prior distribution for the next period), if new practices/policies were introduced in the bank that affect the frequency of the loss. That is, if we have a new policy at time k, the expert may reassess the parameters and replace $\widehat{\alpha}_k$ and $\widehat{\beta}_k$ by $\widehat{\alpha}_k^*$ and $\widehat{\beta}_k^*$ respectively.

In Fig. 4.2, we show the posterior best estimate for the arrival rate $\widehat{\lambda}_k = \widehat{\alpha}_k \times \widehat{\beta}_k$, $k = 1, \ldots, 15$ (with the prior distribution as in the above example), when the annual number of events N_k, $k = 1, \ldots, 15$ are simulated from $Poisson(\lambda = 0.6)$ and are given in Table 4.1.

On the same figure, we show the standard maximum likelihood estimate of the arrival rate $\widehat{\lambda}_k^{\mathrm{MLE}} = \frac{1}{k}\sum_{i=1}^{k} n_i$. After approximately 8 years, the estimators are very close to each other. However, for a small number of observed years, the Bayesian estimate is more accurate as it takes the prior information into account. Only after 12 years do both estimators converge to the true value of 0.6 (this is because the bank

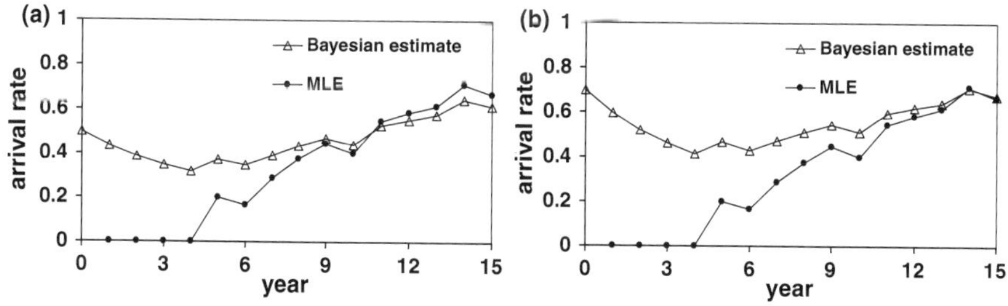

Fig. 4.2 The Bayesian and the standard maximum likelihood estimates of the arrival rate vs the observation year. The Bayesian estimate is a mean of the posterior distribution when the prior distribution is Gamma with: **(a)** $\mathrm{E}[\Lambda] = 0.5$; $\alpha \approx 3.41$ and $\beta \approx 0.15$; **(b)** $\mathrm{E}[\Lambda] = 0.7$ and $\mathrm{Vco}[\Lambda] = 0.5$; $\alpha = 4$ and $\beta = 0.175$. The maximum likelihood estimate is a simple average over the number of observed events. The annual counts were sampled from the Poisson(0.6) and are given in Table 4.1. See Example 4.1 for details

Table 4.1 The annual number of losses simulated from $Poisson(0.6)$

i	1	2	3	4	5	6	7	8	9	10	11	12	13	14	15
n_i	0	0	0	0	1	0	1	1	1	0	2	1	1	2	0

was very lucky to have no events during the first 4 years). Note that for this example we assumed the prior distribution with a mean equal to 0.5, which is different from the true arrival rate. Thus this example shows that an initially incorrect prior estimator is corrected by the observations as they become available. It is interesting to observe that, in year 14, the estimators become slightly different again. This is because the bank was unlucky to experience event counts (1, 1, 2) in the years (12, 13, 14). As a result, the maximum likelihood estimate becomes higher than the true value, while the Bayesian estimate is more stable (smooth) with respect to the unlucky years. If this example is repeated with different sequences of random numbers, then one would observe quite different maximum likelihood estimates (for small k) and more stable Bayesian estimates.

4.3.4 The Lognormal $\mathcal{LN}(\mu, \sigma)$ Severity with Unknown μ

Assume that the loss severity for a risk in one bank is modelled as a random variable from a lognormal distribution, $\mathcal{LN}(\mu, \sigma)$, whose density is

$$f(x|\mu, \sigma) = \frac{1}{x\sqrt{2\pi\sigma^2}} \exp\left(-\frac{(\ln x - \mu)^2}{2\sigma^2}\right). \tag{4.23}$$

This distribution often gives a good fit for operational loss data. Also, it belongs to a class of heavy-tailed distributions that will be considered in Chap. 6. The parameters μ and σ are not known and the Bayesian approach models these as a random variables Θ_μ and Θ_σ respectively. We assume that the losses over the years $t = 1, 2, \ldots, T$ are observed and should be modelled for next year $T + 1$. To simplify notation, we denote the losses over past T years as $X_1, \ldots, X_n$ and the future losses are $X_{n+1}, \ldots$. Then the model can be structured as follows. For simplicity, below we assume that σ is known and μ is unknown. The case where both σ and μ are unknown will be treated in Sect. 4.3.5.

Model Assumptions 4.2

- *Suppose that, given σ and $\Theta_\mu = \mu$, the data $X_1, \ldots, X_n, \ldots$ are independent random variables from $\mathcal{LN}(\mu, \sigma)$. That is, $Y_i = \ln X_i$, $i = 1, 2, \ldots$ are distributed from the normal distribution $\mathcal{N}(\mu, \sigma)$.*
- *Assume that parameter σ is known and the prior distribution for Θ_μ is the normal distribution, $\mathcal{N}(\mu_0, \sigma_0)$. That is the prior density is*

$$\pi(\mu) = \frac{1}{\sigma_0\sqrt{2\pi}} \exp\left(-\frac{(\mu - \mu_0)^2}{2\sigma_0^2}\right). \tag{4.24}$$

Denote the losses over past years as $\mathbf{X} = (X_1, \ldots, X_n)'$ and corresponding log-losses as $\mathbf{Y} = (Y_1, \ldots, Y_n)'$.

Remark 4.3 That is, μ plays the role of $\boldsymbol{\theta}$ in (4.9). The case of a conjugate joint prior for both μ and σ unknown is considered in Sect. 4.3.5.

Posterior. Under the above assumptions, the joint density of the data over past years (conditional on σ and $\Theta_\mu = \mu$) at position $\mathbf{Y} = \mathbf{y}$ is

$$h(\mathbf{y}|\mu, \sigma) = \prod_{i=1}^{n} \frac{1}{\sigma\sqrt{2\pi}} \exp\left(-\frac{(y_i-\mu)^2}{2\sigma^2}\right). \tag{4.25}$$

Then, using formula (4.9), the posterior density can be written as

$$\pi(\mu|\mathbf{y}) \propto \frac{\exp\left(-\frac{(\mu-\mu_0)^2}{2\sigma_0^2}\right)}{\sigma_0\sqrt{2\pi}} \prod_{i=1}^{n} \frac{\exp\left(-\frac{(y_i-\mu)^2}{2\sigma^2}\right)}{\sigma\sqrt{2\pi}}$$
$$\propto \exp\left(-\frac{(\mu-\mu_{0,n})^2}{2\sigma_{0,n}^2}\right), \tag{4.26}$$

that corresponds to a normal distribution, $\mathcal{N}(\mu_{0,n}, \sigma_{0,n})$, i.e. the same as the prior distribution with updated parameters

$$\mu_0 \to \mu_{0,n} = \frac{\mu_0 + \omega \sum_{i=1}^{n} y_i}{1 + n \times \omega}, \tag{4.27}$$

$$\sigma_0^2 \to \sigma_{0,n}^2 = \frac{\sigma_0^2}{1 + n \times \omega}, \quad \text{where } \omega = \sigma_0^2/\sigma^2. \tag{4.28}$$

The expected value of Y_{n+1} (given past observations), $\mathrm{E}[Y_{n+1}|\mathbf{Y} = \mathbf{y}]$, allows for a good interpretation, as follows:

$$\mathrm{E}[Y_{n+1}|\mathbf{Y} = \mathbf{y}] = \mathrm{E}[\Theta_\mu|\mathbf{Y} = \mathbf{y}] = \mu_{0,n} = \frac{\mu_0 + \omega \sum_{i=1}^{n} y_i}{1 + n \times \omega}$$
$$= w_n \overline{y}_n + (1 - w_n)\mu_0, \tag{4.29}$$

where

- $\overline{y}_n = \frac{1}{n} \sum_{i=1}^{n} y_i$ is the estimate of μ using the observed losses only;
- μ_0 is the estimate of μ using a prior distribution only (e.g. specified by expert);
- $w_n = \frac{n}{n+\sigma^2/\sigma_0^2}$ is the credibility weight in [0,1) used to combine μ_0 and $\overline{y}_n$.

Remark 4.4

- As the number of observations increases, the credibility weight w increases and vice versa. That is, the more observations we have the greater weight we assign to the estimator based on the observed counts and the lesser weight is attached to the expert opinion estimate. Also, larger uncertainty in the expert opinion σ_0^2 leads to a higher credibility weight for observations and larger volatility of observations σ^2 leads to a higher credibility weight for expert opinions.

- The posterior distribution can be calculated recursively as follows. Consider the data $Y_1, Y_2, \ldots, Y_k, \ldots$. Assume that the prior distribution, $\mathcal{N}(\mu_0, \sigma_0)$, is specified initially, then the posterior density $\pi(\mu|y_1, \ldots, y_k)$ after the k-th event is the normal distribution $\mathcal{N}(\mu_{0,k}, \sigma_{0,k})$ with

$$\mu_{0,k} = \frac{\mu_0 + \omega \sum_{i=1}^{k} y_i}{1 + k \times \omega}, \quad \sigma_{0,k}^2 = \frac{\sigma_0^2}{1 + k \times \omega},$$

where $\omega = \sigma_0^2/\sigma^2$. It is easy to show that

$$\mu_{0,k} = \frac{\mu_{0,k-1} + \omega_{k-1} y_k}{1 + \omega_{k-1}}, \quad \sigma_{0,k}^2 = \frac{\sigma^2 \omega_{k-1}}{1 + \omega_{k-1}} \tag{4.30}$$

with $\omega_{k-1} = \sigma_{0,k-1}^2/\sigma^2$. That is, calculation of the posterior distribution parameters can be based on the most recent observation and the parameters of the posterior distribution calculated just before this observation.

Estimating prior. Suppose that X, the severity of operational losses, is modelled by the lognormal distribution, $\mathcal{LN}(\mu, \sigma)$ and Model Assumptions 4.2 are satisfied. Then, for given Θ_μ (and σ is known), the expected loss is

$$\Omega = \mathrm{E}[X|\Theta_\mu] = \exp\left(\Theta_\mu + \frac{1}{2}\sigma^2\right) \tag{4.31}$$

and the quantile at level q is

$$Q_q = \exp(\Theta_\mu + \sigma z_q), \tag{4.32}$$

where $z_q = F_N^{-1}(q)$ is the inverse of the standard normal distribution. That is Ω and Q_q are functions of Θ_μ.

Consider the case when the prior distribution for Θ_μ is $\mathcal{N}(\mu_0, \sigma_0)$. In this case, unconditionally, Ω is distributed from $\mathcal{LN}\left(\mu_0 + \frac{1}{2}\sigma^2, \sigma_0\right)$ and the quantile Q_q is distributed from $\mathcal{LN}(\mu_0 + \sigma z_q, \sigma_0)$. Then, the expert may specify 'the best' estimate of the expected loss $\mathrm{E}[\Omega]$ and uncertainty, that is, the interval $[a, b]$ such that the true expected loss is within the interval with a probability $p = \Pr[a \le \Omega \le b]$. Then, the equations

$$p = \Pr[a \le \Omega \le b] = F_N\left(\frac{\ln b - \frac{1}{2}\sigma^2 - \mu_0}{\sigma_0}\right) - F_N\left(\frac{\ln a - \frac{1}{2}\sigma^2 - \mu_0}{\sigma_0}\right),$$

$$\mathrm{E}[\Omega] = \exp\left(\mu_0 + \frac{1}{2}\sigma^2 + \frac{1}{2}\sigma_0^2\right) \tag{4.33}$$

can be solved to find μ_0, σ_0. Here, $F_N(\cdot)$ is the standard normal distribution.

Example 4.2 For example, assume that $\sigma = 2$ and the expert estimates are $\mathrm{E}[\Omega] = 10$ and $p = \Pr[8 \le \Omega \le 12] = 2/3$. Then, solving (4.33) gives $\mu_0 \approx 0.28$ and $\sigma_0 \approx 0.21$. Finally, using (4.27) we can calculate the posterior parameters $\mu_{0,k}, \sigma_{0,k}$ as observations $X_k, k = 1, 2, \ldots$ become available.

One can also try to fit parameters μ_0 and σ_0 using estimates for some quantile and uncertainty by solving

$$p = \Pr[a \le Q_q \le b] = F_N\left(\frac{\ln b - \sigma z_q - \mu_0}{\sigma_0}\right) - F_N\left(\frac{\ln a - \sigma z_q - \mu_0}{\sigma_0}\right),$$

$$\mathrm{E}[Q_q] = \exp\left(\mu_0 + \sigma z_q + \frac{1}{2}\sigma_0^2\right). \tag{4.34}$$

Remark 4.5 If the uncertainty for Ω or Q_q in (4.33) and (4.34) is measured using the coefficient of variation $\mathrm{Vco}[X] = \sqrt{\mathrm{Var}[X]}/\mathrm{E}[X]$, then μ_0, σ_0 are easily expressed in the closed form. In the insurance industry Vco is often provided by regulators.

4.3.5 The Lognormal $\mathcal{LN}(\mu, \sigma)$ Severity with Unknown μ and σ

As in the previous section, assume that the loss severity for a risk in one bank is modelled as a random variable from a lognormal distribution, $\mathcal{LN}(\mu, \sigma)$. However, now consider the case of both μ and σ unknown and modelled by random variables Θ_μ and Θ_σ respectively.

Model Assumptions 4.3

- *Suppose that, given $\Theta_\mu = \mu$ and $\Theta_\sigma = \sigma$, the data $X_1, \ldots, X_n, \ldots$ are independent random variables from $\mathcal{LN}(\mu, \sigma)$, i.e. $Y_i = \ln X_i \sim \mathcal{N}(\mu, \sigma)$.*
- *Assume that the prior distribution of Θ_σ^2 is the inverse Chi-squared distribution, $InvChiSq(\nu, \beta)$, and the prior distribution of Θ_μ (given $\Theta_\sigma = \sigma$) is $\mathcal{N}(\theta, \sigma/\sqrt{\phi})$, i.e. the corresponding densities are:*

$$\pi(\sigma^2) = \frac{(\sigma^2/\beta)^{-1-\nu/2}}{\Gamma(\nu/2)\beta 2^{\nu/2}} \exp\left(-\frac{\beta}{2\sigma^2}\right), \tag{4.35}$$

$$\pi(\mu|\sigma^2) = \frac{1}{\sqrt{2\pi\sigma^2/\phi}} \exp\left(-\frac{(\mu-\theta)^2}{2\sigma^2/\phi}\right). \tag{4.36}$$

Denote the losses over past years as $\mathbf{X} = (X_1, \ldots, X_n)'$ and corresponding log-losses as $\mathbf{Y} = (Y_1, \ldots, Y_n)'$.

Posterior. Under the above assumptions, the joint prior density is

$$\begin{aligned}\pi(\mu,\sigma^2) &= \frac{1}{\sqrt{2\pi\sigma^2/\phi}}\exp\left(-\frac{(\mu-\theta)^2}{2\sigma^2/\phi}-\frac{\beta}{2\sigma^2}\right)\times\frac{2^{-\nu/2}}{\beta\Gamma(\nu/2)}\left(\frac{\sigma^2}{\beta}\right)^{-\frac{\nu}{2}-1}\\ &\propto(\sigma^2)^{-\frac{\nu+1}{2}-1}\exp\left(-\frac{1}{2\sigma^2}(\beta+\phi(\mu-\theta)^2)\right).\end{aligned}\tag{4.37}$$

It is easy to see that the marginal prior distribution of Θ_μ is shifted t-distribution with ν degrees of freedom:

$$\begin{aligned}\pi(\mu) = \int\pi(\mu,\sigma^2)d\sigma^2 &\propto \int x^{-\frac{\nu+1}{2}-1}\exp\left(-\frac{1}{2x}[\beta+\phi(\mu-\theta)^2]\right)dx\\ &\propto\int y^{\frac{\nu+1}{2}-1}\exp\left(-\frac{y}{2}[\beta+\phi(\mu-\theta)^2]\right)dy\\ &\propto[\beta+\phi(\mu-\theta)^2]^{-\frac{\nu+1}{2}}\int z^{\frac{\nu+1}{2}-1}\exp(-z)dz\\ &\propto\left(1+\frac{\phi\nu(\mu-\theta)^2}{\nu\beta}\right)^{-\frac{\nu+1}{2}}.\end{aligned}\tag{4.38}$$

Denote $\Psi_\sigma = (\sigma^2)^{-\frac{\nu+1+n}{2}-1}$, $\overline{y}=\frac{1}{n}\sum_{i=1}^{n}y_i$ and $\overline{y^2}=\frac{1}{n}\sum_{i=1}^{n}y_i^2$. Then, given $\mathbf{Y}=\mathbf{y}$, the joint posterior density

$$\begin{aligned}\pi(\mu,\sigma^2|\mathbf{y}) &\propto \Psi_\sigma\exp\left(-\frac{1}{2\sigma^2}\left(\beta+\phi(\mu-\theta)^2+\sum_{i=1}^{n}(y_i-\mu)^2\right)\right)\\ &\propto\Psi_\sigma\exp\left(-\frac{1}{2\sigma^2}\left(\beta+(\phi+n)\mu^2+\phi\theta^2-2\mu(\phi\theta+n\overline{y})+n\overline{y^2}\right)\right)\\ &\propto\Psi_\sigma\exp\left(-\frac{1}{2\sigma^2}\left(\beta+\phi\theta^2+n\overline{y^2}-\frac{(\phi\theta+n\overline{y})^2}{\phi+n}\right.\right.\\ &\quad\left.\left.+(\phi+n)\left(\mu-\frac{\phi\theta+n\overline{y}}{\phi+n}\right)^2\right)\right)\\ &\propto(\sigma^2)^{-\frac{\nu_n+1}{2}-1}\exp\left(-\frac{1}{2\sigma^2}\left(\beta_n+\phi_n(\mu-\theta_n)^2\right)\right)\end{aligned}$$

has the same form as the joint prior density (4.37) with parameters updated as:

$$\begin{aligned}
\nu &\to \nu_n = \nu + n, \\
\beta &\to \beta_n = \beta + \phi\theta^2 + n\overline{y^2} - \frac{(\phi\theta + n\overline{y})^2}{\phi + n}, \\
\theta &\to \theta_n = \frac{\phi\theta + n\overline{y}}{\phi + n}, \\
\phi &\to \phi_n = \phi + n.
\end{aligned} \tag{4.39}$$

Improper constant prior. It is easy to see that if the prior is constant (improper prior), i.e. $\pi(\mu, \sigma|\mathbf{y}) \propto h(\mathbf{y}|\mu, \sigma)$, then the posteriors densities $\pi(\sigma^2|\mathbf{y})$ and $\pi(\mu|\sigma^2, \mathbf{y})$ correspond to the $InvChiSq(\nu_n, \beta_n)$ and $\mathcal{N}(\theta_n, \sigma_n/\sqrt{\phi_n})$ respectively with

$$\nu_n = n - 3, \quad \beta_n = n\overline{y^2} - n(\overline{y})^2, \quad \theta_n = \overline{y}, \quad \phi_n = n. \tag{4.40}$$

In this case, the mode of the posterior density $\pi(\mu, \sigma|\mathbf{y})$ is

$$\widehat{\mu}^{\text{MAP}} = \overline{y}, \quad (\widehat{\sigma}^2)^{\text{MAP}} = \overline{y^2} - (\overline{y})^2 \tag{4.41}$$

which are the same as MLEs of μ and σ^2.

Estimating prior for both μ and σ. For given Θ_μ and Θ_σ, the loss quantile at the level q is

$$Q_q = \exp(\Theta_\mu + \Theta_\sigma z_q); \tag{4.42}$$

see (4.32). Thus one can find Θ_σ via two quantiles Q_{q_2} and Q_{q_1} as

$$\Theta_\sigma = \frac{\ln(Q_{q_2}/Q_{q_1})}{z_{q_2} - z_{q_1}}. \tag{4.43}$$

Then, one can try to fit the prior distribution for Θ_σ using the expert opinions on $\mathrm{E}[\ln(Q_{q_2}/Q_{q_1})]$ and $\Pr[a \le Q_{q_2}/Q_{q_1} \le b]$ or the opinions involving several pairs of quantiles. Given σ, the prior distribution for μ can be estimated using Eqs. (4.33) or (4.34).

4.3.6 Pareto Severity

Another important example of the severity distribution, which is very useful to fit heavy-tailed losses, for a given threshold $L > 0$, is the Pareto distribution, $Pareto(\xi, L)$, with a density

$$f(x|\xi) = \frac{\xi}{L}\left(\frac{x}{L}\right)^{-\xi-1}. \tag{4.44}$$

It is defined for $x \geq L$ and $\xi > 0$. If $\xi > 1$, then the mean is $L\xi/(\xi - 1)$, otherwise the mean does not exist. The tail parameter ξ is unknown and modelled by a random variable Θ_ξ.

Model Assumptions 4.4

- *Suppose that conditionally, given $\Theta_\xi = \xi$, the data $X_1, \ldots, X_n, \ldots$ are independent random variables from $Pareto(\xi, L)$;*
- *The prior distribution for the tail parameter Θ_ξ is $Gamma(\alpha, \beta)$, i.e. the prior density is*

$$\pi(\xi) \propto \xi^{\alpha-1} \exp(-\xi/\beta). \tag{4.45}$$

Denote the losses over past years as $\mathbf{X} = (X_1, \ldots, X_n)'$.

Posterior. Given $\mathbf{X} = \mathbf{x}$, under the above assumptions, the posterior density (using (4.9))

$$\begin{aligned} \pi(\xi|\mathbf{x}) &= \xi^n \exp\left(-(\xi+1)\sum_{i=1}^{n} \ln(x_i/L)\right) \times \xi^{\alpha-1} \exp(-\xi/\beta) \\ &\propto \xi^{\alpha_n - 1} \exp(-\xi/\beta_n) \end{aligned} \tag{4.46}$$

is $Gamma(\alpha_n, \beta_n)$, i.e. the same as the prior distribution with updated parameters

$$\alpha \to \alpha_n = \alpha + n, \quad \beta^{-1} \to \beta_n^{-1} = \beta^{-1} + \sum_{i=1}^{n} \ln(x_i/L). \tag{4.47}$$

The mean of the posterior distribution for Θ_ξ allows for a good interpretation, as follows:

$$\begin{aligned} \widehat{\xi} = \mathrm{E}[\Theta_\xi|\mathbf{X} = \mathbf{x}] = \alpha_n \beta_n &= \frac{\alpha + n}{\beta^{-1} + \sum_{i=1}^{n} \ln(x_i/L)} \\ &= w_n \widehat{\xi}_n^{\mathrm{MLE}} + (1 - w_n)\xi_0, \end{aligned} \tag{4.48}$$

where

- $\widehat{\xi}_n^{\mathrm{MLE}} = \frac{1}{n} \sum_{i=1}^{n} \ln(x_i/L)$ is the maximum likelihood estimate of ξ using the observed losses;
- $\xi_0 = \alpha\beta$ is the estimate of ξ using a prior distribution only (e.g. specified by expert);
- $w_n = \left[\sum_{i=1}^{n} \ln(x_i/L)\right] \times \left[\sum_{i=1}^{n} \ln(x_i/L) + 1/\beta\right]^{-1}$ is the weight in [0,1) combining ξ_0 and $\widehat{\xi}_n^{\mathrm{MLE}}$.

The posterior distribution can be easily calculated recursively. Consider the observed losses $x_1, x_2, \ldots, x_k, \ldots$. Assume that the prior, $Gamma(\alpha, \beta)$, is specified initially, then the posterior $\pi(\xi|x_1, \ldots, x_k)$ after the k-th event is $Gamma(\alpha_k, \beta_k)$ with

$$\alpha_k = \alpha + k, \quad \beta_k^{-1} = \beta^{-1} + \sum_{i=1}^{k} \ln(x_i/L).$$

It is easy to show that

$$\alpha_k = \alpha_{k-1} + 1, \quad \beta_k^{-1} = \beta_{k-1}^{-1} + \ln(x_k/L). \tag{4.49}$$

Again, this leads to a very efficient recursive scheme, where the calculation of the posterior distribution parameters is based on the most recent observation and parameters of the posterior distribution calculated just before this observation.

Remark 4.6 It is important to note that the prior and posterior distributions of Θ_ξ are gamma distributions formally defined for $\xi > 0$. Thus there is a finite probability that $\Pr[\Theta_\xi \leq 1] > 0$, which leads to infinite means of predicted distributions, that is, $\mathrm{E}[X_i] = \infty$ and $\mathrm{E}[X_{n+1}|\mathbf{X}] = \infty$. If we do not want to allow for infinite mean behaviour, then ξ should be restricted to $\xi > 1$. See Sect. 2.9.4 on how to deal with this.

Improper constant prior. It is easy to see that if the prior is constant (improper prior), i.e. $\pi(\xi|\mathbf{x}) \propto h(\mathbf{x}|\xi)$, then the posterior is $Gamma(\alpha_n, \beta_n)$ with

$$\alpha_n = n + 1, \quad \beta_n^{-1} = \sum_{i=1}^{n} \ln(x_i/L). \tag{4.50}$$

In this case, the mode of the posterior density $\pi(\xi|\mathbf{x})$ is

$$\widehat{\xi}^{\mathrm{MAP}} = \frac{n}{\sum_{i=1}^{n} \ln(x_i/L)}, \tag{4.51}$$

which is the same as MLE of ξ.

Estimating prior. Suppose that X, the severity of operational losses exceeding threshold L, is modelled by the Pareto distribution, $Pareto(\xi, L)$. Then, conditionally on $\Theta_\xi = \xi$, the expected loss

$$\mathrm{E}[X|\Theta_\xi = \xi] = \mu(\xi) = \frac{L\xi}{\xi - 1}, \quad \text{if} \quad \xi > 1 \tag{4.52}$$

and the loss quantile at level q is

$$f_q(\xi) = L \exp\left(-\frac{\ln(1-q)}{\xi}\right), \quad \xi > 0, \tag{4.53}$$

The mean and quantile of the loss are functions of ξ and thus, unconditionally, are random variables

$$\Omega = \mu(\Theta_\xi) \quad \text{and} \quad Q_q = f_q(\Theta_\xi)$$

respectively. If there is a reason to believe that, unconditionally, expected loss is finite, then the tail parameter ξ should satisfy $\xi \geq B > 1$. Now, assume that we choose the prior distribution for Θ_ξ to be $Gamma(\alpha, \beta)$ distribution truncated below B, i.e. to have a density:

$$\pi(\xi) = \frac{\xi^{\alpha-1}\exp(-\xi/\beta)}{(1 - F^{(G)}_{\alpha,\beta}(B))\Gamma(\alpha)\beta^\alpha} 1_{\{\xi \geq B\}}, \quad \xi \geq B, \alpha > 0, \beta > 0, \tag{4.54}$$

where $F^{(G)}_{\alpha,\beta}(\cdot)$ is a gamma distribution $Gamma(\alpha, \beta)$. If the expert estimates $\mathrm{E}[\Theta_\xi]$ and the uncertainty $\Pr[a \leq \Theta_\xi \leq b] = p$, then the following two equations

$$\begin{aligned} \mathrm{E}[\Theta_\xi] &= \alpha\beta \frac{1 - F^{(G)}_{\alpha+1,\beta}(B)}{1 - F^{(G)}_{\alpha,\beta}(B)}, \\ \Pr[a \leq \Theta_\xi \leq b] &= \frac{F^{(G)}_{\alpha,\beta}(b) - F^{(G)}_{\alpha,\beta}(a)}{1 - F^{(G)}_{\alpha,\beta}(B)} \end{aligned} \tag{4.55}$$

can be solved to estimate the structural parameters α and β.

Example 4.3 Assume that, the lower bound for the tail parameter is B=2 and the expert estimates are $\mathrm{E}[\Theta_\xi] = 5$, $\Pr[4 \leq \Theta_\xi \leq 6] = 2/3$. Then we can fit $\alpha \approx 23.086$, $\beta \approx 0.217$ and can calculate the posterior distribution parameters α_k, β_k, when observations $x_1, x_2, \ldots$ become available, using (4.21). In Fig. 4.3a, we show the subsequent posterior best estimates for the tail parameter

$$\xi_k = \alpha_k\beta_k \frac{1 - F^{(G)}_{\alpha+1,\beta}(B)}{1 - F^{(G)}_{\alpha,\beta}(B)}, \quad k = 1, 2, \ldots, \tag{4.56}$$

when the losses X_k are simulated from $Pareto(4, 1)$. The actual simulated loss values are presented in Table 4.2. On the same figure, we show the standard maximum likelihood estimate of the tail parameter

$$\widehat{\xi}_k^{\mathrm{MLE}} = \left(\frac{1}{k}\sum_{i=1}^{k} \ln(x_i/L)\right)^{-1}.$$

It is easy to see that the Bayesian estimates are more stable while the maximum likelihood estimates are quite volatile when the number of observations is small.

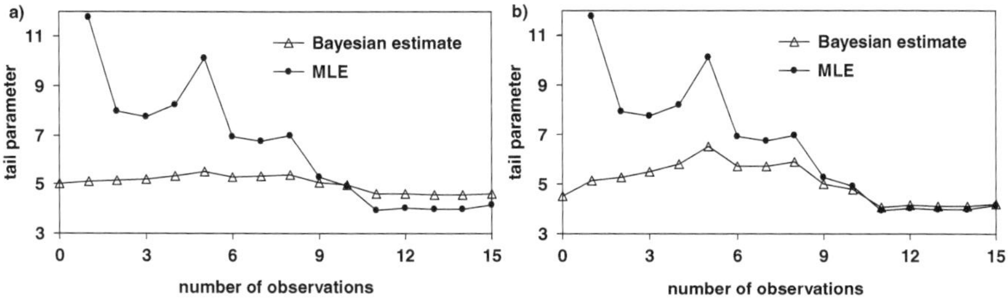

Fig. 4.3 The Bayesian and the standard maximum likelihood estimates (MLE) of the Pareto tail parameter vs the number of observations. The losses were sampled from $Pareto(4, 1)$. The prior distribution is gamma: **(a)** $Gamma(23.1,\ 0.22)$, truncated below $B = 2$; **(b)** $Gamma(4,\ 1.125)$. See Example 4.3 for details

Table 4.2 Loss severities $x_i, i = 1, 2, \ldots, 15$ sampled from a $Pareto(4, 1)$ distribution

i	1	2	3	4	5	6	7	8	9	10	11	12	13	14	15
x_i	1.089	1.181	1.145	1.105	1.007	1.451	1.187	1.116	1.753	1.383	2.167	1.180	1.334	1.272	1.123

As the number of observations increases, two estimators become almost the same. As another example, Fig. 4.3b compares the Bayesian estimate and MLE when the gamma prior is specified by expert who says that $\mathrm{E}[\Theta_\xi] = 4.5$ and $\mathrm{Vco}[\Theta_\xi] = 0.5$. This gives the parameters of the prior $\alpha = 4$ and $\beta = 1.125$.

If it is difficult to express opinions on ξ directly, then the expert may try to estimate the expected loss, quantile or their uncertainties. It might be difficult numerically to fit α, β if the expert specifies unconditional expected loss or expected quantile

$$\mathrm{E}[\Omega] = \mathrm{E}[\mu(\Theta_\xi)] = \int_B^\infty \mu(\xi)\pi(\xi)d\xi,$$

$$\mathrm{E}[Q_q] = \mathrm{E}[f_q(\Theta_\xi)] = L \int_B^\infty f_q(\xi)\pi(\xi)d\xi, \tag{4.57}$$

respectively, as these are not easily expressed. Nevertheless, there is no problem in principle. Fitting opinions on uncertainties might be easier. For example, if the expert estimates the interval $[a, b]$ such that the true expected loss is within the interval with the probability $\Pr[a \le \Omega \le b] = p$, then it leads to the equation

$$\Pr[a \le \Omega \le b] = p = \int_{\widetilde{b}}^{\widetilde{a}} \pi(\xi)d\xi = \frac{F^{(G)}_{\alpha,\beta}(\widetilde{a}) - F^{(G)}_{\alpha,\beta}(\widetilde{b})}{1 - F^{(G)}_{\alpha,\beta}(B)}, \tag{4.58}$$

where $\widetilde{a} = \dfrac{a}{a-L}, \quad \widetilde{b} = \dfrac{b}{b-L}.$

Here, the interval bounds should satisfy $L < a < b \leq B \times L/(B-1)$. The estimation of the interval $[a, b]$, $L < a < b$, such that the true quantile is within the interval with the probability $\Pr[a \leq Q_q \leq b] = p$ leads to the equation

$$\Pr[a \leq Q_q \leq b] = L \int_{C_1}^{C_2} \pi(\xi) d\xi = \frac{F_{\alpha,\beta}^{(G)}(C_2) - F_{\alpha,\beta}^{(G)}(C_1)}{1 - F_{\alpha,\beta}^{(G)}(B)}, \tag{4.59}$$

$$C_1 = -\frac{\ln(1-q)}{\ln(b/L)}, \quad C_2 = -\frac{\ln(1-q)}{\ln(a/L)},$$

where the interval bounds should satisfy

$$L < a < b \leq L \exp\left(-\frac{\ln(1-q)}{B}\right).$$

Equations (4.58) and (4.59) or similar ones can be used to fit α and β. If the expert specifies more than two quantities, then one can use, for example, a nonlinear least square procedure to fit the structural parameters.

4.4 Estimation of the Prior Using Data

The prior distribution can be estimated using a marginal distribution of observations. The data can be collective industry data, collective data in the bank, etc. This approach is referred to as *empirical Bayes*; see Sect. 4.3.1 and Fig. 4.1.

4.4.1 The Maximum Likelihood Estimator

Consider, for example, J *similar* risk cells with the data $\{X_k^{(j)}, k = 1, 2, \ldots, j = 1, \ldots, J\}$. This can be, for example, a specific business line/event type risk cell in J banks. Denote the data over past years as $\mathbf{X}^{(j)} = (X_1^{(j)}, \ldots, X_{K_j}^{(j)})'$, that is, K_j is the number of observations in bank j over past years. Assume that $X_1^{(j)}, \ldots, X_{K_j}^{(j)}$ are conditionally independent and identically distributed from the density $f(\cdot|\boldsymbol{\theta}^j)$, for given $\boldsymbol{\Theta}^{(j)} = \boldsymbol{\theta}^{(j)}$. That is, the risk cells have different risk profiles $\boldsymbol{\Theta}^j$. Assume now that the risks are similar, in a sense that $\boldsymbol{\Theta}^{(1)}, \ldots, \boldsymbol{\Theta}^{(J)}$ are independent and identically distributed from the same density $\pi(\boldsymbol{\theta})$. That is, it is assumed that the risk cells are the same a priori (before we have any observations); see Fig. 4.1. Then the joint density of all observations can be written as

$$f(\mathbf{x}^{(1)}, \ldots, \mathbf{x}^{(J)}) = \prod_{j=1}^{J} \int \left[\prod_{k=1}^{K_j} f\left(x_k^{(j)} | \boldsymbol{\theta}^{(j)}\right) \right] \pi\left(\boldsymbol{\theta}^{(j)}\right) d\boldsymbol{\theta}^{(j)}. \tag{4.60}$$

The parameters of $\pi(\boldsymbol{\theta})$ can be estimated using the maximum likelihood method by maximising (4.60). The density $\pi(\boldsymbol{\theta})$ is a prior density for the j-th cell. Using internal data of the j-th risk cell, its posterior density is calculated from (4.9) as

$$\pi(\boldsymbol{\theta}^{(j)}|\mathbf{x}^{(j)}) = \prod_{k=1}^{K_j} f\left(x_k^{(j)}|\boldsymbol{\theta}^{(j)}\right)\pi\left(\boldsymbol{\theta}^{(j)}\right), \tag{4.61}$$

where $\pi(\boldsymbol{\theta})$ was fitted with MLE using (4.60). The basic idea here is that the estimates based on observations from all banks are better then those obtained using smaller number of observations available in the risk cell of a particular bank.

4.4.2 Poisson Frequencies

It is not difficult to include a priori known differences (exposure indicators, expert opinions on the differences, etc) between the risk cells from the different banks. As an example, we consider the case when the annual frequency of the events is modelled by the Poisson distribution with the gamma prior and estimate structural parameters using the industry data with differences between the banks taken into account.

Model Assumptions 4.5 *Consider J risk cell with the loss frequencies $\{N_{j,k}$, $k = 1, 2, \ldots, j = 1, \ldots, J\}$, where $N_{j,k}$ is the annual number of events in the j-th risk cell in the k-th year. Denote the data over past years in risk cell j as $\mathbf{N}_j = (N_{j,1}, \ldots, N_{j,K_j})$ and the data over past years in all risk cells as $\mathbf{N}_{1:J} = (\mathbf{N}_1, \ldots, \mathbf{N}_J)$. Assume that:*

- *Given $\Lambda_j = \lambda_j$, $N_{j,k}$ are independent random variables from $Poisson(\lambda_j V_{j,k})$, with probability mass function denoted as $f(\cdot|\lambda_j)$. Here, $V_{j,k}$ is the known constant (i.e. the gross income or the volume or combination of several exposure indicators) and λ_j is a risk profile of the cell in the j-th bank.*
- *$\Lambda_1, \ldots, \Lambda_J$ are independent and identically distributed from $Gamma(\alpha, \beta)$ with the density denoted as $\pi(\cdot)$.*
- *Denote $N_j = \sum_{k=1}^{K_j} N_{j,k}$ and $V_j = \sum_{k=1}^{K_j} V_{j,k}$.*

Given the Model Assumptions 4.5, the joint density of all data (over all J risk cells) can be written as

$$\begin{aligned} f(\mathbf{n}_{1:J}) &= \prod_{j=1}^{J} \int \left[\prod_{k=1}^{K_j} f(n_{j,k}|\lambda_j)\right] \pi(\lambda_j) d\lambda_j \\ &= \prod_{j=1}^{J} \int \left[\prod_{k=1}^{K_j} e^{-\lambda_j V_{j,k}} \frac{(V_{j,k}\lambda_j)^{n_{j,k}}}{(n_{j,k})!}\right] \frac{\lambda_j^{\alpha-1} e^{-\lambda_j/\beta}}{\Gamma(\alpha)\beta^\alpha} d\lambda_j \\ &= \left[\prod_{j=1}^{J}\prod_{k=1}^{K_j} \frac{(V_{j,k})^{n_{j,k}}}{(n_{j,k})!}\right] \prod_{j=1}^{J} \frac{\Gamma(\alpha + n_j)}{\Gamma(\alpha)\beta^\alpha (V_j + 1/\beta)^{\alpha+n_j}}. \end{aligned} \tag{4.62}$$

The parameters α and β can now be estimated using the maximum likelihood method by maximising

$$\ln f(\mathbf{n}_{1:J}) \propto \sum_{j=1}^{J} \left\{ \ln \Gamma(\alpha + n_j) - \ln \Gamma(\alpha) - \alpha \ln \beta - (\alpha + n_j) \ln \left(\frac{1}{\beta} + V_j \right) \right\} \tag{4.63}$$

over α and β. To avoid the use of numerical optimisation required for maximising (4.63), one could also use a method of moments; see Proposition 4.1. Once the prior distribution parameters α and β are estimated, then, using (4.9), the posterior distribution of λ_j for the j-th risk cell has a density

$$\begin{aligned} \pi(\lambda_j|\mathbf{n}_j) &\propto \frac{(\lambda_j/\beta)^{\alpha-1}}{\Gamma(\alpha)\beta} e^{-\lambda_j/\beta} \prod_{k=1}^{K_j} e^{-\lambda_j V_{j,k}} \frac{(V_{j,k}\lambda_j)^{n_{j,k}}}{n_{j,k}!} \\ &\propto \lambda^{n_j+\alpha-1} \exp\left(-\lambda_j V_j - \frac{\lambda_j}{\beta} \right), \end{aligned} \tag{4.64}$$

which is $Gamma(\widehat{\alpha}, \widehat{\beta})$ with

$$\widehat{\alpha} = \alpha + \sum_{k=1}^{K_j} n_{j,k}, \; \widehat{\beta} = \beta \left(1 + \beta \sum_{k=1}^{K_j} V_{j,k} \right)^{-1}. \tag{4.65}$$

Assume that the exposure indicator of the cell in the j-th bank for the next year is $V_{j,K_j+1} = V$. Then, the predictive distribution for the annual number of events in the cell (conditional on the past internal data) is negative binomial, $NegBin(\widehat{\alpha}, \widehat{p} = 1/(1 + V\widehat{\beta}))$:

$$\begin{aligned} \Pr[N_{K_j+1} = n|\mathbf{N}_j = \mathbf{n}_j] &= \int e^{-\lambda V} \frac{(V\lambda)^n}{n!} \frac{\lambda^{\widehat{\alpha}-1}}{\Gamma(\widehat{\alpha})\widehat{\beta}^{\widehat{\alpha}}} e^{-\lambda/\widehat{\beta}} d\lambda \\ &= \frac{\Gamma(n + \widehat{\alpha})}{\Gamma(\widehat{\alpha}) n!} (1 - \widehat{p})^n \widehat{p}^{\widehat{\alpha}}. \end{aligned} \tag{4.66}$$

Remark 4.7 Observe that we have scaled the parameters for considering a priori differences. This leads to a linear volume relation for the variance function. To obtain different functional relations, it might be better to scale the actual observations. For example, given observations $X_{j,k}$, $j = 1, \dots, J$, $k = 1, \dots, K_j$ (these could be frequencies or severities), consider variables $Y_{j,k} = X_{j,k}/V_{j,k}$. Assume that, for given $\mathbf{\Theta}_j = \boldsymbol{\theta}_j$, $\{Y_{j,k}, k = 1, \dots, K_j\}$ are independent and identically distributed from $f(\cdot|\boldsymbol{\theta}_j)$. Also, assume that $\mathbf{\Theta}_1, \dots, \mathbf{\Theta}_J$ are independent and identically distributed from $\pi(\cdot)$. Then one can construct the likelihood of $Y_{j,k}$ using (4.60) to fit parameters of $\pi(\cdot)$ or try to use the method of moments.

Estimating prior using method of moments. To avoid the use of numerical optimisation required for maximising (4.63), one could also use a method of moments utilising the following proposition.

Proposition 4.1 *Given Model Assumptions 4.5, denote $\lambda_0 = \mathrm{E}[\Lambda_j] = \alpha\beta$, $\sigma_0^2 = \mathrm{Var}[\Lambda_j] = \alpha\beta^2$. Then the estimates $\widehat{\lambda}_0$ and $\widehat{\sigma}_0^2$ for λ_0 and σ_0^2 respectively are*

$$\widehat{\lambda}_0 = \frac{1}{J}\sum_{j=1}^{J}\widehat{\lambda}_j,\ \widehat{\lambda}_j = \frac{1}{K_j}\sum_{k=1}^{K_j}\frac{n_{j,k}}{V_{j,k}},\ j = 1,\ldots,J,$$

$$\widehat{\sigma}_0^2 = \max\left[\frac{1}{J-1}\sum_{j=1}^{J}(\widehat{\lambda}_j - \widehat{\lambda}_0)^2 - \frac{\widehat{\lambda}_0}{J}\sum_{j=1}^{J}\frac{1}{K_j^2}\sum_{k=1}^{K_j}\frac{1}{V_{j,k}},\ 0\right].$$

These can easily be used to estimate α and β as $\widehat{\alpha} = \widehat{\lambda}_0/\widehat{\beta}$ and $\widehat{\beta} = \widehat{\sigma}_0^2/\widehat{\lambda}_0$ correspondingly[2].

Proof Consider the standardised frequencies $F_{j,k} = N_{j,k}/V_{j,k}$. It is easy to observe that,

$$\begin{aligned}\mathrm{E}[N_{j,k}|\Lambda_j] &= \Lambda_j V_{j,k},\ \mathrm{Var}[N_{j,k}|\Lambda_j] = \Lambda_j V_{j,k},\\ \mathrm{E}[F_{j,k}|\Lambda_j] &= \Lambda_j,\quad \mathrm{Var}[F_{j,k}|\Lambda_j] = \Lambda_j/V_{j,k}\end{aligned}$$

and

$$\begin{aligned}\mathrm{E}[F_{j,k}] &= \mathrm{E}[\mathrm{E}[F_{j,k}|\Lambda_j]] = \mathrm{E}[\Lambda_j] = \lambda_0,\\ \mathrm{Var}[F_{j,k}] &= \mathrm{E}[\mathrm{Var}[F_{j,k}|\Lambda_j]] + \mathrm{Var}(\mathrm{E}[F_{j,k}|\Lambda_j])\\ &- \mathrm{E}[\Lambda_j/V_{j,k}] + \mathrm{Var}[\Lambda_j] = \frac{\lambda_0}{V_{j,k}} + \sigma_0^2.\end{aligned}$$

Note that, given Λ_j, $F_{j,k}$ are independent. Consider estimators

$$\widehat{\Lambda}_j = \frac{1}{K_j}\sum_{k=1}^{K_j}F_{j,k},\ j = 1,\ldots,J.$$

These estimators are independent and

$$\mathrm{E}[\widehat{\Lambda}_j] = \frac{1}{K_j}\sum_{k=1}^{K_j}\mathrm{E}[F_{j,k}] = \lambda_0,$$

$$\mathrm{Var}[\widehat{\Lambda}_j] = \mathrm{E}[\mathrm{Var}[\widehat{\Lambda}_j|\Lambda_j]] + \mathrm{Var}[\mathrm{E}[\widehat{\Lambda}_j|\Lambda_j]] = \frac{\lambda_0}{K_j^2}\sum_{k=1}^{K_j}\frac{1}{V_{j,k}} + \sigma_0^2.$$

[2] Alternative unbiased moment estimators are given in Bühlmann and Gisler ([44], section 4.10).

Thus

$$\widehat{\Lambda}_0 = \frac{1}{J}\sum_{j=1}^{J}\widehat{\Lambda}_j$$

is an unbiased estimator for λ_0. In the case of the same number of observations per company and the same weights $V_{j,k}$, this estimator would have a minimal variance among all linear combinations of $\widehat{\Lambda}_1, \ldots, \widehat{\Lambda}_J$. Next, calculate

$$\begin{aligned}
\sum_{j=1}^{J}\mathrm{E}[(\widehat{\Lambda}_j - \widehat{\Lambda}_0)^2] &= \sum_{j=1}^{J}\mathrm{E}[(\widehat{\Lambda}_j - \lambda_0 + \lambda_0 - \widehat{\Lambda}_0)^2] \\
&= \sum_{j=1}^{J}\left(\mathrm{Var}[\widehat{\Lambda}_j] + \mathrm{Var}[\widehat{\Lambda}_0] - 2\mathrm{Cov}[\widehat{\Lambda}_j, \widehat{\Lambda}_0]\right) \\
&= \sum_{j=1}^{J}\left(\mathrm{Var}[\widehat{\Lambda}_j] + \frac{1}{J^2}\sum_{i=1}^{J}\mathrm{Var}[\widehat{\Lambda}_i] - \frac{2}{J}\mathrm{Var}[\widehat{\Lambda}_j]\right) \\
&= \frac{J-1}{J}\sum_{j=1}^{J}\mathrm{Var}[\widehat{\Lambda}_j] \\
&= \lambda_0\frac{J-1}{J}\sum_{j=1}^{J}\frac{1}{K_j^2}\sum_{k=1}^{K_j}\frac{1}{V_{j,k}} + \sigma_0^2(J-1).
\end{aligned} \tag{4.67}$$

Thus

$$\widetilde{\Xi}_0^2 = \frac{1}{J-1}\sum_{j=1}^{J}(\widehat{\Lambda}_j - \widehat{\Lambda}_0)^2 - \frac{\widehat{\Lambda}_0}{J}\sum_{j=1}^{J}\frac{1}{K_j^2}\sum_{k=1}^{K_j}\frac{1}{V_{j,k}} \tag{4.68}$$

is an unbiased estimator for σ_0^2. Observe that $\widetilde{\Xi}_0^2$ is not necessarily positive, hence we take $\widehat{\Xi}_0^2 = \max\{\widetilde{\Xi}_0^2,\ 0\}$ as the final estimator for σ_0^2. This completes the proof. □

4.5 Combining Expert Opinions with External and Internal Data

In the above sections we showed how to combine two data sources: expert opinions and internal data; or external data and internal data. In order to estimate the risk capital of a bank and to fulfil the Basel II requirements, risk managers have to take into account internal data, relevant external data (industry data) and expert opinions. The aim of this section is to provide an example of methodology to be used to combine these three sources of information. Here, we follow the approach suggested in Lambrigger, Shevchenko and Wüthrich [141]. As in the previous section, we consider one risk cell only. In terms of methodology we go through the following steps:

- In any risk cell, we model the loss frequency and the loss severity by parametric distributions (e.g. Poisson for the frequency or Pareto, lognormal, etc. for the severity). For the considered bank, the unknown parameter vector $\boldsymbol{\theta}$ (for example, the Poisson parameter or the Pareto tail index) of these distributions has to be quantified.
- A priori, before we have any company specific information, only industry data are available. Hence, the best prediction of our bank specific parameter $\boldsymbol{\theta}$ is given by the belief in the available external knowledge such as the provided industry data. This unknown parameter of interest is modelled by a prior distribution (structural distribution) corresponding to a random vector $\boldsymbol{\Theta}$. The parameters of the prior distribution (hyper-parameters) are estimated using data from the whole industry by, for example, maximum likelihood estimation, as described in Sect. 4.4. If no industry data are available, the prior distribution could come from a "super expert" that has an overview over all banks.
- The true bank specific parameter $\boldsymbol{\theta}_0$ is treated as a realisation of $\boldsymbol{\Theta}$. The prior distribution of a random vector $\boldsymbol{\Theta}$ corresponds to the whole banking industry sector, whereas $\boldsymbol{\theta}$ stands for the unknown underlying parameter set of the bank being considered. Due to the variability amongst banks, it is natural to model $\boldsymbol{\theta}$ by a probability distribution. Note that $\boldsymbol{\Theta}$ is random with known distribution, whereas $\boldsymbol{\theta}_0$ is deterministic but unknown.
- As time passes, internal data $\mathbf{X} = (X_1, \ldots, X_K)'$ as well as expert opinions $\boldsymbol{\Delta} = (\Delta_1, \ldots, \Delta_M)'$ about the underlying parameter $\boldsymbol{\theta}$ become available. This affects our belief in the distribution of $\boldsymbol{\Theta}$ coming from external data only and adjust the prediction of $\boldsymbol{\theta}_0$. The more information on $\mathbf{X}$ and $\boldsymbol{\Delta}$ we have, the better we are able to predict $\boldsymbol{\theta}_0$. That is, we replace the prior density $\pi(\boldsymbol{\theta})$ by a conditional density of $\boldsymbol{\Theta}$ given $\mathbf{X}$ and $\boldsymbol{\Delta}$.

In order to determine the posterior density $\pi(\boldsymbol{\theta}|\mathbf{x}, \boldsymbol{\delta})$, consider the joint conditional density of observations and expert opinions (given the parameter vector $\boldsymbol{\theta}$):

$$h(\mathbf{x}, \boldsymbol{\delta}|\boldsymbol{\theta}) = h_1(\mathbf{x}|\boldsymbol{\theta}) h_2(\boldsymbol{\delta}|\boldsymbol{\theta}), \tag{4.69}$$

where h_1 and h_2 are the conditional densities (given $\boldsymbol{\Theta} = \boldsymbol{\theta}$) of $\mathbf{X}$ and $\boldsymbol{\Delta}$, respectively. Thus $\mathbf{X}$ and $\boldsymbol{\Delta}$ are assumed to be conditionally independent given $\boldsymbol{\Theta}$.

Remark 4.8

- Notice that, in this way, we naturally combine external data information, $\pi(\boldsymbol{\theta})$, with internal data $\mathbf{X}$ and expert opinion $\boldsymbol{\Delta}$.
- In classical Bayesian inference (as it is used, for example, in actuarial science), one usually combines only two sources of information as described in the previous sections. Here, we combine three sources simultaneously using an appropriate structure, that is, Eq. (4.69).
- Equation (4.69) is quite a reasonable assumption. Assume that the true bank specific parameter is $\boldsymbol{\theta}_0$. Then, (4.69) says that the experts in this bank estimate $\boldsymbol{\theta}_0$

(by their opinion $\mathbf{\Delta}$) independently of the internal observations. This makes sense if the experts specify their opinions regardless of the data observed. Otherwise we should work with the joint distribution $h(\mathbf{x}, \boldsymbol{\delta}|\boldsymbol{\theta})$.

We further assume that observations as well as expert opinions are conditionally independent and identically distributed, given $\mathbf{\Theta} = \boldsymbol{\theta}$, so that

$$h_1(\mathbf{x}|\boldsymbol{\theta}) = \prod_{k=1}^{K} f_1(x_k|\boldsymbol{\theta}), \tag{4.70}$$

$$h_2(\boldsymbol{\delta}|\boldsymbol{\theta}) = \prod_{m=1}^{M} f_2(\delta_m|\boldsymbol{\theta}), \tag{4.71}$$

where f_1 and f_2 are the marginal densities of a single observation and a single expert opinion, respectively. We have assumed that all expert opinions are identically distributed, but this can be generalised easily to expert opinions having different distributions.

Here, the unconditional parameter density $\pi(\boldsymbol{\theta})$ is the *prior* density, whereas the conditional parameter density $\pi(\boldsymbol{\theta}|\mathbf{x}, \boldsymbol{\delta})$ is the *posterior* density. Let $h(\mathbf{x}, \boldsymbol{\delta})$ denote the unconditional joint density of the data $\mathbf{X}$ and expert opinions $\mathbf{\Delta}$. Then, it follows from Bayes's theorem that

$$h(\mathbf{x}, \boldsymbol{\delta}|\boldsymbol{\theta})\pi(\boldsymbol{\theta}) = \pi(\boldsymbol{\theta}|\mathbf{x}, \boldsymbol{\delta})h(\mathbf{x}, \boldsymbol{\delta}). \tag{4.72}$$

Note that the unconditional density $h(\mathbf{x}, \boldsymbol{\delta})$ does not depend on $\boldsymbol{\theta}$ and thus the posterior density is given by

$$\pi(\boldsymbol{\theta}|\mathbf{x}, \boldsymbol{\delta}) \propto \pi(\boldsymbol{\theta}) \prod_{k=1}^{K} f_1(x_k|\boldsymbol{\theta}) \prod_{m=1}^{M} f_2(\delta_m|\boldsymbol{\theta}). \tag{4.73}$$

For the purposes of operational risk, it should be used to estimate the predictive distribution of future losses.

Hereafter, in this section, we assume that the parameters of the prior distribution are known and we look at a single risk cell in one bank. Therefore, the index representing bank or risk cell is not introduced.

4.5.1 Conjugate Prior Extension

Equation (4.73) can be used in a general setup, but it is convenient to find some *conjugate* prior distributions such that the prior and the posterior distribution have a similar type, or where at least the posterior distribution can be calculated analytically. This type of distribution has been treated in Sect. 4.3 when two data sources have to be combined. For the case of (4.73), the standard definition of the conjugate prior distributions, Definition 2.22, can be extended as follows.

Definition 4.1 (Conjugate Prior Distribution) Let F denote the class of density functions $h(\mathbf{x}, \boldsymbol{\delta}|\boldsymbol{\theta})$, indexed by $\boldsymbol{\theta}$. A class U of prior densities $\pi(\boldsymbol{\theta})$ is said to be a *conjugate family* for F if the posterior density $\pi(\boldsymbol{\theta}|\mathbf{x}, \boldsymbol{\delta}) \propto \pi(\boldsymbol{\theta})h(\mathbf{x}, \boldsymbol{\delta}|\boldsymbol{\theta})$ also belongs to the class U for all $h \in F$ and $\pi \in U$.

Again, in general, the posterior distribution cannot be calculated analytically but can be estimated numerically – for instance by the Markov chain Monte Carlo methods described in Sect. 2.11.

4.5.2 Modelling Frequency: Poisson Model

To model the loss frequency for operational risk in a risk cell, consider the following model.

Model Assumptions 4.6 (Poisson-gamma-gamma) *Assume that a risk cell in a bank has a scaling factor V for the frequency in a specified risk cell (it can be the product of several economic factors such as the gross income, the number of transactions or the number of staff).*

(a) *Let $\Lambda \sim Gamma(\alpha_0, \beta_0)$ be a gamma distributed random variable with shape parameter $\alpha_0 > 0$ and scale parameter $\beta_0 > 0$, which are estimated from (external) market data. That is, the density of $Gamma(\alpha_0, \beta_0)$, plays the role of $\pi(\boldsymbol{\theta})$ in (4.73).*

(b) *Given $\Lambda = \lambda$, the annual frequencies, $N_1, \ldots, N_T, N_{T+1}$, where $T+1$ refers to next year, are assumed to be independent and identically distributed with $N_t \sim Poisson(V\lambda)$. That is, $f_1(\cdot|\lambda)$ in (4.73) corresponds to the probability mass function of a $Poisson(V\lambda)$ distribution.*

(c) *A financial company has M expert opinions Δ_m, $1 \leq m \leq M$, about the intensity parameter Λ. Given $\Lambda = \lambda$, Δ_m and N_t are independent for all t and m, and $\Delta_1, \ldots, \Delta_M$ are independent and identically distributed with $\Delta_m \sim Gamma(\xi, \lambda/\xi)$, where ξ is a known parameter. That is, $f_2(\cdot|\lambda)$ in (4.73) corresponds to the density of a $Gamma(\xi, \lambda/\xi)$ distribution.*

Remark 4.9

- The parameters α_0 and β_0 in Model Assumptions 4.6 are hyper-parameters (parameters for parameters) and can be estimated using the maximum likelihood method or the method of moments; see for instance Sect. 4.4.
- In Model Assumptions 4.6 we assume

$$\mathrm{E}[\Delta_m|\Lambda] = \Lambda, \quad 1 \leq m \leq M, \tag{4.74}$$

  that is, expert opinions are unbiased. A possible bias might only be recognised by the regulator, as he alone has the overview of the whole market.

Note that the *coefficient of variation* of the conditional expert opinion $\Delta_m|\Lambda$ is

$$\mathrm{Vco}[\Delta_m|\Lambda] = (\mathrm{Var}[\Delta_m|\Lambda)])^{1/2}/\mathrm{E}[\Delta_m|\Lambda] = 1/\sqrt{\xi},$$

and thus is independent of Λ. This means that ξ, which characterises the uncertainty in the expert opinions, is independent of the true bank specific Λ. For simplicity, we have assumed that all experts have the same conditional coefficient of variation and thus have the same credibility. Moreover, this allows for the estimation of ξ as

$$\widehat{\xi} = (\widehat{\mu}/\widehat{\sigma})^2, \tag{4.75}$$

where

$$\widehat{\mu} = \frac{1}{M}\sum_{m=1}^{M}\delta_m \quad \text{and} \quad \widehat{\sigma}^2 = \frac{1}{M-1}\sum_{m=1}^{M}(\delta_m - \widehat{\mu})^2, \quad M \geq 2.$$

In a more general framework the parameter ξ can be estimated, for example, by maximum likelihood method.

In the insurance practice ξ is often specified by the regulator denoting a lower bound for expert opinion uncertainty; e.g. Swiss Solvency Test, see Swiss Financial Market Supervisory Authority ([230], appendix 8.4). If the credibility differs among the experts, then $\mathrm{Vco}[\Delta_m|\Lambda]$ should be estimated for all m, $1 \leq m \leq M$. Admittedly, this may often be a challenging issue in practice.

Remark 4.10 This model can be extended to a model where one allows for more flexibility in the expert opinions. For convenience, it is preferred that experts are conditionally independent and identically distributed, given Λ. This has the advantage that there is only one parameter, ξ, that needs to be estimated.

Using the notation from Sect. 4.5, the posterior density of Λ, given the losses up to year K and the expert opinions of M experts, can be calculated. Denote the data over past years as follows:

$$\mathbf{N} = (N_1, \ldots, N_T)',$$
$$\boldsymbol{\Delta} = (\Delta_1, \ldots, \Delta_M)'.$$

Also, denote the arithmetic means by

$$\overline{N} = \frac{1}{T}\sum_{t=1}^{T} N_t, \quad \overline{\Delta} = \frac{1}{M}\sum_{m=1}^{M}\Delta_m, \quad \text{etc.} \tag{4.76}$$

Then, the posterior density is given by the following theorem.

Theorem 4.2 *Under Model Assumptions 4.6, given loss information* $\mathbf{N} = \mathbf{n}$ *and expert opinion* $\boldsymbol{\Delta} = \boldsymbol{\delta}$, *the posterior density of* Λ *is*

$$\pi(\lambda|\mathbf{n}, \boldsymbol{\delta}) = \frac{(\omega/\phi)^{(\nu+1)/2}}{2K_{\nu+1}(2\sqrt{\omega\phi})}\lambda^{\nu} e^{-\lambda\omega - \lambda^{-1}\phi}, \tag{4.77}$$

with

$$\begin{aligned}\nu &= \alpha_0 - 1 - M\xi + T\overline{n},\\ \omega &= VT + \frac{1}{\beta_0},\\ \phi &= \xi M\overline{\delta},\end{aligned} \tag{4.78}$$

and

$$K_{\nu+1}(z) = \frac{1}{2}\int_0^\infty u^\nu e^{-z(u+1/u)/2}du. \tag{4.79}$$

Here, $K_\nu(z)$ is a modified Bessel function of the third kind; see for instance Abramowitz and Stegun ([3], p. 375).

Proof Model Assumptions 4.6 applied to (4.73) yield

$$\begin{aligned}\pi(\lambda|\mathbf{n},\boldsymbol{\delta}) &\propto \lambda^{\alpha_0-1}e^{-\lambda/\beta_0}\prod_{t=1}^{T} e^{-V\lambda}\frac{(V\lambda)^{n_t}}{n_t!}\prod_{m=1}^{M}\frac{(\delta_m)^{\xi-1}}{(\lambda/\xi)^{\xi}}e^{-\delta_m\xi/\lambda}\\ &\propto \lambda^{\alpha_0-1}e^{-\lambda/\beta_0}\prod_{t=1}^{T} e^{-V\lambda}\lambda^{n_t}\prod_{m=1}^{M}(\xi/\lambda)^{\xi}e^{-\delta_m\xi/\lambda}\\ &\propto \lambda^{\alpha_0-1-M\xi+T\overline{n}}\exp\left(-\lambda\left(VT+\frac{1}{\beta_0}\right)-\frac{1}{\lambda}\xi M\overline{\delta}\right).\end{aligned}$$

Remark 4.11

- A distribution with density (4 77) is known as the generalised inverse Gaussian distribution GIG(ω, ϕ, ν). This is a well-known distribution with many applications in finance and risk management; see McNeil, Frey and Embrechts ([157], pp. 75, 497). The GIG has been analysed by many authors; see a discussion in Jørgensen [129]. The GIG belongs to the popular class of subexponential (heavy-tailed) distributions; see Embrechts [80] for a proof and Sect. 6.7 for a detailed treatment of subexponential distributions. The GIG with $\nu \leq 1$ is a distribution of the first hitting time for certain time-homogeneous processes; see for instance Jørgensen ([129], chapter 6). In particular, the standard inverse Gaussian (i.e. the GIG with $\nu = -3/2$) is known by financial practitioners as the distribution function determined by the first passage time of a Brownian motion. The algorithm for generating realisations from a GIG is provided in Appendix B.1.
- In comparison with the classical Poisson-gamma case of combining two sources of information (considered in Sect. 4.3.3), where the posterior is a gamma distribution, the posterior $\pi(\lambda|\cdot)$ in (4.80) is more complicated. In the exponent, it involves both λ and $1/\lambda$. Note that expert opinions enter via the term $1/\lambda$ only.
- Observe that the classical exponential dispersion family with associated conjugates (see chapter 2.5 in Bühlmann and Gisler [44]) allows for a natural extension to GIG-like distributions. In this sense the GIG distributions enlarge the classical Bayesian inference theory on the exponential dispersion family.

For our purposes it is interesting to observe how the posterior density transforms when new data from a newly observed year arrive. Let ν_k, ω_k and ϕ_k denote the parameters for the data $(N_1, \ldots, N_k)$ after k accounting years. Implementation of the update processes is then given by the following equalities (assuming that expert opinions do not change).

Information update process. Year $k \to$ year $k+1$:

$$\begin{aligned} \nu_{k+1} &= \nu_k + n_{k+1}, \\ \omega_{k+1} &= \omega_k + V, \\ \phi_{k+1} &= \phi_k. \end{aligned} \tag{4.80}$$

Obviously, the information update process has a very simple form and only the parameter ν is affected by the new observation n_{k+1}. The posterior density does not change its type every time new data arrive and hence, is easily calculated.

The moments of a GIG are not available in a closed form through elementary functions but can be expressed in terms of Bessel functions; see Appendix A.2.11. In particular, the posterior expected number of losses is

$$\mathrm{E}[\Lambda|\mathbf{N} = \mathbf{n}, \boldsymbol{\Delta} = \delta] = \sqrt{\frac{\phi}{\omega}} \frac{K_{\nu+2}(2\sqrt{\omega\phi})}{K_{\nu+1}(2\sqrt{\omega\phi})}. \tag{4.81}$$

The mode of a GIG has a simple expression (see Appendix A.2.11) that gives the posterior mode

$$\mathrm{mode}\,[\Lambda|\mathbf{N} = \mathbf{n}, \boldsymbol{\Delta} = \delta] = \frac{1}{2\omega}\left(\nu + \sqrt{\nu^2 + 4\omega\phi}\right). \tag{4.82}$$

It can be used as an alternative point estimator instead of the mean. Also, the mode of a GIG differs only slightly from the expected value for large $|\nu|$.

We are clearly interested in robust prediction of the bank specific Poisson parameter and thus the Bayesian estimator (4.81) is a promising candidate within this operational risk framework. The examples below show that, in practice, (4.81) outperforms other classical estimators. To interpret (4.81) in more detail, we make use of asymptotic properties. Using properties of Bessel functions, it is easy to show that

$$R_{\nu^2}(2\nu) \to \nu \quad \text{as} \quad \nu \to \infty, \tag{4.83}$$

where

$$R_\nu(z) = \frac{K_{\nu+1}(z)}{K_\nu(z)};$$

see Lambrigger, Shevchenko and Wüthrich ([141], lemma B.1 in appendix B). Using this result, a full asymptotic interpretation of the Bayesian estimator (4.81) can be found as follows.

Theorem 4.3 *Under Model Assumptions 4.6, the following asymptotic relations hold:*

(a) If $T \to \infty$ *then* $\mathrm{E}[\Lambda|\mathbf{N}, \mathbf{\Delta}] \to \mathrm{E}[N_t|\Lambda = \lambda]/V = \lambda$.
(b) If $\mathrm{Vco}[\Delta_m|\Lambda] \to 0$ *then* $\mathrm{E}[\Lambda|\mathbf{N}, \mathbf{\Delta}] \to \Delta_m$, $m = 1, \ldots, M$.
(c) If $M \to \infty$ *then* $\mathrm{E}[\Lambda|\mathbf{N}, \mathbf{\Delta}] \to \mathrm{E}[\Delta_m|\Lambda = \lambda] = \lambda$.
(d) If $\mathrm{Vco}[\Delta_m|\Lambda] \to \infty$, $m = 1, \ldots, M$ *then*

$$\mathrm{E}[\Lambda|\mathbf{N}, \mathbf{\Delta}] \to \frac{1}{VT\beta_0 + 1}\mathrm{E}[\Lambda] + \frac{1}{V}\left(1 - \frac{1}{VT\beta_0 + 1}\right)\overline{N}.$$

(e) If $\mathrm{E}[\Lambda] =$ *constant and* $\mathrm{Vco}[\Lambda] \to 0$ *then* $\mathrm{E}[\Lambda|\mathbf{N}, \mathbf{\Delta}] \to \mathrm{E}[\Lambda]$.

Proof The proof is given in Lambrigger, Shevchenko and Wüthrich ([141], appendix C). These asymptotic relations should be understood in a probability sense, that is, true with probability 1 (the so-called P-*almost surely*).

Remark 4.12 The GIG mode and mean are asymptotically the same for $\nu \to \infty$; also $4\omega\phi/\nu^2 \to 0$ for $T \to \infty$, $M \to \infty$, $M \to 0$ or $\xi \to 0$. Then, one can approximate the posterior mode as

$$\text{mode}\,[\Lambda|\mathbf{N} = \mathbf{n}, \mathbf{\Delta} = \boldsymbol{\delta}] \approx \frac{\nu}{2\omega}1_{\{\nu \geq 0\}} + \frac{\phi}{|\nu|} \tag{4.84}$$

and obtain the results of Theorem 4.3 in an elementary manner avoiding Bessel functions.

Theorem 4.3 yields a natural interpretation of the posterior density (4.77) and its expected value (4.81):

- As the number of observations increases, we give more weight to them and in the limit $T \to \infty$ (case a) we completely believe in the observations N_k and neglect a priori information and expert opinion.
- On the other hand, the more the coefficient of variation of the expert opinions decreases, the more weight is given to them (case b).
- In Model Assumptions 4.6, we assume experts to be conditionally independent. In practice, however, even for $\mathrm{Vco}[\Delta_m|\Lambda] \to 0$, the variance of $\overline{\Delta}|\Lambda$ cannot be made arbitrarily small when increasing the number of experts, as there is always a positive covariance term due to positive dependence between experts. Since we predict random variables, we never have "perfect diversification", that is, in practical applications we would probably question property c.
- Conversely, if experts become less credible in terms of having an increasing coefficient of variation, our model behaves as if the experts do not exist (case d). The

Bayes estimator is then a weighted sum of prior and posterior information with appropriate credibility weights. This is the classical credibility result obtained from Bayesian inference on the exponential dispersion family with two sources of information; see (4.20).

- Of course, if the coefficient of variation of the prior distribution (i.e. of the whole banking industry) vanishes, the external data are not affected by internal data and expert opinion (case e).

The above interpretation shows that the model behaves as we would expect and require in practice. Thus there are good reasons to believe that it provides an adequate model to combine internal observations with relevant external data and expert opinions, as required by many risk managers. One can even go further and generalise the results from this section in a natural way to a Poisson-gamma-GIG model, that is, where the prior distribution is a GIG. Then, the posterior distribution is again a GIG (see also Model Assumptions 4.16 below).

Example 4.4 A simple example, taken from Lambrigger, Shevchenko and Wüthrich ([141], example 3.7), illustrates the above methodology combining three data sources. It also extends Example 4.1 displayed in Fig. 4.2, where two data sources are combined using classical Bayesian inference approach. Assume that:

- External data (for example, provided by external databases or regulator) estimate the intensity of the loss frequency (i.e. the Poisson parameter Λ), which has a prior gamma distribution $\Lambda \sim Gamma(\alpha_0, \beta_0)$, as $\mathrm{E}[\Lambda] = \alpha_0\beta_0 = 0.5$ and $\Pr[0.25 \leq \Lambda \leq 0.75] = 2/3$. Then, the parameters of the prior are $\alpha_0 \approx 3.407$ and $\beta_0 \approx 0.147$; see Example 4.1.
- One expert gives an estimate of the intensity as $\delta = 0.7$. For simplicity, we consider in this example one single expert only and hence, the coefficient of variation is not estimated using (4.75), but given a priori (e.g. by the regulator): $\mathrm{Vco}[\Delta|\Lambda] = \sqrt{\mathrm{Var}[\Delta|\Lambda]}/\mathrm{E}[\Delta|\Lambda] = 0.5$, i.e. $\xi = 4$.
- The observations of the annual number of losses $n_1, n_2, \ldots$ are sampled from $Poisson(0.6)$ and are the same as in the Example 4.1 (i.e. given in Table 4.1).

This means that a priori we have a frequency parameter distributed as $Gamma(\alpha_0, \beta_0)$ with mean $\alpha_0\beta_0 = 0.5$. The true value of the parameter λ for this risk in a bank is 0.6, that is, it does worse than the average institution. However, our expert has an even worse opinion of his institution, namely $\delta = 0.7$. Now, we compare:

- the pure maximum likelihood estimate

$$\widehat{\lambda}_k^{\mathrm{MLE}} = \frac{1}{k}\sum_{i=1}^{k} n_i;$$

- the Bayesian estimate (4.20),

$$\widehat{\lambda}_k^{(2)} = \mathrm{E}[\Lambda|N_1 = n_1, \ldots, N_k = n_k], \tag{4.85}$$

without expert opinion; and

- the Bayesian estimate derived in formula (4.81)

$$\widehat{\lambda}_k^{(3)} = \mathrm{E}[\Lambda | N_1 = n_1, \ldots, N_k = n_k, \Delta = \delta], \tag{4.86}$$

that combines internal data and expert opinions with the prior.

The results are plotted in Fig. 4.4a. The estimator $\widehat{\lambda}_k^{(3)}$ shows a much more stable behaviour around the true value $\lambda = 0.6$, due to the use of the prior information (market data) and the expert opinions. Given adequate expert opinions, $\widehat{\lambda}_k^{(3)}$ clearly outperforms the other estimators, particularly if only a few data points are available.

One could think that this is only the case when the experts' estimates are appropriate. However, even if experts fairly under- (or over-) estimate the true parameter λ, the method presented here performs better for our dataset than the other mentioned methods, when a few data points are available. Figure 4.4b displays the same estimators, but where the expert's opinion is $\delta = 0.4$, which clearly underestimates the true expected value 0.6.

In Fig. 4.4a, $\widehat{\lambda}_k^{(3)}$ gives better estimates when compared to $\lambda_k^{(2)}$. Observe also that in Fig. 4.4b, $\widehat{\lambda}_k^{(3)}$ gives more appropriate estimates than $\lambda_k^{(2)}$. Though the expert is too optimistic, $\widehat{\lambda}_k^{(3)}$ manages to correct $\widehat{\lambda}_k^{\mathrm{MLE}}$ ($k \leq 10$), which is clearly too low.

The above example yields a typical picture observed in numerical experiments that demonstrates that the Bayes estimator (4.81) is often more suitable and stable than maximum likelihood estimators based on internal data only.

Remark 4.13 Note that in this example the prior distribution as well as the expert opinion do not change over time. However, as soon as new information is available or when new risk management tools are in place, the corresponding parameters may be easily adjusted.

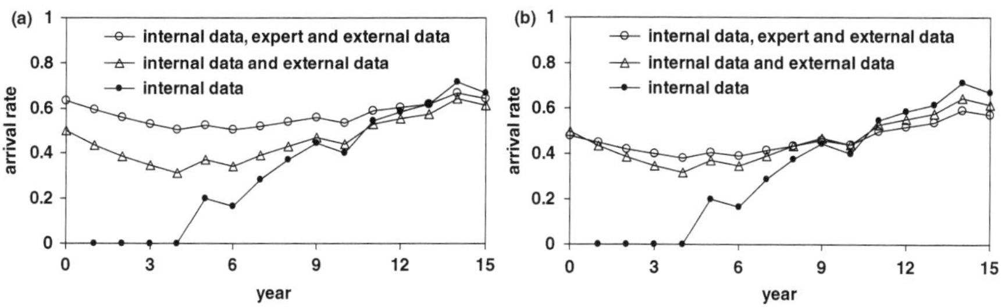

Fig. 4.4 (∘) The Bayes estimate $\widehat{\lambda}_k^{(3)}$, $k = 1, \ldots, 15$, combines the internal data simulated from $Poisson(0.6)$, external data giving $\mathrm{E}[\Lambda] = 0.5$, and expert opinion δ. It is compared with the Bayes estimate $\widehat{\lambda}_k^{(2)}$ (△), that combines external data and internal data, and the classical maximum likelihood estimate $\widehat{\lambda}_k^{\mathrm{MLE}}$ (•). **(a)** is the case of expert opinion $\delta = 0.7$ and **(b)** is the case of expert opinion $\delta = 0.4$. See Example 4.4 for details

4.5.3 Modelling Frequency: Poisson with Stochastic Intensity

Consider the annual number of events for a risk in one bank in year t modelled as random variable from the Poisson distribution $Poisson(\Lambda_t = \lambda_t)$. Conditional on Λ_t, the expected number of events per year is Λ_t. The latter is not only different for different banks and different risks but also may change from year to year for a risk in the same bank. In the previous section, we considered the situation where Λ_t is the same for all years $t = 1, 2, \ldots$. However, in general, the evolution of Λ_t, can be modelled as having deterministic (trend, seasonality) and stochastic components, that is, we consider a sequence $\Lambda_1, \Lambda_2, \ldots, \Lambda_T, \Lambda_{T+1}$, where $T + 1$ corresponds to the next year. In actuarial mathematics this is called a mixed Poisson model.

For simplicity, assume that Λ_t is purely stochastic and distributed according to a gamma distribution. In the context of operational risk, this case was considered in Peters, Shevchenko and Wüthrich [187]. The frequency risk profile Λ_t is characterised by a risk characteristic Θ_Λ. This Θ_Λ represents a vector of unknown distribution parameters of risk profile Λ_t. The true value of Θ_Λ is not known. Then, under the Bayesian approach, it is modelled as a random variable. A priori, before having any company specific information, the prior distribution of Θ_Λ is based on external data only. Our aim then is to find the distribution of Θ_Λ when we have company specific information about risk cell such as observed losses and expert opinions. This can be achieved by developing the following Bayesian model.

Model Assumptions 4.7 *Assume that a risk cell has a fixed, deterministic volume V (i.e. number of transactions, etc.).*

1. *The risk characteristics Θ_Λ of a risk cell has a gamma prior distribution:*

$$\Theta_\Lambda \sim Gamma(a, 1/b), \quad a > 0,\ b > 0.$$

2. *Given $\Theta_\Lambda = \theta_\Lambda$, $(\Lambda_1, N_1), \ldots, (\Lambda_{T+1}, N_{T+1})$ are independent and identically distributed, and the intensity of events of year $t \in \{1, \ldots, T+1\}$ has conditional marginal distribution*

$$\Lambda_t \sim Gamma(\alpha, \theta_\Lambda/\alpha)$$

   for a given parameter $\alpha > 0$.
3. *Given $\Theta_\Lambda = \theta_\Lambda$ and $\Lambda_t = \lambda_t$, the frequencies*

$$N_t \sim Poisson(V\lambda_t).$$

4. *The financial company has M expert opinions Δ_m, $m = 1, \ldots, M$ about Θ_Λ. Given $\Theta_\Lambda = \theta_\Lambda$, Δ_m and (Λ_t, N_t) are independent for all m and t, and $\Delta_1, \ldots, \Delta_M$ are independent and identically distributed with*

$$\Delta_m \sim Gamma(\xi, \theta_\Lambda/\xi).$$

Remark 4.14

- Given that $\Theta_\Lambda \sim Gamma(a, 1/b)$, $\mathrm{E}[\Theta_\Lambda] = a/b$ and $\mathrm{Var}[\Theta_\Lambda] = a/b^2$. These are the prior two moments of the underlying risk characteristics Θ_Λ. The prior is determined by external data (or the regulator). In general parameters a and b can be estimated by the maximum likelihood method using the data from all banks.
- The first moments are

$$\begin{aligned}
\mathrm{E}[\,\Lambda_t|\,\Theta_\Lambda] &= \Theta_\Lambda, \\
\mathrm{E}[\Lambda_t] &= \frac{a}{b}, \\
\mathrm{E}[\,N_t|\,\Theta_\Lambda, \Lambda_t] &= V\ \Lambda_t, \\
\mathrm{E}[\,N_t|\,\Theta_\Lambda] &= V\ \Theta_\Lambda, \\
\mathrm{E}[N_t] &= V\ \frac{a}{b}.
\end{aligned}$$

The second moments are given by

$$\begin{aligned}
\mathrm{Var}[\,\Lambda_t|\,\Theta_\Lambda] &= \alpha^{-1}\ \Theta_\Lambda^2, \\
\mathrm{Var}[\Lambda_t] &= \alpha^{-1}\ \frac{a^2}{b^2} + (\alpha^{-1}+1)\ \frac{a}{b^2}, \\
\mathrm{Var}[\,N_t|\,\Theta_\Lambda, \Lambda_t] &= V\ \Lambda_t, \\
\mathrm{Var}[\,N_t|\,\Theta_\Lambda] &= V\ \Theta_\Lambda + V^2\,\alpha^{-1}\ \Theta_\Lambda^2, \\
\mathrm{Var}[N_t] &= V\ \frac{a}{b} + V^2\,\alpha^{-1}\ \frac{a^2}{b^2} + V^2\,(\alpha^{-1}+1)\ \frac{a}{b^2}.
\end{aligned}$$

Note that if we measure diversification in terms of the variational coefficient (Vco) we obtain

$$\lim_{V\to\infty} \mathrm{Vco}^2[\,N_t|\,\Theta_\Lambda] = \lim_{V\to\infty} \frac{\mathrm{Var}[\,N_t|\,\Theta_\Lambda]}{\mathrm{E}^2[\,N_t|\,\Theta_\Lambda]} = \alpha^{-1} > 0 \tag{4.87}$$

and

$$\lim_{V\to\infty} \mathrm{Vco}^2[N_t] = \lim_{V\to\infty} \frac{\mathrm{Var}[N_t]}{\mathrm{E}^2[N_t]} = \alpha^{-1} + \left(\alpha^{-1}+1\right)\ a^{-1} > 0. \tag{4.88}$$

That is, the model makes perfect sense from a practical perspective. Namely, as volume increases, $V \to \infty$, there always remains a non-diversifiable element; see (4.87) and (4.88). This is exactly what has been observed in practice and what regulators require from internal models. Note that if we model Λ_t as constant and known, then $\mathrm{Vco}^2[N_t|\Lambda_t] \to 0$ as $V \to \infty$.
- Contrary to the developments in the previous section, where the intensity Λ_t was constant over time, now Λ_t is a stochastic process. From a practical point of view, it is not plausible that the intensity of the annual counts is constant over time. In such a setting parameter risks completely vanish if we have infinitely many

observed years or infinitely many expert opinions, respectively (see theorem 3.6 (a) and (c) in Lambrigger, Shevchenko and Wüthrich [141]). This is because Λ_t can then be perfectly forecasted. In the present model, parameter risks will also decrease with increasing information. As we gain information the posterior standard deviation of Θ_Λ will converge to 0. However, since Λ_{T+1} viewed from time T is always random, the posterior standard deviation for Λ_{T+1} will be finite.

- Note that conditionally given $\Theta_\Lambda = \theta_\Lambda$, N_t has a negative binomial distribution with probability weights for $n \geq 0$,

$$\Pr[N_t = n|\theta_\Lambda] = \binom{\alpha+n-1}{n}\left(\frac{\alpha}{\alpha+\theta_\Lambda V}\right)^{\alpha}\left(\frac{\theta_\Lambda V}{\alpha+\theta_\Lambda V}\right)^{n}. \tag{4.89}$$

  That is, we could directly work with a negative binomial distribution, instead of introducing stochastic intensity Λ_t explicitly. In Sect. 7.12, we extend this model to the case of many risks with dependence induced by the dependence between risk profiles Λ_t of different risks.

- Δ_m denotes the expert opinion of expert m which predicts the true risk characteristics Θ_Λ of his company. We have

$$\begin{aligned} \mathrm{E}[\Delta_m|\,\Theta_\Lambda] &= \mathrm{E}\left[\Lambda_j|\,\Theta_\Lambda\right] = \mathrm{E}\left[N_j/V|\,\Theta_\Lambda\right] = \Theta_\Lambda, \\ \mathrm{Var}[\,\Delta_m|\,\Theta_\Lambda] &= \Theta_\Lambda^2/\xi,\ \ \mathrm{Vco}[\,\Delta_m|\,\Theta_\Lambda] = \xi^{-1/2}. \end{aligned} \tag{4.90}$$

  That is, the relative uncertainty Vco in the expert opinion does not depend on the value of Θ_Λ. That means that ξ can be given externally, for example, by the regulator, who is able to give a lower bound to the uncertainty. Moreover, we see that the expert predicts the average frequency for his company. Alternatively, ξ can be estimated using a method of moments as in (4.75).

Denote $\mathbf{\Lambda} = (\Lambda_1, \ldots, \Lambda_T)'$, $\mathbf{N} = (N_1, \ldots, N_T)'$ and $\mathbf{\Delta} = (\Delta_1, \ldots, \Delta_M)'$. Then, using Bayes's theorem, the joint posterior density of the random vector $(\Theta_\Lambda, \mathbf{\Lambda})$ given observations $\mathbf{N} = \mathbf{n}$ and $\mathbf{\Delta} = \boldsymbol{\delta}$ is

$$\pi(\theta_\Lambda, \boldsymbol{\lambda}|\mathbf{n}, \boldsymbol{\delta}) \propto \pi(\mathbf{n}|\theta_\Lambda, \boldsymbol{\lambda})\,\pi(\boldsymbol{\lambda}|\theta_\Lambda)\,\pi(\boldsymbol{\delta}|\theta_\Lambda)\,\pi(\theta_\Lambda).$$

Here, the explicit expressions for the likelihood terms and the prior are

$$\begin{aligned} \pi(\mathbf{n}|\theta_\Lambda, \boldsymbol{\lambda})\,\pi(\boldsymbol{\lambda}|\theta_\Lambda) &= \prod_{t=1}^{T} \frac{(V\lambda_t)^{n_t}}{n_t!}\frac{(\alpha/\theta_\Lambda)^{\alpha}}{\Gamma(\alpha)}\lambda_t^{\alpha-1} \\ &\quad \times \exp\{-\lambda_t(V+\alpha/\theta_\Lambda)\}, \end{aligned} \tag{4.91}$$

$$\pi(\boldsymbol{\delta}|\theta_\Lambda) = \prod_{m=1}^{M} \frac{(\xi/\theta_\Lambda)^{\xi}}{\Gamma(\xi)}\,\delta_m^{\xi-1}\,\exp\{-\delta_m\xi/\theta_\Lambda\}, \tag{4.92}$$

$$\pi(\theta_\Lambda) = \frac{b^a}{\Gamma(a)}\,\theta_\Lambda^{a-1}\,\exp\{-\theta_\Lambda b\}. \tag{4.93}$$

Note that the intensities $\Lambda_1, \ldots, \Lambda_T$ are non-observable. Therefore we take the integral over their densities to obtain the posterior distribution of the random variable Θ_Λ, given $\mathbf{N} = \mathbf{n}$ and $\boldsymbol{\Delta} = \boldsymbol{\delta}$,

$$
\begin{aligned}
\pi(\theta_\Lambda | \mathbf{n}, \boldsymbol{\delta}) &\propto \prod_{t=1}^{T} \binom{\alpha + n_t - 1}{n_t} \left(\frac{\alpha}{\alpha + \theta_\Lambda V}\right)^{\alpha} \left(\frac{\theta_\Lambda V}{\alpha + \theta_\Lambda V}\right)^{n_t} \\
&\quad \times \prod_{m=1}^{M} \frac{(\xi/\theta_\Lambda)^{\xi}}{\Gamma(\xi)} \delta_m^{\xi - 1} \exp\{-\delta_m \xi / \theta_\Lambda\} \frac{b^a}{\Gamma(a)} \theta_\Lambda^{a-1} \exp\{-\theta_\Lambda b\} \\
&\propto \left(\frac{1}{\alpha + \theta_\Lambda V}\right)^{T\alpha + \sum_{t=1}^{T} n_t} \theta_\Lambda^{a - M\xi + \sum_{t=1}^{T} n_t - 1} \\
&\quad \times \exp\left\{-\theta_\Lambda b - \frac{\xi}{\theta_\Lambda} \sum_{m=1}^{M} \delta_m \right\}.
\end{aligned}
\quad (4.94)
$$

Given Θ_Λ, the distribution of the number of losses N_t is negative binomial. Hence, one could start with a negative binomial model for N_t. The reason for the introduction of the random intensities Λ_t is that in Sect. 7.12 we will utilise them to model dependence between different risk cells, by introducing dependence between $\Lambda_t^{(1)}, \ldots, \Lambda_t^{(J)}$, where superscript refers to the risk cell.

Typically, a closed-form expression for the marginal posterior density of Θ_Λ, given $(\mathbf{N}, \boldsymbol{\Lambda})$, cannot be obtained. In general, one can integrate out the latent variables $\Lambda_1, \ldots, \Lambda_T$ numerically through a MCMC approach, as will be done in Sect. 7.13, to obtain an empirical distribution for the posterior of $\pi(\theta_\Lambda | \mathbf{n}, \boldsymbol{\delta})$. This empirical posterior distribution then allows for the simulation of Λ_{T+1} and N_{T+1}, respectively, conditional on data $(\mathbf{N}, \boldsymbol{\Delta})$.

4.5.4 Lognormal Model for Severities

In general, one can use the methodology summarised by Eq. (4.73) to develop a model combining external data, internal data and expert opinion for estimation of the severity. For illustration purposes, this section considers the lognormal severity model; the Pareto severity model is developed in the next section.

Consider modelling severities $X_1, \ldots, X_K, \ldots$ using the lognormal distribution $\mathcal{LN}(\mu, \sigma)$, where $\mathbf{X} = (X_1, \ldots, X_K)'$ are the losses over past T years. Here, we take an approach considered in Sect. 4.3.4, where μ is unknown and σ is known. The unknown μ is treated under the Bayesian approach as a random variable Θ_μ. Then combining external data, internal data and expert opinions can be accomplished using the following model.

Model Assumptions 4.8 (Lognormal-normal-normal) *Let us assume the following severity model for a risk cell in one bank:*

(a) Let $\Theta_\mu \sim \mathcal{N}(\mu_0, \sigma_0)$ be a normally distributed random variable with parameters μ_0, σ_0, which are estimated from (external) market data, i.e. $\pi(\boldsymbol{\theta})$ in (4.73) is the density of $\mathcal{N}(\mu_0, \sigma_0)$.

(b) Given $\Theta_\mu = \mu$, the losses $X_1, X_2, \ldots$ are conditionally independent with a common lognormal distribution:

$$X_k \sim \mathcal{LN}(\mu, \sigma),$$

where σ is assumed known. That is, $f_1(\cdot|\mu)$ in (4.73) corresponds to the density of a $\mathcal{LN}(\mu, \sigma)$ distribution.

(c) The financial company has M experts with opinions Δ_m, $1 \le m \le M$, about Θ_μ. Given $\Theta_\mu = \mu$, Δ_m and X_k are independent for all m and k, and $\Delta_1, \ldots, \Delta_M$ are independent with a common normal distribution:

$$\Delta_m \sim \mathcal{N}(\mu, \xi),$$

where ξ is a parameter estimated using expert opinion data. That is, $f_2(\cdot|\mu)$ corresponds to the density of a $\mathcal{N}(\mu, \xi)$ distribution.

Remark 4.15

- For $M \ge 2$, the parameter ξ can be estimated by the standard deviation of δ_m:

$$\widehat{\xi} = \left(\frac{1}{M-1} \sum_{m=1}^{M} (\delta_m - \overline{\delta})^2 \right)^{1/2}. \tag{4.95}$$

- The hyper-parameters μ_0 and σ_0 are estimated from market data, for example, by maximum likelihood estimation or by the method of moments.
- In practice one often uses an ad-hoc estimate for σ, which usually is based on expert opinion only. However, one could think of a Bayesian approach for σ, but then an analytical formula for the posterior distribution in general does not exist and the posterior needs then to be calculated numerically, for example, by MCMC methods.

Under Model Assumptions 4.8, the posterior density is given by

$$\pi(\mu|\mathbf{x}, \boldsymbol{\delta}) \propto \frac{1}{\sigma_0\sqrt{2\pi}} \exp\left(-\frac{(\mu-\mu_0)^2}{2\sigma_0^2}\right) \prod_{k=1}^{K} \frac{1}{\sigma\sqrt{2\pi}} \exp\left(-\frac{(\ln x_k - \mu)^2}{2\sigma^2}\right)$$
$$\prod_{m=1}^{M} \frac{1}{\xi\sqrt{2\pi}} \exp\left(-\frac{(\delta_m - \mu)^2}{2\xi^2}\right)$$

$$\propto \exp\left[-\left(\frac{(\mu-\mu_0)^2}{2\sigma_0^2}+\sum_{k=1}^{K}\frac{(\ln x_k-\mu)^2}{2\sigma^2}+\sum_{m=1}^{M}\frac{(\delta_m-\mu)^2}{2\xi^2}\right)\right]$$
$$\propto \exp\left[-\frac{(\mu-\widehat{\mu})^2}{2\widehat{\sigma}^2}\right], \tag{4.96}$$

with

$$\widehat{\sigma}^2=\left(\frac{1}{\sigma_0^2}+\frac{K}{\sigma^2}+\frac{M}{\xi^2}\right)^{-1},$$

and

$$\widehat{\mu}=\widehat{\sigma}^2\times\left(\frac{\mu_0}{\sigma_0^2}+\frac{1}{\sigma^2}\sum_{k=1}^{K}\ln x_k+\frac{1}{\xi^2}\sum_{m=1}^{M}\delta_m\right).$$

In summary, we derived the following theorem (also see Lambrigger, Shevchenko and Wüthrich [141]).

Theorem 4.4 *Under Model Assumptions 4.8, the posterior distribution of Θ_μ, given loss information $\mathbf{X}=\mathbf{x}$ and expert opinion $\boldsymbol{\Delta}=\boldsymbol{\delta}$, is a normal distribution $\mathcal{N}(\widehat{\mu},\widehat{\sigma})$ with*

$$\widehat{\sigma}^2=\left(\frac{1}{\sigma_0^2}+\frac{K}{\sigma^2}+\frac{M}{\xi^2}\right)^{-1}$$

and

$$\widehat{\mu}=\mathrm{E}[\Theta_\mu|\mathbf{X}=\mathbf{x},\boldsymbol{\Delta}=\boldsymbol{\delta}]=\omega_1\mu_0+\omega_2\overline{\ln x}+\omega_3\overline{\delta}, \tag{4.97}$$

where $\overline{\ln x}=\frac{1}{K}\sum_{k=1}^{K}\ln x_k$ and the credibility weights are

$$\omega_1=\widehat{\sigma}^2/\sigma_0^2,\ \omega_2=\widehat{\sigma}^2K/\sigma^2,\ \omega_3=\widehat{\sigma}^2M/\xi^2.$$

This theorem yields a natural interpretation of the considered model. The estimator $\widehat{\mu}$ in (4.97) weights the internal and external data as well as the expert opinion in an appropriate manner. Observe that under Model Assumptions 4.8, the mean of the posterior distribution can be calculated explicitly. This is different from the frequency model in Sect. 4.5.2, where asymptotic calculations (Theorem 4.3) were required for the interpretation of the terms. However, interpretation of the terms is exactly the same as in Theorem 4.3. The more credible the information, the higher is the credibility weight in (4.97) – as expected from an appropriate model for combining internal observations, relevant external data and expert opinions.

4.5.5 Pareto Model

Consider modelling severities $X_1, \ldots, X_K, \ldots$ using $Pareto(\gamma, L)$ with a density

$$f(x) = \frac{\gamma}{L}\left(\frac{x}{L}\right)^{-\gamma-1}, \quad x \geq L, \quad \xi > 0, \tag{4.98}$$

where $\mathbf{X} = (X_1, \ldots, X_K)'$ are the losses over past T years. Note that if $\xi > 1$, then the mean is $L\xi/(\xi - 1)$, otherwise the mean does not exist. Here, we take an approach considered in Sect. 4.3.6, where γ is unknown and the threshold L is known. The unknown γ is treated under the Bayesian approach as a random variable Θ_γ. Then, combining external data, internal data and expert opinions can be accomplished using the following model.

Model Assumptions 4.9 (Pareto-gamma-gamma) *Let us assume the following severity model for a risk cell in one bank:*

(a) *Let $\Theta_\gamma \sim Gamma(\alpha_0, \beta_0)$ be a gamma distributed random variable with parameters α_0 and β_0, which are estimated from (external) market data, i.e. $\pi(\boldsymbol{\theta})$ in (4.73) is the density of a $Gamma(\alpha_0, \beta_0)$ distribution.*
(b) *Given, $\Theta_\gamma = \gamma$, the losses $X_1, X_2, \ldots$ in the risk cell are assumed to be conditionally independent and Pareto distributed:*

$$X_k \sim Pareto(\gamma, L),$$

where the threshold $L \geq 0$ is assumed to be known and fixed. That is, $f_1(\cdot|\gamma)$ in (4.73) corresponds to the density of a Pareto(γ, L) distribution.
(c) *A financial company has M experts with opinions Δ_m, $1 \leq m \leq M$, about the parameter Θ_γ. Given $\Theta_\gamma = \gamma$, Δ_m and X_k are independent for all m and k, and $\Delta_1, \ldots, \Delta_M$ are independent and identically distributed with*

$$\Delta_m \sim Gamma(\xi, \gamma/\xi),$$

where ξ is a parameter estimated using expert opinion data; see (4.75). That is, $f_2(\cdot|\gamma)$ corresponds to the density of a $Gamma(\xi, \gamma/\xi)$ distribution.

Theorem 4.5 *Under Model Assumptions 4.9, given loss information $\mathbf{X} = \mathbf{x}$ and expert opinion $\boldsymbol{\Delta} = \boldsymbol{\delta}$, the posterior distribution of Θ_γ is $GIG(\omega, \phi, \nu)$ with the density*

$$\pi(\gamma|\mathbf{x}, \boldsymbol{\delta}) = \frac{(\omega/\phi)^{(\nu+1)/2}}{2K_{\nu+1}(2\sqrt{\omega\phi})}\gamma^{\nu} e^{-\gamma\omega-\gamma^{-1}\phi}, \tag{4.99}$$

where

$$\begin{aligned}\nu &= \alpha_0 - 1 - M\xi_i + K,\\ \omega &= \frac{1}{\beta_0} + \sum_{k=1}^{K} \ln \frac{x_k}{L},\\ \phi &= \xi M \overline{\delta}.\end{aligned} \tag{4.100}$$

Proof This is straightforward from the calculation of the posterior density

$$\begin{aligned}\pi(\gamma|\mathbf{x},\boldsymbol{\delta}) &\propto \gamma^{\alpha_0-1} e^{-\gamma/\beta_0} \prod_{k=1}^{K} \frac{\gamma}{L}\left(\frac{x_k}{L}\right)^{-\gamma-1} \prod_{m=1}^{M} \frac{(\delta_m)^{\alpha-1}}{\beta^{\alpha}} e^{-\delta_m/\beta}\\ &\propto \gamma^{\alpha_0-1-M\xi+K} \exp\left[-\gamma\left(\frac{1}{\beta_0} + \sum_{k=1}^{K} \ln \frac{x_k}{L}\right) - \frac{\xi M \overline{\delta}}{\gamma}\right].\end{aligned} \tag{4.101}$$

Hence, as in Theorem 4.2 for modelling Poisson frequencies, the posterior distribution is a GIG with the nice property that the term γ in the exponent in (4.101) is only affected by the internal observations, whereas the term $1/\gamma$ is driven by the expert opinions.

Remark 4.16 It seems natural to generalise this result to the case of the GIG prior distribution. In particular, changing the assumption a) in Model Assumptions 4.9 to $\Theta_\gamma \sim GIG(\omega_0, \phi_0, \nu_0)$, with the parameters ν_0, ω_0, ϕ_0, the posterior density $\pi(\gamma|\mathbf{x},\boldsymbol{\delta})$ is $GIG(\omega, \phi, \nu)$ with

$$\begin{aligned}\nu &= \nu_0 - M\xi + K,\\ \omega &= \omega_0 + \sum_{k=1}^{K} \ln(x_k/L),\\ \phi &= \phi_0 + \xi M \overline{\delta}.\end{aligned} \tag{4.102}$$

Note that for $\phi_0 = 0$, the prior GIG is a gamma distribution and hence we are in the Pareto-gamma-gamma situation of Model Assumptions 4.9.

The posterior mean (that can be used as a Bayesian point estimator for γ) can be calculated as

$$\mathrm{E}[\Theta_\gamma|\mathbf{X}=\mathbf{x},\boldsymbol{\Delta}=\boldsymbol{\delta}] = \sqrt{\frac{\phi}{\omega}} \frac{K_{\nu+2}(2\sqrt{\omega\phi})}{K_{\nu+1}(2\sqrt{\omega\phi})}; \tag{4.103}$$

see Appendix A.2.11. The maximum likelihood estimator of the Pareto tail index γ is also easily calculated as

$$\widehat{\gamma}^{\mathrm{MLE}} = \frac{K}{\sum_{k=1}^{K} \ln(x_k/L)}. \tag{4.104}$$

Then, completely analogous to Theorem 4.3, the following theorem gives a natural interpretation of the Bayesian (posterior mean) estimator.

Theorem 4.6 *Under Model Assumptions 4.9, the following asymptotic relations hold* P*-almost surely:*

(a) If $K \to \infty$ *then* $\mathrm{E}[\Theta_\gamma|\mathbf{X}, \boldsymbol{\Delta}] \to \mathrm{E}[X_k|\Theta_\gamma = \gamma]/V = \gamma$.
(b) If $\mathrm{Vco}[\Delta_m|\Delta_\gamma] \to 0$ *then* $\mathrm{E}[\Theta_\gamma|\mathbf{X}, \boldsymbol{\Delta}] \to \Delta_m,\ m = 1, \ldots, M$.
(c) If $M \to \infty$ *then* $\mathrm{E}[\Theta_\gamma|\mathbf{X}, \boldsymbol{\Delta}] \to \mathrm{E}[\Delta_m|\Theta_\gamma = \gamma] = \gamma$.
(d) If $\mathrm{Vco}[\Delta_m|\Theta_\gamma] \to \infty,\ m = 1, \ldots, M$ *then*

$$\mathrm{E}[\Theta_\gamma|\mathbf{X}, \boldsymbol{\Delta}] \to (1 - w)\,\mathrm{E}[\Theta_\gamma] + w\widehat{\gamma}^{\mathrm{MLE}},$$

where $w = K\beta_0/(\widehat{\gamma}^{\mathrm{MLE}} + K\beta_0)$.
(e) If $\mathrm{E}[\Theta_\gamma] = constant$ *and* $\mathrm{Vco}[\Theta_\gamma] \to 0$ *then* $\mathrm{E}[\Theta_\gamma|\mathbf{X}, \boldsymbol{\Delta}] \to \mathrm{E}[\Theta_\gamma]$.

Remark 4.17

- Theorem 4.6 basically says that the higher the precision of a particular source of risk information, the higher its corresponding credibility weight. This means that we obtain the same interpretations as for Theorem 4.3 and formula (4.97).
- Observe that Model Assumptions 4.9 lead to an infinite mean model because the Pareto parameter Θ_γ can be less than one with positive probability. For finite mean models, the range of possible γ has to be restricted to $\gamma > 1$. This does not impose difficulties; see Sect. 2.9.4.

The update process of (4.100) and (4.102) has again a simple linear form when new information arrives. The posterior density (4.99) does not change its type every time a new observation arrives. In particular, only the parameter ω is affected by a new observation.

Information update process. Loss $k \to$ loss $k + 1$:

$$\begin{aligned} \nu_{k+1} &= \nu_k + 1, \\ \omega_{k+1} &= \omega_k + \ln(x_{k+1}/L), \\ \phi_{k+1} &= \phi_k. \end{aligned} \tag{4.105}$$

The following example illustrates the simplicity and robustness of the posterior mean estimator.

Example 4.5 Assume that a bank would like to model its risk severity by a Pareto distribution with tail index Θ_γ. The regulator provides external prior data, saying that $\Theta_\gamma \sim Gamma(\alpha_0, \beta_0)$ with $\alpha_0 = 4$ and $\beta_0 = 9/8$, i.e. $\mathrm{E}[\Theta_\gamma] = 4.5$ and $\mathrm{Vco}[\Theta_\gamma] = 0.5$. The bank has one expert opinion $\delta = 3.5$ with $\mathrm{Vco}[\Delta|\Theta_\gamma = \gamma] = 0.5$, i.e. $\xi = 4$. We then observe the losses given in Table 4.2 (sampled from a $Pareto(4, 1)$ distribution). In Fig. 4.5, the following estimators are compared:

- the classical maximum likelihood estimate

$$\widehat{\gamma}_k^{\mathrm{MLE}} = \frac{k}{\sum_{i=1}^{k} \ln(x_i/L)}; \tag{4.106}$$

- the Bayesian posterior mean estimate (4.48)

$$\gamma_k^{(2)} = \mathrm{E}[\Theta_\gamma | X_1 = x_1, \ldots, X_k = x_k], \tag{4.107}$$

  that does not account for expert opinions; and
- the Bayesian posterior mean estimate

$$\widehat{\gamma}_k^{(3)} = \mathrm{E}[\Theta_\gamma | X_1 = x_1, \ldots, X_k = x_k, \Delta = \delta], \tag{4.108}$$

  given by (4.103).

Figure 4.5 shows the high volatility of the maximum likelihood estimator for small numbers k. It is very sensitive to newly arriving losses. The estimator $\widehat{\gamma}_k^{(3)}$ shows a much more stable behaviour around the true value $\alpha = 4$, most notably when a few data points are available. This example also shows that consideration of the relevant external data and well-specified expert opinions stabilises and smoothes the estimator in an appropriate way.

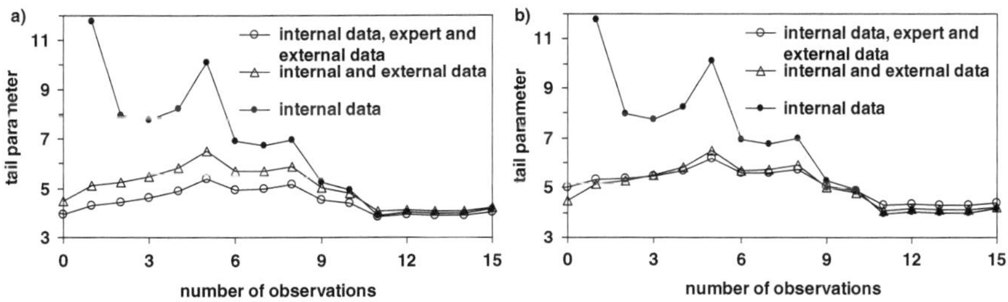

Fig. 4.5 (∘) The Bayes estimate $\widehat{\gamma}_k^{(3)}$, $k = 1, \ldots, 15$, combines the internal data simulated from *Pareto*(4, 1), external data giving $\mathrm{E}[\Theta_\gamma] = 4.5$, and expert opinion δ. It is compared with the Bayes estimate $\widehat{\gamma}_k^{(2)}$ (△), that combines external data and internal data, and the classical maximum likelihood estimate $\widehat{\gamma}_k^{\mathrm{MLE}}$ (•). **(a)** is the case of expert opinion $\delta = 3$ and **(b)** is the case of expert opinion $\delta = 3$. See Example 4.5 for details

4.6 Combining Data Sources Using Credibility Theory

Quantification of the frequency and severity distributions of the low-frequency/high-severity losses (that typically account for most of the operational risk capital) is a challenging task. The data are so limited that often full quantification of frequency, severity and related prior distributions is problematic. In this situation, methods of credibility theory are very useful as they require less information. Credibility theory approach has been successfully used in the insurance industry and actuarial sciences for many decades. It can be used to estimate frequency and severity distributions of the low frequency large losses in each risk cell by taking into account bank internal data, expert opinions and industry data. An excellent textbook on credibility theory is Bühlmann and Gisler [44]; also see Kaas, Goovaerts, Dhaene and Denuit ([130], section 7.2).

Consider a model parameterised by θ that generates data $X_1, \ldots, X_n, \ldots$. In general we are interested in estimation of some function of θ (e.g. $\mu(\theta)$) given past data $\mathbf{X} = (X_1, \ldots, X_n)'$. Under the Bayesian approach, θ is modelled by random variable Θ. Let $\widehat{\mu(\Theta)}$ be some estimator of $\mu(\Theta)$. Then the unconditional MSEP (mean square error of prediction) of an estimator $\widehat{\mu(\Theta)}$ is

$$\begin{aligned}\text{MSEP} &= \text{E}[(\mu(\Theta) - \widehat{\mu(\Theta)})^2] \\ &= \text{E}\left[\text{E}[(\mu(\Theta) - \text{E}[\mu(\Theta)|\mathbf{X}] + \text{E}[\mu(\Theta)|\mathbf{X}] - \widehat{\mu(\Theta)})^2]|\mathbf{X}\right] \\ &= \text{E}[(\mu(\Theta) - \text{E}[\mu(\Theta)|\mathbf{X}])^2] + \text{E}[\text{E}[(\mu(\Theta)|\mathbf{X}] - \widehat{\mu(\Theta)})^2]\end{aligned}$$

It is easy to see that the posterior mean

$$\widehat{\mu(\Theta)} = \text{E}[\mu(\Theta)|\mathbf{X}]$$

minimises MSEP and thus is the best estimator with respect to the quadratic loss function; also see Sect. 2.10.

In general, the posterior mean cannot be found in closed form. The prior and conditional distributions should also be specified which is certainly a problem in the case of small datasets. The credibility theory initiated by Bühlmann [42] considers estimators which are linear in observations $X_1, X_2, \ldots$ and minimise quadratic loss function. This allows for simple calculation of the estimators, referred to as *credibility estimators* or linear Bayes estimators.

The credibility estimators have already appeared in the above sections. For example, the estimator for the expected intensity of events (4.20), when frequencies are modelled by $Poisson(\Lambda = \lambda)$ and the prior for Λ is $Gamma(\alpha, \beta)$, is

$$\widehat{\Lambda} = \text{E}[\Lambda|N_1, \ldots, N_T] = w\overline{N} + (1 - w)\lambda_0,$$

where

- $\overline{N} = \frac{1}{T}\sum_{t=1}^{T} N_t$ is the estimator of λ using the observed counts only;
- $\lambda_0 = \alpha\beta$ is the estimate of λ using a prior distribution only (e.g. specified by expert or from external data); and
- $w = \frac{T}{T+1/\beta}$ is the credibility weight in [0,1) used to combine λ_0 and $\overline{N}$.

The estimator $\widehat{\Lambda}$ is linear in data $N_1, \ldots, N_T$ and minimises the mean square error of prediction

$$\text{E}[(\widehat{\Lambda} - \Lambda)^2].$$

Of course, the estimator $\widehat{\Lambda}$ was derived assuming a specific prior distribution. The credibility theory avoids this assumption and requires the knowledge of the first and second moments only. To demonstrate the idea, consider a simplistic credibility model.

Model Assumptions 4.10 (Simple credibility model)

- *Given, $\Theta = \theta$, random variables $X_1, X_2, \ldots$ are independent and identically distributed with*

$$\mu(\theta) = \mathrm{E}[X_j|\Theta = \theta], \quad \sigma^2(\theta) = \mathrm{Var}[X_j|\Theta = \theta].$$

- *Θ is a random variable with*

$$\mu_0 = \mathrm{E}[\mu(\Theta)], \quad \tau^2 = \mathrm{Var}[\mu(\Theta)].$$

The aim of credibility estimators is to find an estimator of $\mu(\Theta)$ which is linear in $X_1, \ldots, X_n$, i.e.

$$\widehat{\mu(\Theta)} = \widehat{a}_0 + \widehat{a}_1 X_1 + \cdots + \widehat{a}_n X_n$$

and minimise quadratic loss function, i.e.

$$(\widehat{a}_0, \ldots, \widehat{a}_n) = \min_{a_0,\ldots,a_n} \mathrm{E}\left[(\mu(\Theta) - a_0 - a_1 X_1 - \cdots - a_n X_n)^2\right]$$

The invariance of the distribution of $X_1, \ldots, X_n$ under permutations of X_j, gives $\widehat{a}_1 = \widehat{a}_2 = \cdots = \widehat{a}_n := \widehat{b}$. Then, by solving the minimisation problem for two parameters a_0 and b by setting corresponding partial derivatives with respect to a_0 and b to zero, obtain

$$\widehat{\mu(\Theta)} = w\overline{X} + (1 - w)\mu_0,$$

where

$$w = \frac{n}{n + \sigma^2/\tau^2}, \quad \overline{X} = \frac{1}{n}\sum_{i=1}^{n} X_i.$$

For details of the proof and discussion, see Bühlmann and Gisler ([44], section 3.1).

4.6.1 Bühlmann-Straub Model

In operational risk we are interested in the LDA model for the annual loss (4.1). That is, for a risk cell, the annual number of events $N_1, N_2, \ldots$ are modelled as random variables from some discrete distribution $P(\cdot|\boldsymbol{\lambda})$ and the severities of the events $X_1, X_2, \ldots$ are modelled as random variables from a continuous distribution

$F(\cdot|\boldsymbol{\theta})$. Under the Bayesian approach, $\boldsymbol{\lambda}$ and $\boldsymbol{\theta}$ are distribution parameters which are not known and are modelled by random variables $\boldsymbol{\Lambda}$ and $\boldsymbol{\Theta}$ respectively. Often, the credibility approach takes the empirical Bayes setup (see Sect. 4.3.1). That is, it considers a group of risks, where $\boldsymbol{\Lambda}$ are different for different risks but are drawn from the same distribution (the prior distribution) common across the risks (and similar for $\boldsymbol{\Theta}$). In this framework we do not consider risks individually but regard each risk as embedded in a group of *similar* risks (*collective*). If a pure Bayesian setup is taken then the prior distribution is specified by the expert.

Usually, the credibility estimators are used to estimate expected number of events or expected loss. However, in general, they can be applied to estimate any square integrable valued random variable Z based on some known random vector $\mathbf{Y}$. For example, the elements of $\mathbf{Y}$ can be the maximum likelihood estimators, transformed data, quantiles, etc. In particular, the credibility estimators for the severity and frequency distribution parameters can be calculated using the model developed in Bühlmann and Straub [46]; also see Bühlmann and Gisler ([44], Model Assumptions 4.1 and Theorems 4.2, 4.4).

Model Assumptions 4.11 (Bühlmann-Straub model) *Consider a portfolio of J risks modelled by random variables $Y_{j,k}$: $k = 1, 2, \dots, j = 1, \dots, J$. Assume that, for known weights $w_{j,k}$, the j-th risk is characterised by an individual risk profile θ_j, which is itself the realisation of a random variable Θ_j, and*

- *Given Θ_j, the data $Y_{j,1}, Y_{j,2}, \dots$ are independent with*

$$\mathrm{E}[Y_{j,k}|\Theta_j] = \mu(\Theta_j), \quad \mathrm{Var}[Y_{j,k}|\Theta_j] = \sigma^2(\Theta_j)/w_{j,k} \tag{4.109}$$

  for all $j = 1, \dots, J$.
- *The pairs $(\Theta_1, Y_{1,k};\ k \geq 1), \dots, (\Theta_J, Y_{J,k};\ k \geq 1)$ are independent.*
- *$\Theta_1, \dots, \Theta_J$ are independent and identically distributed with*

$$\mu_0 = \mathrm{E}[\mu(\Theta_j)], \quad \sigma^2 = \mathrm{E}[\sigma^2(\Theta_j)], \quad \tau^2 = \mathrm{Var}[\mu(\Theta_j)].$$

  for all j.

Theorem 4.7 (Bühlmann-Straub credibility estimators) *Under the Model Assumptions 4.11, given the available data $\mathbf{Y}_j = (Y_{j,1}, \dots, Y_{j,K_j})'$, $j = 1, \dots, J$, the inhomogeneous and homogeneous credibility estimators of $\mu(\Theta_j)$ are given as follows:*

- *The inhomogeneous credibility estimator is*

$$\widehat{\widehat{\mu(\Theta_j)}} = \alpha_j \overline{Y}_j + (1 - \alpha_j)\mu_0. \tag{4.110}$$

- *The homogeneous credibility estimator is*

$$\widehat{\widehat{\mu(\Theta_j)}} = \alpha_j \overline{Y}_j + (1 - \alpha_j)\widehat{\mu_0}. \tag{4.111}$$

Here:

$$\widehat{\mu_0} = \sum_{j=1}^{J} \frac{\alpha_j}{\alpha_0} \overline{Y}_j, \quad \overline{Y}_j = \sum_{k=1}^{K_j} \frac{w_{j,k}}{\widetilde{w}_j} Y_{j,k}, \quad \alpha_j = \frac{\widetilde{w}_j}{\widetilde{w}_j + \sigma^2/\tau^2},$$

$$\alpha_0 = \sum_{j=1}^{J} \alpha_j, \quad \widetilde{w}_j = \sum_{k=1}^{K_j} w_{j,k}.$$

Remark 4.18

- Note that K_j may vary between the risks.
- Structural parameters μ_0, σ^2 and τ^2 can be determined using expert opinions (pure Bayes) or using data of all risks (empirical Bayes).
- The difference between inhomogeneous and homogeneous credibility estimators is that the latter estimates μ_0 by $\widehat{\mu_0}$ using the data for all risks.

Using the above credibility estimators, Bühlmann, Shevchenko and Wüthrich [45] suggested a "toy" model for operational risk, where the Pareto and Poisson distributions were used for modelling severity and frequency respectively. Although the model might be simple, it is a very good illustration of a consistent credibility approach for estimating low-frequency/high-severity operational risks. Below we illustrate the use of the model in a simple case of J risks without considering a full hierarchical model.

4.6.2 Modelling Frequency

Consider a collection of J similar risk cells; see Fig. 4.6. Let $N_{j,k}$ be the annual number of events, with the event losses exceeding some high threshold L, in the j-th risk cell ($j = 1, \ldots, J$) in the k-th year. That is, the same threshold L is used across all risk cells in a collection (for example, one can choose the threshold equal to the threshold in the database of external data).

Model Assumptions 4.12 (Poisson frequency) *Assume that:*

(a) Given, $\Lambda_j = \lambda_j$, $N_{j,k}$ are independent and distributed from $Poisson(\nu_j \lambda_j)$, i.e.

$$\Pr\left[N_{j,k} = n | \Lambda_j = \lambda_j\right] = \frac{\left(\nu_j \lambda_j\right)^n}{n!} \exp\left(-\nu_j \lambda_j\right) \tag{4.112}$$

and moments

$$\mathrm{E}[N_{j,k}|\Lambda_j] = \nu_j \Lambda_j, \;\; \mathrm{Var}[N_{j,k}|\Lambda_j] = \nu_j \Lambda_j. \tag{4.113}$$

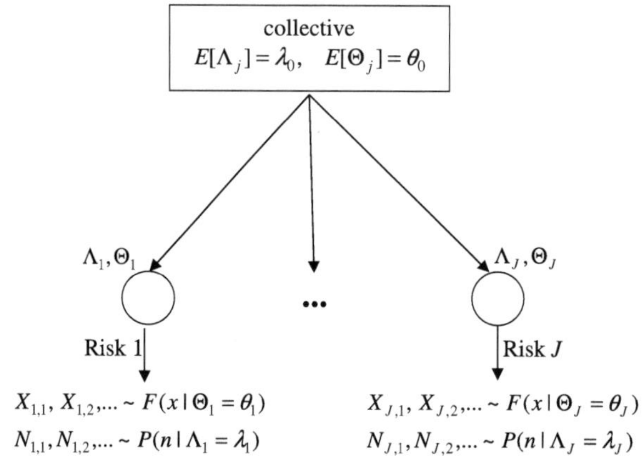

Fig. 4.6 Example of the credibility model for operational risk. Given $\Theta_j = \theta_j$ and $\Lambda_j = \lambda_j$, $X_{j,k} \sim Pareto(a_j\theta_j, L_j)$ and $N_{j,k} \sim Poisson(\nu_j\lambda_j)$ are the losses (above threshold L_j and their annual frequencies in risk cells $j = 1, \dots, J$ respectively. The risk profiles Λ_j are drawn from common distribution with $\mathrm{E}[\Lambda_j] = \lambda_0$,$\mathrm{Var}[\Lambda_j] = (\omega_0)^2$; risk profiles Θ_j are from common distribution with $\mathrm{E}[\Theta_j] = \theta_0$, $\mathrm{Var}[\Theta_j] = (\tau_0)^2$. Scaling factors a_j and ν_j for the relative differences between the risks can be specified using expert opinions or known factors

The arrival rate parameter is defined as $\nu_j\Lambda_j$, where ν_j are the known a priori constants and Λ_j are the risk profiles of the bank cells. The constants ν_j are scaling factors, reflecting differences in frequencies across the risks, discussed below.

(b) Assume that $\Lambda_1, \dots, \Lambda_J$ are independent and identically distributed with

$$\mathrm{E}[\Lambda_j] = \lambda_0 \quad \textit{and} \quad \mathrm{Var}[\Lambda_j] = (\omega_0)^2,$$

and $(\Lambda_1, N_{1,k};\ k \geq 1), \dots, (\Lambda_J, N_{J,k}; k \geq 1)$ are independent.

(c) The available data are $\{N_{j,1}, \dots, N_{j,K_j}\ :\ j = 1, \dots, J\}$.

4.6.2.1 The Arrival Rate MLE Using Data in a Risk Cell

Under the first assumption in Model Assumptions 4.12, $N_{j,k}$, $k = 1, \dots, K_j$ in the j-th risk cell are conditionally independent. Thus, given $\Lambda_j = \lambda_j$, the standard MLE of λ_j is

$$\widehat{\Lambda_j} = \frac{1}{\widetilde{\nu}_j}\sum_{k=1}^{K_j} N_{j,k}, \quad \widetilde{\nu}_j = \nu_j K_j \tag{4.114}$$

with

$$\mathrm{E}[\widehat{\Lambda_j}|\Lambda_j = \lambda_j] = \lambda_j,$$
$$\mathrm{Var}[\widehat{\Lambda_j}|\Lambda_j = \lambda_j] = \lambda_j/\widetilde{\nu}_j.$$

Again, a common situation in operational risk is that only few large losses are observed for some risk cells, so the standard MLEs of parameters λ_j will not be reliable. The idea is to use data from a collection of risks to improve the estimates of the arrival rate parameter.

4.6.2.2 The Arrival Rate Estimator Improved by Bank Data

Under the second assumption in Model Assumptions 4.12, $\Lambda_1, \Lambda_2, \ldots$ are independent and identically distributed with $\mathrm{E}[\Lambda_j] = \lambda_0$ and $\mathrm{Var}[\Lambda_j] = (\omega_0)^2$. Observe that the standardised frequencies $F_{j,k} = N_{j,k}/\nu_j$ satisfy

$$\mathrm{E}[F_{j,k}|\Lambda_j] = \Lambda_j \quad \text{and} \quad \mathrm{Var}[F_{j,k}|\Lambda_j] = \Lambda_j/\nu_j. \tag{4.115}$$

Thus $F_{j,k}$ satisfy the Bühlmann-Straub model (4.109), (4.110), and (4.111) and the credibility estimator for Λ_j is given by

$$\widehat{\widehat{\Lambda_j}} = \gamma_j \widehat{\Lambda_j} + (1 - \gamma_j)\lambda_0, \tag{4.116}$$

where

$$\gamma_j = \frac{\widetilde{\nu}_j}{\widetilde{\nu}_j + \lambda_0/(\omega_0)^2}. \tag{4.117}$$

The structural parameters λ_0 and ω_0 can be estimated using all data from a collection of J risks by solving two nonlinear equations (using, for example, an iterative procedure; see Bühlmann and Gisler [44], pp. 102–103):

$$(\widehat{\omega}_0)^2 = \max\left[c \times \left\{A - \frac{J\widehat{\lambda}_0}{\nu_0}\right\}, 0\right], \quad \widehat{\lambda}_0 = \frac{1}{\widetilde{\gamma}}\sum_j \gamma_j \widehat{\Lambda}_j, \tag{4.118}$$

where

$$\nu_0 = \sum_{j=1}^{J} \widetilde{\nu}_j, \; A = \frac{J}{J-1}\sum_{j=1}^{J}\frac{\widetilde{\nu}_j}{\nu_0}(\widehat{\Lambda_j} - \overline{F})^2, \; \widetilde{\gamma} = \sum_j \gamma_j,$$

$$\overline{F} = \frac{1}{J}\sum_{j=1}^{J}\widehat{\Lambda_j}, \quad c = \frac{J}{J-1}\left\{\sum_{j=1}^{J}\frac{\widetilde{\nu}_j}{\nu_0}\left(1 - \frac{\widetilde{\nu}_j}{\nu_0}\right)\right\}^{-1}.$$

Here, the coefficients γ_j are given in (4.117) with λ_0 and ω_0 replaced by $\widehat{\lambda}_0$ and $\widehat{\omega}_0$ respectively.

Remark 4.19

- Based on the cell data and all data in a collection of J risks, the best credibility estimator of the arrival rate parameter in the j-th cell is $\nu_j \widehat{\widehat{\Lambda_j}}$.

- We assumed that the constants ν_j are known a priori. Note that these constants are defined up to a constant factor, that is, the coefficients γ_j (and the final estimates of the arrival rate parameters) will not change if all ν_j are changed by the same factor. Hence, only relative differences between risks play a role. These constants have the interpretation of a priori differences and can be fixed by the expert opinions on expected annual number of losses exceeding threshold L for each risk cell. For example, the expert may estimate the expected annual number of events (exceeding threshold L_j) n_j in the j-th cell as $\widehat{n}_j$ and estimate ν_j as $\widehat{n}_j/\lambda_0$. Only relative differences play a role here, thus (without loss of generality) λ_0 can be set equal to 1. For an example of using expert opinions for quantification of frequency and severity distributions, see Alderweireld, Garcia and Léonard [6], and Shevchenko and Wüthrich [218].

4.6.3 Modelling Severity

Again, consider a collection of J similar risk cells; see Fig. 4.6.

Model Assumptions 4.13 (Pareto severity) *Assume that:*

- *Given,* $\Theta_j = \theta_j$, *the losses* $X_{j,k}$, $k \geq 1$ *above threshold* L_j *in the* j*-th risk cell* $(j = 1, \ldots, J)$ *are independent and Pareto distributed,* $Pareto(a_j\theta_j, L)$, *with the density*

$$f(x) = \frac{a_j\theta_j}{L}\left(\frac{x}{L}\right)^{-a_j\theta_j-1} \tag{4.119}$$

  respectively, for $x \geq L$ *and* $a_j\theta_j > 0$. *It is assumed that the threshold* L *is known and the same across risk cells. Here* a_j *are known a priori constants (differences) and* θ_j *are the risk profiles of the cells in the bank. The constants* a_j *are scaling factors, reflecting differences in severities across the risks, that can be fixed by experts as discussed below.*
- *Assume that* $\Theta_1, \ldots, \Theta_J$ *are independent and identically distributed with*

$$\mathrm{E}[\Theta_j] = \theta_0 \quad and \quad \mathrm{Var}[\Theta_j] = (\tau_0)^2,$$

  and $(\Theta_1, X_{1,k};\ k \geq 1), \ldots, (\Theta_J, X_J, k;\ k \geq 1)$ *are independent. Here,* θ_0 *is a risk profile of the collection.*
- *The available data are denoted as* $\{X_{j,1}, \ldots, X_{j,\widetilde{K}_j} : j = 1, \ldots, J\}$.

Remark 4.20

- Note that the number of available losses in the j-th risk cell, denoted as $\widetilde{K}_j$, is the number of events over K_j years. The latter is the number of observed years for modelling annual frequencies in the previous section.
- The results in this section are valid if thresholds are different for different risk cells, although in the previous section for modelling frequencies we assumed the same threshold across risk cells.

- The Pareto distribution is often used in the insurance industry to model large claims and is a good candidate for modelling large operational risk losses. It is interesting to note that the conditional distribution of the losses exceeding any higher level $\widetilde{L}$ is also a Pareto distribution with parameters $a_j\theta_j$ and $\widetilde{L}$.

4.6.3.1 The Tail Parameter MLE Using Data in a Risk Cell

Under the first assumption in Model Assumptions 4.13, the losses $X_{j,k}$, $k \geq 1$ in the j-th risk cell are conditionally (given Θ_j) independent and Pareto distributed. Thus MLE of θ_j is

$$\widehat{\Psi}_j = \left[\frac{a_j}{\widetilde{K}_j}\sum_{k=1}^{\widetilde{K}_j}\ln\left(\frac{X_{j,k}}{L}\right)\right]^{-1}. \tag{4.120}$$

It is easy to show (see Rytgaard [206]) that an unbiased estimator of θ_j is

$$\widehat{\Theta}_j = \frac{\widetilde{K}_j - 1}{\widetilde{K}_j}\widehat{\Psi}_j, \tag{4.121}$$

with

$$\mathrm{E}[\widehat{\Theta}_j|\Theta_j = \theta_j] = \theta_j, \quad \mathrm{Var}[\widehat{\Theta}_j|\Theta_j = \theta_j] = \frac{(\theta_j)^2}{\widetilde{K}_j - 2}. \tag{4.122}$$

A common situation in operational risk is that only a few losses are observed for certain risk cells. Thus the standard MLE $a_j\widehat{\Theta}_j$ (based on the data in the j-th risk cell only) for the Pareto tail parameters will not be reliable (this is easy to see from the variance in (4.122)). The idea is to use the collective losses (from bank, industry, etc) to improve the estimates of the Pareto parameters in the risk cells.

4.6.3.2 The Tail Parameter Estimator Improved by Collective Data

The tail parameter estimator $a_j\widehat{\Theta}_j$ can be improved using all data in the collection of J risks as follows. Under the second assumption in Model Assumptions 4.13, $\Theta_1, \ldots, \Theta_J$ are independent and identically distributed random variables with $\mathrm{E}[\Theta_j] = \theta_0$ and $\mathrm{Var}[\Theta_j] = (\tau_0)^2$, where θ_0 is a risk profile for the whole collective. Observe that the unbiased estimators $\widehat{\Theta}_j$, see (4.122), satisfy the assumptions of the Bühlmann-Straub model (4.109), (4.110), and (4.111) and thus the credibility estimator is given by

$$\widehat{\widehat{\Theta}}_j = \alpha_j\widehat{\Theta}_j + (1 - \alpha_j)\theta_0, \tag{4.123}$$

where

$$\alpha_j = \frac{\widetilde{K}_j - 2}{\widetilde{K}_j - 1 + (\theta_0/\tau_0)^2}.$$

The structural parameters θ_0 and $(\tau_0)^2$ can be estimated using data across all risk cells in the bank by solving two nonlinear equations (using for example an iterative procedure; see Bühlmann and Gisler [44], pp. 116–117):

$$(\widehat{\tau}_0)^2 = \frac{1}{J-1}\sum_{j=1}^{J}\alpha_j(\widehat{\Theta}_j - \widehat{\theta}_0)^2, \quad \widehat{\theta}_0 = \frac{1}{W}\sum_{j=1}^{J}\alpha_j\widehat{\Theta}_j, \tag{4.124}$$

$$\text{where} \quad W = \sum_{j=1}^{J}\alpha_j.$$

Here, the coefficients α_j are given in (4.123), with θ_0 and $(\tau_0)^2$ replaced by $\widehat{\theta}_0$ and $(\widehat{\tau}_0)^2$ respectively. If the solution for $(\widehat{\tau}_0)^2$ is negative, then we set $\alpha_j = 0$ and

$$\widehat{\theta}_0 = \frac{1}{W}\sum_{j=1}^{J} w_j\widehat{\Theta}_j, \quad w_j = K_j - 2, \quad W = \sum_{j=1}^{J} w_j.$$

The best credibility estimate for the tail parameter in the j-th cell (based on the cell data and all data in the collection) is $a_j\widehat{\widehat{\Theta}}_j$. We assumed that constants a_j are known a priori. Note that these constants are defined up to a constant factor, that is, coefficients α_j (and final estimates of tail parameters) will not change if all $a_j, j = 1, \ldots, J$, are changed/scaled by the same factor. Hence, only relative differences between risks play a role. These constants have the interpretation of a priori differences and can be fixed by expert using opinions on, for example, quantiles of losses exceeding L_j. For example, the expert may estimate the probability q_j, that the loss in the j-th cell will exceed level H_j, as $\widehat{q}_j$ and use relations

$$a_j\theta_j = -\ln q_j/\ln(H_j/L_j) \quad \text{and} \quad \mathrm{E}[\Theta_j] = \theta_0$$

to estimate a_j as $-\ln\widehat{q}_j/[\theta_0\ln(H_j/L_j)]$. Only relative differences play a role, so here (without loss of generality) θ_0 can be set equal to 1. Experts may specify several quantiles, then a_j can be estimated using, for example, a least square method. Ideally, the expert specifying constants a_j has a complete overview over all risk cells in the bank, as only relative differences between risks are important. However, in practice, opinions from experts with special knowledge of business specifics within a risk cell are required. Combining opinions from different experts is one of the problems to be resolved by a practitioner.

4.6.4 Numerical Example

To illustrate the above procedures consider an example where losses (exceeding USD 1 million) observed across 10 risk cells are given in Table 4.3 and all risk cells are the same a priori, $a_1 = \cdots = a_{10} = 1$. Using these losses the MLEs for the tail parameters $\widehat{\Theta}_j$, presented in Table 4.3, are calculated by (4.121). Then, using (4.123) and (4.124), we estimate the structural parameters $(\widehat{\tau}_0)^2 \approx 1.116$ and $\widehat{\theta}_0 \approx 3.157$, and credibility coefficients $\alpha_j \approx 0.446$ (the coefficients are the same because equal number of losses is observed in the cells).

The credibility estimators $\widehat{\widehat{\Theta}}_j$, shown in Table 4.3, are calculated using (4.123). In this example, the MLEs are quite volatile as the number of observations is small. For example, cell 7 has no large losses and thus its MLE is high; cell 10 has one large loss and thus its MLE is smaller. One could easily calculate cell MLEs vs the number of observations in a cell and observe that MLEs are highly volatile for small number of observations. One large observation may lead to a substantial change in MLE. The credibility estimators (based on data in the bank) are smoother in comparison with MLEs. This is because a credibility estimator is a weighted average, according to credibility theory, between a risk cell MLE and the estimator of the structural parameter $\widehat{\theta}_0$ based on all data in the collection. The credibility weights α_j are approximately 0.45 which means that a risk cell MLE (based on observations in a cell) $\widehat{\Theta}_j$ and the a priori estimate $\widehat{\theta}_0 \approx 3.157$ are weighted with 0.45 and 0.55 respectively.

Table 4.3 Losses (in millions USD) exceeding USD 1 million observed across 10 risk cells; and corresponding maximum likelihood and credibility estimators for the Pareto tail parameter in the risk cells

Cell 1	Cell 2	Cell 3	Cell 4	Cell 5	Cell 6	Cell 7	Cell 8	Cell 9	Cell 10
Losses (in millions USD) exceeding USD 1 million observed in risk cells									
1.557	9.039	1.166	1.548	1.578	1.201	1.006	1.741	1.364	1.074
1.079	2.138	1.037	1.040	1.282	2.815	1.169	1.165	2.036	1.103
1.047	1.008	1.136	1.045	1.092	3.037	1.215	1.010	1.014	1.664
1.199	1.761	2.104	1.774	1.658	1.001	1.116	1.096	1.217	1.049
1.395	1.654	1.774	1.045	2.025	1.114	1.010	1.060	1.202	1.104
1.060	1.073	1.161	1.856	1.129	1.422	1.560	1.352	1.095	2.924
3.343	2.435	1.080	1.636	1.946	2.397	1.059	1.044	1.348	1.265
2.297	4.357	1.154	1.403	1.831	1.241	1.059	1.678	1.191	1.333
1.297	1.576	1.257	2.522	1.478	1.522	1.050	1.882	1.161	1.424
1.180	1.113	1.231	1.113	1.208	1.243	1.231	1.401	1.017	1.435
Maximum likelihood estimators (MLEs) $\widehat{\Theta}_j,\ j = 1, \ldots, 10$									
2.499	1.280	3.688	2.487	2.264	1.992	6.963	3.335	4.194	2.870
Credibility estimators $\widehat{\widehat{\Theta}}_j,\ j = 1, \ldots, 10$ disregarding industry data									
2.863	2.319	3.394	2.858	2.759	2.637	4.855	3.236	3.620	3.029

4.6.5 Remarks and Interpretation

The credibility formulas (4.116) and (4.123) for the frequency and severity parameter estimators, based on a cell and collective data, have a simple interpretation.

- As the number of observations in the j-th cell increases, the larger credibility weights γ_j and α_j are assigned to the estimators $\widehat{\Lambda}_j$ and $\widehat{\Theta}_j$ (based on the cell observations) and the lesser weights are assigned to the estimators $\widehat{\theta}_0$ and $\widehat{\lambda}_0$ (based on all observations in a collection of risks) respectively.
- Also, the larger τ_0 and ω_0 (variance across risk cells in a collection), the larger weights are assigned to $\widehat{\Theta}_j$ and $\widehat{\Lambda}_j$ correspondingly. For a detailed discussion on the credibility parameters, refer to Bühlmann and Gisler ([44], section 4.4).

It is not difficult to consider a hierarchical model, where the collection of risks is part of another larger collection. For example, one can consider the collection of similar risks in the bank and then consider a collection of banks (i.e. the banking industry). This will further improve the estimates of arrival rate $\nu_j\lambda_j$ and the tail parameter $a_j\theta_j$. This can be done using a hierarchical credibility model; see Bühlmann and Gisler ([44], chapter 6). In particular, one can consider M banks with bank specific parameters $\lambda_0^{(m)}$ and $\theta_0^{(m)}$ modelled by random variables $\Lambda_0^{(m)}$ and $\Theta_0^{(m)}$, $m = 1, \ldots, M$ respectively. Then assume that:

(a) $\Lambda_0^{(m)}$ are independent and identically distributed random variables with

$$\mathrm{E}[\Lambda_0^{(m)}] = \lambda_{coll} \quad \text{and} \quad \mathrm{Var}[\Lambda_0^{(m)}] = \omega_{coll}^2.$$

(b) $\Theta_0^{(m)}$ are independent and identically distributed random variables with

$$\mathrm{E}[\vartheta_0^{(m)}] = \vartheta_{coll} \quad \text{and} \quad \mathrm{Var}[\vartheta_0^{(m)}] = \tau_{coll}^2.$$

The credibility weights and estimators in such a hierarchical model can be calculated as described in Bühlmann, Shevchenko and Wüthrich [45].

The Capital Calculations. For the purposes of the regulatory capital calculations of operational risk, the annual loss distribution, in particular its 0.999 quantile (VaR) as a risk measure, should be quantified for each Basel II risk cell in the matrix of eight business lines times seven risk types and for the whole bank. The credibility model presented in the above is for modelling low-frequency/high-severity losses exceeding some large threshold L. Given the credibility estimates for the model parameters the annual loss distribution can be calculated as usual using methods listed in Chap. 3; also see Bühlmann, Shevchenko and Wüthrich ([45], section 5). Of course, modelling of the high-frequency/low-severity losses (below threshold L) should be added to the model before the final operational risk capital charge is estimated. For a related actuarial literature on this topic, see Sandström [207] and Wüthrich [239]. That is, one can model the losses above threshold L using credibility theory as described in the above, while the losses below the threshold are modelled separately.

Note that typically the low-frequency/high-severity losses give the largest contribution to the final capital charge. The number of high-frequency/low-impact losses recorded in the bank internally is usually large enough to obtain reliable estimates by a standard fitting of the frequency and severity distributions without the use of the external data.

The important assumption in calculation of the credibility estimates is that the risk cells are independent. While it is an important (and quite realistic) assumption of the proposed model that the low-frequency/high-severity losses from different risk cells are independent, dependence can be considered between the high frequency low impact losses from different risk cells. Accurate quantification of the dependencies between the risks is a difficult task; this is an open field for future research. The dependence can be introduced using different methods (for example, copula methods, common shocks, etc.) that will be discussed in Chap. 7.

4.7 Capital Charge Under Parameter Uncertainty

According to the Basel II requirements (BCBS [17]) the final bank capital should be calculated as a sum of the risk measures in the risk cells if the bank's model cannot account for correlations between risks accurately. If this is the case, then one needs to calculate VaR for each risk cell separately and sum VaRs over risk cells to estimate the total bank capital. It is equivalent to the assumption of perfect dependence between risks. Modelling of dependence between risks and aggregation issues will be discussed in Chap. 7. In this section, we consider one risk cell, but note that adding quantiles over the risk cells to find the quantile of the total loss distribution is not necessarily conservative. In fact it can underestimate the capital in the case of heavy-tailed distribution as will be discussed in Chap. 7.

Under the LDA model, the annual loss in a risk cell over the next year $T+1$ is modelled as a random variable Z_{T+1} with some density $f(z_{T+1}|\boldsymbol{\theta})$, where $\boldsymbol{\theta}$ are model parameters. Given data $\mathbf{Y}$ over past T years (frequencies and severities) generated from some distributions parameterised by $\boldsymbol{\theta}$, the main task is to estimate the distribution of Z_{T+1}. The MLE $\widehat{\boldsymbol{\theta}}^{\text{MLE}}$ is often used as the "best fit" point estimate for $\boldsymbol{\theta}$. Then, a typical industry practice is to estimate the annual loss distribution for the next year as $f(z_{T+1}|\widehat{\boldsymbol{\theta}}^{\text{MLE}})$ and its 0.999 quantile, $Q_{0.999}(\widehat{\boldsymbol{\theta}}^{\text{MLE}})$, is used for the capital charge calculation.

However, the parameters $\boldsymbol{\theta}$ are unknown and it is important to account for this uncertainty when the capital charge is estimated especially for risks with small datasets. The Bayesian inference approach is an elegant and convenient way to accomplish this task.

4.7.1 Predictive Distributions

Under the Bayesian approach, the unknown parameters are modelled by random variables $\boldsymbol{\Theta}$ and their posterior density $\pi(\boldsymbol{\theta}|\mathbf{y})$ is calculated. Then, the predictive density of Z_{T+1}, given data $\mathbf{Y}=\mathbf{y}$, is defined as follows.

Definition 4.2 (Predictive density for annual loss) Suppose that:

(a) Given $\boldsymbol{\Theta} = \boldsymbol{\theta}$, the conditional density of the annual loss Z_{T+1} is $f(z_{T+1}|\boldsymbol{\theta})$.
(b) Given data $\mathbf{Y} = \mathbf{y}$, the posterior density of $\boldsymbol{\Theta}$ is $\pi(\boldsymbol{\theta}|\mathbf{y})$.
(c) Given $\boldsymbol{\Theta}$, Z_{T+1} and $\mathbf{Y}$ are independent.

Then the predictive density of Z_{T+1} is

$$f(z_{T+1}|\mathbf{y}) = \int f(z_{T+1}|\boldsymbol{\theta})\pi(\boldsymbol{\theta}|\mathbf{y})d\boldsymbol{\theta}. \quad (4.125)$$

Remark 4.21

- The predictive distribution accounts for both process and parameter uncertainties.
- It is assumed that, given $\boldsymbol{\Theta}$, Z_{T+1} and $\mathbf{Y}$ are independent. If they are not independent, then $f(z_{T+1}|\boldsymbol{\theta})$ should be replaced by $f(z_{T+1}|\boldsymbol{\theta}, \mathbf{y})$.
- If a frequentist approach is taken to estimate the parameters, then $\boldsymbol{\theta}$ should be replaced by the point estimators $\widehat{\boldsymbol{\theta}}$ and the integration should be done with respect to the density of $\widehat{\boldsymbol{\theta}}$.

The ultimate goal in capital charge calculation is to estimate the 0.999 quantile of the annual loss distribution. It is important to realise that there are two ways to define the required quantile to account for parameter uncertainty.

Definition 4.3 (Quantile of the predictive density $f(z_{T+1}|\mathbf{y})$**)** The quantile of a random variable with the predictive density (4.125) is

$$Q_q^P = F^{-1}_{Z_{T+1}|\mathbf{Y}}(q) = \inf\{z \in \mathbb{R} : \Pr[Z_{T+1} > z|\mathbf{Y}] \leq 1 - q\}, \quad (4.126)$$

where $q \in (0, 1)$ is a quantile level and $F^{-1}_{Z_{T+1}|\mathbf{Y}}(q)$ is the inverse of the distribution corresponding to the density (4.125).

Then, $Q^P_{0.999}$ can be used as a risk measure for capital calculations. Here, "P" in the upper script is used to emphasise that this is a quantile of the full predictive distribution.

Another approach under a Bayesian framework to account for parameter uncertainty is to consider a quantile of the annual loss density $f(z_{T+1}|\boldsymbol{\theta})$ conditional on parameter $\boldsymbol{\Theta} = \boldsymbol{\theta}$, defined in a standard way as follows.

Definition 4.4 (Quantile of the conditional density $f(z|\boldsymbol{\theta})$**)** The quantile of a random variable with the density $f(z|\boldsymbol{\theta})$ is

$$Q_q(\boldsymbol{\theta}) = F^{-1}_{Z_{T+1}|\boldsymbol{\Theta}}(q) = \inf\{z \in \mathbb{R} : \Pr[Z_{T+1} > z|\boldsymbol{\Theta} = \boldsymbol{\theta}] \leq 1 - q\}, \quad (4.127)$$

where $q \in (0, 1)$ is a quantile level and $F^{-1}_{Z_{T+1}|\boldsymbol{\Theta}}(q)$ is the inverse of the distribution corresponding to the density $f(z_{T+1}|\boldsymbol{\theta})$.

That is, the quantile $Q_q(\boldsymbol{\theta})$ is a function of $\boldsymbol{\theta}$ and thus $Q_q(\boldsymbol{\Theta})$ is a random variable with some distribution. Given that $\boldsymbol{\Theta}$ is distributed with the density $\pi(\boldsymbol{\theta}|\mathbf{y})$, one can find the *predictive distribution* of $Q_q(\boldsymbol{\Theta})$ and its characteristics. In particular, the mean of this distribution can be used as a point estimator:

$$\widehat{Q_q(\boldsymbol{\Theta})}^{\text{MMSE}} = \int Q_q(\boldsymbol{\theta})\pi(\boldsymbol{\theta}|\mathbf{y})d\boldsymbol{\theta}. \tag{4.128}$$

Other standard point estimators are the mode and median. A predictive interval $[L, U]$ can be formed to contain the true value with a probability α:

$$\Pr\left[L \leq Q_q(\boldsymbol{\Theta}) \leq U\right] = \alpha \tag{4.129}$$

or one-sided predictive interval

$$\Pr\left[Q_q(\boldsymbol{\Theta}) \leq U\right] = \alpha. \tag{4.130}$$

As before, for capital charge calculations we are interested in $q = 0.999$. Then one can argue that the conservative estimate of the capital charge accounting for parameter uncertainty should be based on the upper bound of the constructed predictive interval.

Remark 4.22

- Specification of the confidence level α is required to form a conservative interval for $Q_q(\boldsymbol{\Theta})$. It might be difficult to justify a particular choice of α. For example, it might be difficult to argue that the commonly used confidence level $\alpha = 0.95$ is good enough for estimation of the 0.999 quantile.
- This is similar to forming a confidence interval in the frequentist approach using the distribution of $Q_{0.999}(\widehat{\boldsymbol{\theta}}^{\text{MLE}})$, where $\widehat{\boldsymbol{\theta}}^{\text{MLE}}$ is treated as random.

In operational risk, it seems that the objective should be to estimate the full predictive distribution (4.125) for the annual loss Z_{T+1} over next year conditional on all available information. The capital charge should then be estimated as a quantile of this distribution, i.e. $Q^P_{0.999}$ given by (4.126).

4.7.2 Calculation of Predictive Distributions

Consider a risk cell in the bank. Assume that the frequency $p(\cdot|\boldsymbol{\alpha})$ and severity $f(\cdot|\boldsymbol{\beta})$ densities for the cell are chosen. Also, suppose that the posterior density $\pi(\boldsymbol{\theta}|\mathbf{y})$, $\boldsymbol{\theta} = (\boldsymbol{\alpha}, \boldsymbol{\beta})$ is estimated. Then, the predictive annual loss density (4.125) in the cell can be calculated using Monte Carlo procedure with the following logical steps.

Algorithm 4.1 (Full predictive loss distribution via MC)

1. For $k = 1, \ldots, K$
   a. For a given risk simulate the risk parameters $\boldsymbol{\theta} = (\boldsymbol{\alpha}, \boldsymbol{\beta})$ from the posterior $\pi(\boldsymbol{\theta}|\mathbf{y})$. If the posterior is not known in closed form then this simulation can be done using MCMC (see Sect. 2.11). For example, one can run MCMC for K iterations (after burn-in) beforehand and simply take the k-th iteration parameter values.
   b. Given $\boldsymbol{\theta} = (\boldsymbol{\alpha}, \boldsymbol{\beta})$, simulate the annual number of events N from $p(\cdot|\boldsymbol{\alpha})$ and severities $X^{(1)}, \ldots, X^{(N)}$ from $f(\cdot|\boldsymbol{\beta})$, then calculate the annual loss $Z^{(k)} = \sum_{n=1}^{N} X^{(n)}$.
2. Next k

Obtained annual losses $Z^{(1)}, \ldots, Z^{(K)}$ are samples from the predictive density (4.125). Extending the above procedure to the case of many risks is easy but requires specification of the dependence model; see Chap. 7. In this case, in general, all model parameters (including the dependence parameters) should be simulated from their joint posterior in Step (a). Then, given these parameters, Step (b) should simulate all risks with a chosen dependence structure. In general, sampling from the joint posterior of all model parameters can be accomplished via MCMC; see Peters, Shevchenko and Wüthrich [187] and Dalla Valle [68]. The 0.999 quantile $Q^{P}_{0.999}$ and other distribution characteristics can be estimated using the simulated samples in the usual way; see Sect. 3.2.

The above procedure is easily adapted to calculate the predictive distribution of $Q_{0.999}(\boldsymbol{\Theta})$. In particular, in Step (b) one can calculate the quantile $Q_{0.999}(\boldsymbol{\theta})$ of the conditional density $f(z|\boldsymbol{\theta})$, using for example FFT; see Chap. 3. Then the obtained K samples of the quantile can be used to estimate the distribution of $Q_{0.999}(\boldsymbol{\Theta})$ implied by the posterior $\pi(\boldsymbol{\theta}|\mathbf{y})$. To summarise, the logical steps of Monte Carlo procedure are as follows.

Algorithm 4.2 (Posterior distribution of quantile via MC)

1. For $k = 1, \ldots, K$
   a. For a given risk simulate the risk parameters $\boldsymbol{\theta} = (\boldsymbol{\alpha}, \boldsymbol{\beta})$ from the posterior $\pi(\boldsymbol{\theta}|\mathbf{y})$. If the posterior is not known in closed form then this simulation can be done using MCMC (see Sect. 2.11). For example, one can run MCMC for K iterations beforehand and simply take the k-th iteration parameter values.
   b. Given $\boldsymbol{\theta} = (\boldsymbol{\alpha}, \boldsymbol{\beta})$, calculate the quantile $Q_q^{(k)}(\boldsymbol{\theta})$ of $f(z|\boldsymbol{\theta})$ using FFT or other methods described in Chap. 3.
2. Next k

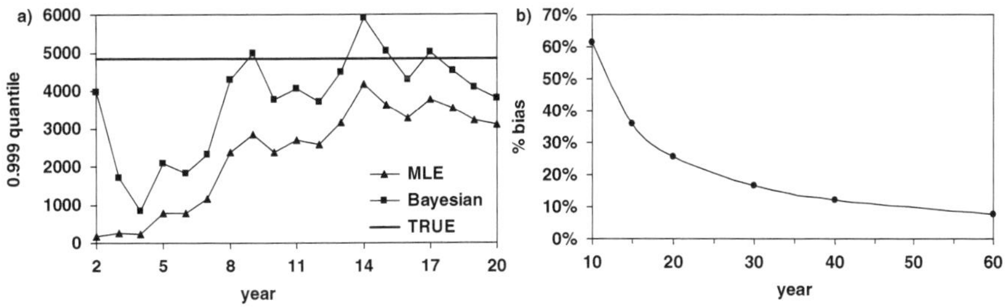

Fig. 4.7 Comparison of the estimators of the 0.999 annual loss quantile vs number of observation years. Losses were simulated from $Poisson(10)$ and $\mathcal{LN}(1, 2)$. Parameter uncertainty is ignored by $Q_{0.999}(\widehat{\theta}^{\text{MLE}})$ (MLE) but taken into account by $Q^P_{0.999}$ (Bayesian). **(a)** The quantile estimators for a one specific data realisation; **(b)** Relative bias $\mathrm{E}[Q^P_{0.999} - Q_{0.999}(\widehat{\boldsymbol{\theta}}^{\text{MLE}})]/Q^{(0)}_{0.999}$ calculated as an average over 100 realisations

Note that in the above Monte Carlo procedures the risk profile $\boldsymbol{\Theta}$ is sampled from its posterior for each simulation $k = 1, \ldots, K$. Thus we model both the process uncertainty, which comes from the fact that frequencies and severities are random variables, and the parameter risk (parameter uncertainty), which comes from the fact that we do not know the true values of $\boldsymbol{\theta}$.

Example 4.6 The parameter uncertainty is ignored by the estimator $Q_{0.999}(\widehat{\boldsymbol{\theta}}^{\text{MLE}})$ but is taken into account by $Q^P_{0.999}$. The following illustrative example is taken from Shevchenko ([215], section 8). Figure 4.7 presents results for the relative bias (averaged over 100 realisations) $\mathrm{E}[Q^P_{0.999} - Q_{0.999}(\widehat{\boldsymbol{\theta}}^{\text{MLE}})]/Q^{(0)}_{0.999}$, where $\widehat{\boldsymbol{\theta}}^{\text{MLE}}$ is MLE, $Q^{(0)}_{0.999}$ is the quantile of $f(\cdot|\boldsymbol{\theta}_0)$ and $\boldsymbol{\theta}_0$ is the true value of the parameter. The frequencies and severities are simulated from $Poisson(\lambda_0 = 10)$ and $\mathcal{LN}(\mu_0 = 1, \sigma_0 = 2)$ respectively. Also, constant priors are used for the parameters so that there are closed form expressions for the posterior; see Sects. 4.3.3 and 4.3.5. In this example, the bias induced by parameter uncertainty is large: it is approximately 10% after 40 years (i.e. approximately 400 data points) and converges to zero as the number of losses increases. A similar analysis for a multivariate case was performed in Dalla Valle [68] with real data. For high-frequency/low-severity risks, where a large amount of data is available, the impact is certainly expected to be small. However, for low-frequency/high-severity risks, where the data are very limited, the impact can be significant.

4.8 General Remarks

This chapter described how the parameters of the frequency and severity distributions are estimated using internal data, external data and expert opinion. Then calculation of VaR (accounting for parameter uncertainty) for each risk cell can easily be done using a simulation approach as described in Sect. 4.7. The approaches and issues related to modelling dependence and aggregation over many risks will be discussed in Chap. 7.

The main motivation for the use of the Bayesian approach is that, typically, the bank's internal data of the large losses in risk cells are so limited that the standard maximum likelihood estimates are not reliable. Overall, the use of the Bayesian inference method for the quantification of the frequency and severity distributions of operational risks is very promising. The method is based on specifying the prior distributions for the parameters of the frequency and severity distributions using expert opinions or industry data. Then, the prior distributions are weighted with the actual observations in the bank to estimate the posterior distributions of the model parameters. These are used to estimate the annual loss distribution for the next accounting year. The estimation of low-frequency risks using this method has several appealing features such as stable estimators, simple calculations (in the case of conjugate priors), and the ability to take into account expert opinions and industry data. The approach allows for combining all three data sources: internal data, external data and expert opinions required by Basel II.

If the data a very limited, it might be difficult to specify the prior distributions. Then one can use a closely related credibility theory approach to estimate parameters of the frequency and severity distributions for the low-frequency/high-severity risks, as described in Sect. 4.6.

The models presented in this chapter give illustrative examples that can be extended to a full scale application. The approach has a simple structure which is beneficial for practical use and can engage the bank risk managers, statisticians and regulators in productive model development and risk assessment.

Several general remarks on the described Bayesian method for operational risk are worth making:

- Validation of the models in the case of small data sets is problematic. Formally, justification of the model assumptions (such as conditional independence between the losses or common distribution for the risk profiles across the risks) can be based on the analysis of the unconditional properties (e.g. unconditional means and covariances) of the losses and should be addressed during model implementation.
- Presented examples have a simplistic dependence on time but can be extended to the case of more realistic time component.
- Adding extra levels to the considered hierarchical structure may be required to model the actual risk cell structure in a bank.
- One of the features of the described method is that the variance of the posterior distribution $\pi(\boldsymbol{\theta}|\cdot)$ will converge to zero for a large number of observations. This means that the true value of the risk profile will be known exactly. However, there are many factors (political, economical, legal, etc.) changing in time that should not allow for the precise knowledge of the risk profiles. One can model this by limiting the variance of the posterior distribution by some lower levels (e.g. 5%). This has been done in many solvency approaches for the insurance industry, for example in the Swiss Solvency Test; see Swiss Financial Market Supervisory Authority ([230], formulas (25) and (26)).

- For convenience, we have assumed that expert opinions are independent and identically distributed. However, all formulas can easily be generalised to the case of expert opinions modelled by different distributions.
- It would be ideal if the industry risk profiles (prior distributions for frequency and severity parameters in risk cells) are calculated and provided by the regulators to ensure consistency across the banks. Unfortunately this may not be realistic at the moment. Banks might thus estimate the industry risk profiles using industry data available through external databases from vendors and consortia of banks. The data quality, reporting and survival biases in external databases are the issues that should be considered in practice.

Finally, in this book we consider modelling operational risk but the use of similar Bayesian models is also useful in other areas (such as credit risk, insurance, environmental risk and ecology) where, mainly due to lack of internal observations, a combination of internal data with external data and expert opinions is required.

Problems[3]

4.1 (⋆⋆) Prove the Theorem 4.1.

4.2 (⋆) Assume that, given $\Theta = \theta$, the counts $N_1, N_2, \ldots, N_T$ are independent and binomial distributed: $N_j \sim Bin(V_j, \theta)$. Also, assume that the prior distribution of Θ is $Beta(\alpha, \beta)$. Find the posterior density of Θ given $\mathbf{N} = \mathbf{n}$, i.e. $\pi(\theta|\mathbf{n})$.

4.3 (⋆) Consider the annual number of losses N_i, $i = 1, \ldots, 10$ given in Table 4.4, that were observed for a risk over 10 years. Assume that N_i are independent and distributed from $Poisson(\lambda)$. Using noninformative constant prior, find the posterior for parameter λ.

Table 4.4 The annual number of losses $N_i, i = 1, 2, \ldots, 10$; see Problem 4.3 for details

i	1	2	3	4	5	6	7	8	9	10
N_i	1	1	4	1	1	0	1	2	0	4

4.4 (⋆) Consider the losses X_i, $i = 1, \ldots, 15$ given in Table 4.5, that were observed for a risk. Assume that X_i are independent and distributed from $\mathcal{LN}(\mu, \sigma)$. Using noninformative constant priors, find the posterior for parameters μ and σ and estimate the mean of these posteriors.

Table 4.5 Loss severities $X_i, i = 1, 2, \ldots, 15$; see Problem 4.4 for details

1.877	2.050	9.050	0.406	0.210	0.066	2.893	321.668	0.421	0.368
0.196	3.290	12.027	2.701	13.528					

[3] Problem difficulty is indicated by asterisks: (⋆) – low; (⋆⋆) – medium; (⋆ ⋆ ⋆) – high.

4.5 (⋆⋆) Repeat calculations of Example 4.1 (combining expert opinion and internal data), if the prior is determined by expert who specifies $\mathrm{E}[\Lambda] = 0.8$ and $\mathrm{Vco}[\Lambda] = 0.5$.

4.6 (⋆⋆) Repeat calculations of Example 4.3 (combining expert opinion and internal data), if the prior is determined by expert who specifies $\mathrm{E}[\Theta_\xi] = 3$ and $\mathrm{Vco}[\Theta_\xi] = 0.5$.

4.7 (⋆⋆) Repeat calculations of Example 4.4 (combining expert opinions, internal data and external data) using mode of the posterior as a Bayesian point estimate.

4.8 (⋆ ⋆ ⋆) Assume that the annual frequency is $N \sim Poisson(\lambda)$ and independent severities are $X_i \sim \mathcal{LN}(\mu, \sigma)$; assume also that severities and frequency are independent. Suppose that past data imply that the posterior for λ is gamma distribution with mean 15 and standard deviation 5, and that severity parameters are known $\mu = 0$, $\sigma = 2$. Calculate the predictive distribution of the annual loss $Z = \sum_{i=1}^{N} X_i$ and find its 0.999 quantile. Given model parameters $\boldsymbol{\theta} = (\lambda, \mu, \sigma)$, denote the 0.999 quantile of the annual loss as $Q_{0.999}(\boldsymbol{\theta})$. Calculate the predictive distribution of the $Q_{0.999}(\boldsymbol{\theta})$ and find its mean, median, 0.25 and 0.75 quantiles.

4.9 (⋆⋆⋆) Repeat calculations of Problem 4.8, if the severity parameter μ is unknown and past data imply that its posterior is the normal distribution with mean 0 and standard deviation 1. Assume that μ and λ are independent in the posterior. Compare with the results of Problem 4.8.

Chapter 5
Addressing the Data Truncation Problem

Whenever you set out to do something, something else must be done first.

Murphy

Abstract Typically, operational risk losses are reported above some threshold. This chapter studies the impact of ignoring data truncation on the 0.999 quantile of the annual loss distribution. Fitting data reported above a constant threshold is a well-known and studied problem. However, in practice, the losses are scaled for business and other factors before the fitting and thus the threshold varies across the scaled data sample. The reporting level may also change when a bank changes its reporting policy. This chapter considers the issue of thresholds – both constant and time-varying. The maximum likelihood and Bayesian Markov chain Monte Carlo approaches to fit the models are discussed.

5.1 Introduction

Accurate modelling of the severity and frequency distributions is the key to estimating a capital charge. One of the challenges in modelling operational risk is the lack of complete data – often a bank's internal data are not reported below a certain level (typically of the order of Euro 10,000). These data are said to be left-truncated. Generally speaking, missing data increase uncertainty in modelling. Sometimes a threshold level is introduced to avoid difficulties with collection of too many small losses. Industry data in external databases from vendors and consortia of banks are available above some thresholds: Algo OpData provides publicly reported operational risk losses above USD 1 million and ORX provides operational risk losses above Euro 20,000 reported by ORX members. The operational risk data from Loss Data Collection Exercises (LDCE) over many institutions are truncated too. For example, Moscadelli [166] analysed 2002 LDCE and Dutta and Perry [77] analysed 2004 LDCE, where the data were mainly above Euro 10,000 and USD 10,000 respectively.

Often, modelling of missing data is done assuming a parametric distribution for losses below and above the threshold. Then fitting is accomplished using

P. Shevchenko, *Modelling Operational Risk Using Bayesian Inference*,
DOI 10.1007/978-3-642-15923-7_5,

losses reported above the threshold via the maximum likelihood method (Frachot, Moudoulaud and Roncalli [95]) or the expectation maximisation (EM) algorithm (Bee [25], [26]). In practice, often the missing data are ignored completely. This may lead to a significant underestimation or overestimation of the capital. The impact of data truncation in operational risk was discussed in the literature; see Baud, Frachot and Roncalli [20], Chernobai, Menn, Trück and Rachev [53], Mignola and Ugoccioni [163], and Luo, Shevchenko and Donnelly [151]. Typically, the case of a constant threshold is discussed in research studies, though in practice, a threshold level is varying across observations; see Shevchenko and Temnov [217]. One of the reasons for appearing a varying threshold in operational risk loss data is that the losses are scaled for inflation and other factors before fitting to reflect changes in risk over time. The reporting level may also change from time to time within a bank when reporting policy is changed. The problem with multiple thresholds also appears when different companies report losses into the same database using different threshold levels; see Baud, Frachot and Roncalli [20].

Of course, for risks with heavy-tailed severities, the impact of the data threshold should not be important in a limit of high quantiles. However, it should be quantified first before making such a conclusion and to justify a chosen reporting level. Also, for light tailed risks, the impact can be significant.

In this chapter, a single risk cell is considered only, and the following notation and assumptions are used:

- The annual loss in a risk cell in year m is

$$Z_m = \sum_{i=1}^{N_m} X_i(m). \tag{5.1}$$

- N_m is the number of events (frequency) and $X_i(m), i = 1, \dots, N_m$ are the severities of the events in year m.
- If convenient, we may index severities $X_i(m)$ and their event times $T_i(m)$, $i = 1, \dots, N_m$, $m = 1, 2, \dots$ (ordered in time) as X_j and T_j, $j = 1, 2, \dots$ respectively, where $T_1 < T_2 < \cdots$.
- The severities of the events X_j, $j = 1, 2, \dots$ occurring at times T_j, $j = 1, 2, \dots$ respectively are modelled as independent and identically distributed random variables from a continuous distribution $F(x|\boldsymbol{\beta})$, $0 < x < \infty$, whose density is denoted as $f(x|\boldsymbol{\beta})$. Here, $\boldsymbol{\beta}$ are the severity distribution parameters.
- N_m, $m = 1, 2, \dots$ are independent and identically distributed random variables from a discrete frequency distribution with probability mass function $p(n|\lambda) = \Pr[N_m = n]$, where λ is a frequency parameter (or a vector of parameters).
- It is assumed that the severities $X_i(m)$ and frequencies N_m of the events are independent.
- $\boldsymbol{\gamma} = (\lambda, \boldsymbol{\beta})$ is a vector of frequency and severity distribution parameters.

5.2 Constant Threshold – Poisson Process

Often, it is assumed that loss events are modelled by a homogeneous Poisson process with the intensity parameter λ. Then, N_m, $m = 1, 2, \ldots$ are independent and identically distributed random variables from the Poisson distribution, $Poisson(\lambda)$, with

$$\Pr[N_m = n] = p(n|\lambda) = \frac{\lambda^n}{n!} \exp(-\lambda), \quad \lambda > 0, n = 0, 1, \ldots \tag{5.2}$$

and the event inter-arrival times $\delta T_j = T_j - T_{j-1}$, $j = 1, 2, \ldots$ (where $T_0 < T_1 < T_2 < \ldots$ are the event times and $T_0 = t_0$ is the start of the observation period) are independent exponentially distributed random variables with the density and distribution functions

$$g(\tau|\lambda) = \lambda \exp(-\lambda\tau) \quad \text{and} \quad G(\tau|\lambda) = 1 - \exp(-\lambda\tau) \tag{5.3}$$

respectively.

If the losses, originating from severity $f(x|\boldsymbol{\beta})$ and frequency $p(n|\lambda)$ densities, are recorded above a known reporting level (truncation level) L, then the density of the losses above L is left-truncated density

$$f_L(x|\boldsymbol{\beta}) = \frac{f(x|\boldsymbol{\beta})}{1 - F(L|\boldsymbol{\beta})}; \quad L \leq x < \infty. \tag{5.4}$$

The events of the losses above L follow the Poisson process with the intensity

$$\theta(\boldsymbol{\gamma}, L) = \lambda(1 - F(L|\boldsymbol{\beta})), \tag{5.5}$$

the so-called *thinned* Poisson process, and the annual number of events above the threshold is distributed from *Poisson* (θ).

The series of the annual counts or event times can be used for estimating frequency distribution. These cases are considered separately below.

Data for annual counts and severities. Consider a random vector $\mathbf{Y}$ of the events recorded above the threshold L over a period of T years consisting of the annual frequencies $\widetilde{N}_m$, $m = 1, \ldots, T$ and severities $\widetilde{X}_j$, $j = 1, \ldots, J$, $J = \widetilde{N}_1 + \ldots + \widetilde{N}_T$. For given model parameters γ, the joint density of $\mathbf{Y}$ at $\widetilde{N}_m = \widetilde{n}_m$ and $\widetilde{X}_j = \widetilde{x}_j$ can be written as

$$h(\mathbf{y}|\boldsymbol{\gamma}) = \prod_{j=1}^{J} f_L(\widetilde{x}_j|\boldsymbol{\beta}) \prod_{m=1}^{T} p(\widetilde{n}_m|\theta(\boldsymbol{\gamma}, L)). \tag{5.6}$$

That is, the likelihood function for this model is $\ell_{\mathbf{y}}(\boldsymbol{\gamma}) = h(\mathbf{y}|\boldsymbol{\gamma})$.

Data for event times and severities. Similarly, if the data $\mathbf{Y}$ of the events above a constant threshold over the time period $[t_0, t_E]$ consist of the event inter-arrival times $\delta\widetilde{T}_j = \widetilde{T}_j - \widetilde{T}_{j-1}$, $j = 1, \ldots, J$ (where $\widetilde{T}_j$, $j = 1, 2, \ldots$ are the event times and $\widetilde{T}_0 = t_0$) and the severities $\widetilde{X}_j$, $j = 1, \ldots, J$, then the joint density (for given $\boldsymbol{\gamma}$) of $\mathbf{Y}$ at $\delta\widetilde{T}_j = \widetilde{\tau}_j$ and $\widetilde{X}_j = \widetilde{x}_j$ is

$$h(\mathbf{y}|\boldsymbol{\gamma}) = \big(1 - G(t_E - \widetilde{t}_J|\theta(\boldsymbol{\gamma}, L))\big) \prod_{j=1}^{J} f_L(\widetilde{x}_j|\boldsymbol{\beta}) g(\widetilde{\tau}_j|\theta(\boldsymbol{\gamma}, L))$$

$$= \lambda^J \exp(-\theta(\boldsymbol{\gamma}, L)(t_E - t_0)) \prod_{j=1}^{J} f(\widetilde{x}_j|\boldsymbol{\beta}). \tag{5.7}$$

Here, $1 - G(t_E - \widetilde{t}_J|\theta(\boldsymbol{\gamma}, L))$ is the probability that no event will occur within $(\widetilde{t}_J, t_E]$. The likelihood function for this model is $\ell_{\mathbf{y}}(\boldsymbol{\gamma}) = h(\mathbf{y}|\boldsymbol{\gamma})$.

Remark 5.1 If the start and end of the observation period correspond to the beginning and end of the first and last years respectively, then the inferences based on the likelihoods (5.7) and (5.6) are equivalent. This is because the likelihoods, in this case, are different by a factor that does not depend on the model parameters.

5.2.1 Maximum Likelihood Estimation

The maximum likelihood estimators (MLEs) $\widehat{\boldsymbol{\gamma}}$ are the values of frequency and severity parameters $\boldsymbol{\gamma}$ that maximise the likelihood function. That is, maximise (5.6) or (5.7) if the frequency data consists from the annual counts or the event times respectively. Note that the likelihood function corresponding to the joint density (5.6) or (5.7) is $\ell_{\mathbf{y}}(\boldsymbol{\gamma}) = h(\mathbf{y}|\boldsymbol{\gamma})$.

Data for annual counts and severities. From (5.6), the maximum likelihood estimators (MLEs) $\widehat{\boldsymbol{\gamma}}$ can be found as a solution of

$$\frac{\partial \ln \ell_{\mathbf{Y}}(\boldsymbol{\gamma})}{\partial \lambda} = (1 - F(L|\boldsymbol{\beta})) \sum_{m=1}^{T} \frac{\partial}{\partial \theta} \ln p(\widetilde{N}_m|\theta(\boldsymbol{\gamma}, L)) = 0, \tag{5.8}$$

$$\frac{\partial \ln \ell_{\mathbf{Y}}(\boldsymbol{\gamma})}{\partial \boldsymbol{\beta}} = \sum_{j=1}^{J} \frac{\partial}{\partial \boldsymbol{\beta}} \ln f_L(\widetilde{X}_j|\boldsymbol{\beta})$$

$$-\lambda \frac{\partial F(L|\boldsymbol{\beta})]}{\partial \boldsymbol{\beta}} \sum_{m=1}^{T} \frac{\partial}{\partial \theta} \ln p(\widetilde{N}_m|\theta(\boldsymbol{\gamma}, L)) = 0. \tag{5.9}$$

It is easy to see that the MLEs $\widehat{\boldsymbol{\beta}}$ for the severity parameters can be found marginally (independently from frequency) by maximising

$$\sum_{j=1}^{J} \ln f_L(\widetilde{X}_j|\boldsymbol{\beta}) \tag{5.10}$$

and then the Eq. (5.8) gives the MLE for the intensity

$$\widehat{\lambda} = \frac{1}{1 - F(L|\widehat{\boldsymbol{\beta}})} \times \frac{1}{T}\sum_{m=1}^{T} \widetilde{N}_m. \tag{5.11}$$

Data for event times and severities. From (5.7), the MLEs $\widehat{\boldsymbol{\gamma}}$ can be found as a solution of

$$\begin{aligned}
\frac{\partial \ln \ell_{\mathbf{Y}}(\boldsymbol{\gamma})}{\partial \lambda} &= \frac{J}{\lambda} - (1 - F(L|\boldsymbol{\beta}))(t_E - t_0) = 0,\\
\frac{\partial \ln \ell_{\mathbf{Y}}(\boldsymbol{\gamma})}{\partial \boldsymbol{\beta}} &= \lambda(t_E - t_0)\frac{\partial F(L|\boldsymbol{\beta})}{\partial \boldsymbol{\beta}} + \sum_{j=1}^{J} \frac{\partial}{\partial \boldsymbol{\beta}} \ln f(\widetilde{X}_j|\boldsymbol{\beta}) = 0.
\end{aligned} \tag{5.12}$$

This gives the intensity MLE

$$\widehat{\lambda} = \frac{J}{(1 - F(L|\widehat{\boldsymbol{\beta}}))(t_E - t_0)}, \tag{5.13}$$

which is equivalent to (5.11) if the start and end of the observation period correspond to the beginning and end of the first and last years respectively. Substituting $\widehat{\lambda}$ into the second equation in (5.12), it is easy to see that the severity MLEs $\widehat{\boldsymbol{\beta}}$ can be obtained by maximising $\sum_{j=1}^{J} \ln f_L(\widetilde{X}_j|\boldsymbol{\beta})$.

MLE errors. The MLE errors are typically estimated using asymptotic Gaussian approximation via the inverse of the Fisher information matrix; see Sect. 2.8.1. The latter is often estimated by the observed information matrix. That is,

$$\mathrm{Cov}[\widehat{\gamma}_i, \widehat{\gamma}_j] \approx (\widehat{\mathbf{I}}^{-1})_{ij}, \quad (\widehat{\mathbf{I}})_{ij} = -\left.\frac{\partial^2 \ln \ell_{\mathbf{y}}(\boldsymbol{\gamma})}{\partial \gamma_i \partial \gamma_j}\right|_{\boldsymbol{\gamma}=\widehat{\boldsymbol{\gamma}}}. \tag{5.14}$$

Whether the sample size is large enough to use this asymptotic approximation is a difficult question that should be addressed in a practical solution. Also, the regularity conditions required for this approximation are mild but difficult to prove.

A point estimator for a risk measure, for example a quantile $Q_q(\boldsymbol{\gamma})$ of the annual loss distribution, is calculated as $\widehat{Q}_q = Q_q(\widehat{\boldsymbol{\gamma}})$. Its accuracy is typically estimated using the error propagation method by performing the first order Taylor series expansion around the true value

$$Q_q(\widehat{\boldsymbol{\gamma}}) - Q_q(\boldsymbol{\gamma}) \approx \sum_i \frac{\partial Q_q(\boldsymbol{\gamma})}{\partial \gamma_i}(\widehat{\gamma}_i - \gamma_i) \tag{5.15}$$

and calculating the standard deviation $\mathrm{stdev}[Q_q(\widehat{\boldsymbol{\gamma}})] = \sqrt{\mathrm{Var}[Q_q(\widehat{\boldsymbol{\gamma}})]}$, where

$$\mathrm{Var}[Q_q(\widehat{\boldsymbol{\gamma}})] \approx \sum_{i,j} \frac{\partial Q_q(\boldsymbol{\gamma})}{\partial \gamma_i} \frac{Q_q(\boldsymbol{\gamma})}{\gamma_j} \mathrm{Cov}[\widehat{\gamma}_i, \widehat{\gamma}_j]. \tag{5.16}$$

Finally, the unknown true parameter values $\boldsymbol{\gamma}$ are replaced by $\widehat{\boldsymbol{\gamma}}$ and the standard deviation is estimated using the above formula with $\partial Q_q(\boldsymbol{\gamma})/\partial \gamma_i$ replaced by $\partial Q_q(\widehat{\boldsymbol{\gamma}})/\partial \widehat{\gamma}_i$.

Example 5.1 As an illustrative example, consider simulated loss amounts X_j and times T_j from $Poisson(10)$-$GPD(0.3, 6)$ over unrealistic $T = 80$ years; GPD[1] is the generalised Pareto distribution formally defined in Appendix A.2.9. The simulated loss amounts and times over realistic time period of $T = 5$ are given in Table 5.1. Then, the truncated data were simulated and fitted using the following procedure:

- **Step 1**. Simulate the Poisson process event times t_j, $j = 1, 2 \ldots$ covering the period of T years by simulating independent inter-arrival times $\tau_j = t_j - t_{j-1}$ from the exponential distribution with the parameter λ. Find the number of events

Table 5.1 Losses and event times simulated from $Poisson(10)$-$GPD(0.3, 6)$ over 5 years

Index, i	t_i	$t_i - t_{i-1}$	Loss, x_i	Index, i	t_i	$t_i - t_{i-1}$	Loss, x_i
1	0.0257	0.0257	7.5306	26	2.6488	0.1077	4.5207
2	0.2540	0.2284	3.1811	27	2.7531	0.1043	26.7507
3	0.4662	0.2121	2.1633	28	2.9669	0.2139	6.5704
4	0.5784	0.1123	2.0684	29	3.1671	0.2002	1.0533
5	0.7248	0.1463	7.0108	30	3.2638	0.0967	11.5740
6	0.7399	0.0151	15.1171	31	3.2988	0.0350	1.3647
7	0.7803	0.0404	0.4791	32	3.3984	0.0996	17.4227
8	0.8533	0.0731	1.9012	33	3.6000	0.2016	6.1744
9	0.9065	0.0532	9.6585	34	3.7285	0.1284	2.1298
10	1.2136	0.3070	6.5786	35	3.7799	0.0514	7.8412
11	1.2265	0.0129	2.9675	36	3.9074	0.1276	13.9317
12	1.3274	0.1009	0.2208	37	3.9117	0.0043	4.1237
13	1.5192	0.1918	13.9372	38	4.0006	0.0889	13.5370
14	1.5728	0.0536	8.3221	39	4.0628	0.0622	8.1449
15	1.8030	0.2301	62.9923	40	4.1023	0.0395	1.2949
16	1.8641	0.0611	4.0205	41	4.3969	0.2946	11.3031
17	1.8648	0.0008	4.5983	42	4.4120	0.0151	1.7095
18	1.8755	0.0107	3.6437	43	4.4696	0.0576	15.2808
19	1.9202	0.0447	0.6435	44	4.6578	0.1882	0.6268
20	1.9538	0.0336	9.5114	45	4.7437	0.0859	2.2327
21	1.9897	0.0359	4.4838	46	4.7440	0.0003	0.3352
22	2.1843	0.1946	26.9387	47	4.7540	0.0100	13.3572
23	2.2377	0.0535	37.6751	48	4.7738	0.0198	6.7379
24	2.3737	0.1359	24.1384	49	4.8508	0.0770	3.0586
25	2.5410	0.1674	0.4814	50	4.9532	0.1024	0.9756

[1] GPD is a distribution of threshold exceedances in the limit of large threshold; see Sect. 6.3.

n_m occurred during each year $m = 1, \ldots, T$ and the total number of events $n = n_1 + \ldots + n_T$.

- **Step 2**. Simulate independent severities x_j from $GPD(\alpha, \beta)$ for the event times t_j, $j = 1, \ldots, n$ respectively.
- **Step 3**. Remove the events from the simulated sample when the event loss is below a threshold L.
- **Step 4**. Given truncated sample of the severities $\widetilde{x}_1, \ldots, \widetilde{x}_J$ exceeding the threshold and corresponding event times, estimate the parameters (α, β, λ) via the MLE and MCMC procedures using likelihood (5.7).

In total, 50 losses occurred over 5 years. It is easy to see that only 38 of these losses, $\widetilde{x}_j$ exceeding L, will be reported if there is a reporting threshold $L = 2$; for $L = 1$, 43 losses will be reported. The probabilities of the loss to be less than L, i.e. $F(L|\xi, \beta)$, are $(0, 0.150, 0.272)$ for $L = (0, 1, 2)$ respectively. The data over 5-year period correspond to a realistic example if the losses X_j, reporting level L, and scaling parameter β are multiplied by USD 10,000. Note that in the case of $GPD(0.3, 6)$ severity, the skewness and higher moments do not exist.

If we know the true distribution types but do not know the model parameters, then the log-likelihood function for the reported data is found using (5.7):

$$\ln \ell_{\mathbf{y}}(\boldsymbol{\gamma}) = J \ln \lambda - \lambda T \left(1 + \frac{\xi L}{\beta}\right)^{-\frac{1}{\xi}} - J \ln \beta - \left(1 - \frac{1}{\xi}\right) \sum_{j=1}^{J} \ln\left(1 + \frac{\xi \widetilde{x}_j}{\beta}\right). \tag{5.17}$$

Maximising the log-likelihood gives the MLEs $\widehat{\xi}, \widehat{\beta}, \widehat{\lambda}$ shown in Table 5.2. For these MLEs we also calculate the 0.999 quantile $\widehat{Q}_{0.999} = Q_{0.999}(\widehat{\boldsymbol{\gamma}})$ of the annual loss distribution and its standard deviation calculated using FFT, as explained in Sect. 3.4. The standard deviation is obtained using (5.16). To demonstrate the behaviour of the estimators, we present the results for several reporting levels $L = 0, 1, 2$ and several time periods $T = 5, 20, 80$. The results for $L = 0, 1, 2$ with $T = 5$ show that reporting level introduces more uncertainly into the estimators and the uncertainty in the quantile estimator is so large (approximately 50% for $T = 5, 20$) that no reliable conclusion can be made. Increasing the time period (i.e. the data sample size), improves the accuracy of the estimators. Only at $T = 80$, the estimator for the 0.999 quantile is getting some certainty where the standard deviation is approximately 20% of the quantile estimator. Moreover, increasing T from 5 to 20 in the case of $L = 2$ does not reduce the standard deviation (which is approx 50% in both cases) of the 0.999 quantile that indicated that the 1st order expansion (5.15) is not a good approximation for this case. Also note that correlations between $\widehat{\lambda}$ and severity MLEs are zero when $L = 0$ and become significant for $L = 1, 2$.

Table 5.2 MLEs $\widehat{\xi}, \widehat{\beta}, \widehat{\lambda}$ when data are simulated from $Poisson(\lambda = 10)$-$GPD(\xi = 0.3, \beta = 6)$ in the case of different reporting levels $L = 0, 1, 2$ and time horizons $T = 5, 20, 80$. The 5-year data are given in Table 5.1

MLE	$L = 0$ $T = 5$	$L = 1$ $T = 5$	$L = 2$ $T = 5$	$L = 2$ $T = 20$	$L = 2$ $T = 80$
$\widehat{\xi}$	0.214	0.203	0.218	0.321	0.357
$\widehat{\text{stdev}[\widehat{\xi}]}$	0.174	0.185	0.203	0.102	0.051
$\widehat{\beta}$	6.980	7.147	6.913	7.409	5.632
$\widehat{\text{stdev}[\widehat{\beta}]}$	1.551	1.873	2.176	1.159	0.464
$\widehat{\lambda}$	10.000	9.872	10.062	9.975	10.740
$\widehat{\text{stdev}[\widehat{\lambda}]}$	1.414	1.543	1.820	0.883	0.505
$\widehat{\rho}_{\widehat{\beta},\widehat{\xi}}$	−0.649	−0.704	−0.754	−0.699	−0.697
$\widehat{\rho}_{\widehat{\lambda},\widehat{\xi}}$	0.000	0.149	0.314	0.269	0.327
$\widehat{\rho}_{\widehat{\lambda},\widehat{\beta}}$	0.000	−0.220	−0.441	−0.413	−0.512
$\widehat{Q}_{0.999}$	325.7762	319.4020	328.3046	552.6983	528.2161
$\widehat{\text{stdev}[\widehat{Q}_{0.999}]}$	168.5775	161.3365	186.5343	242.3923	125.4532

5.2.2 Bayesian Estimation

Under the Bayesian approach, the parameters are modelled as random variables. Given the vector of all frequency and severity parameters $\boldsymbol{\gamma} = (\lambda, \boldsymbol{\beta})$, denote the density of the annual loss as $h(z|\boldsymbol{\gamma})$. Then, given data $\mathbf{Y}$ (severities and frequencies over T years), the predictive density for the next year annual loss Z_{T+1} is

$$h(z|\mathbf{y}) = \int h(z|\boldsymbol{\gamma})\pi(\boldsymbol{\gamma}|\mathbf{y})d\boldsymbol{\gamma}, \tag{5.18}$$

where $\pi(\boldsymbol{\gamma}|\mathbf{y})$ is the joint posterior density of the parameters given data $\mathbf{Y}$. From Bayes's rule

$$\pi(\boldsymbol{\gamma}|\mathbf{y}) \propto h(\mathbf{y}|\boldsymbol{\gamma})\pi(\boldsymbol{\gamma}), \tag{5.19}$$

where $h(\mathbf{y}|\boldsymbol{\gamma})$ is the joint density of the data and $\pi(\boldsymbol{\gamma})$ is a prior density for the parameters (the prior distribution can be specified by an expert or fitted using external data or can be taken to be noninformative so that inference is implied by data only).

For large sample size (and continuous prior distribution), it is common to approximate $\ln \pi(\boldsymbol{\gamma}|\mathbf{y})$ by a second-order Taylor series expansion around $\widehat{\boldsymbol{\gamma}}$. Then $\pi(\boldsymbol{\gamma}|\mathbf{y})$ is approximately a multivariate normal distribution (see (2.42)) that in the case of improper constant priors (that is, $\pi(\boldsymbol{\gamma}|\mathbf{y}) \propto h(\mathbf{y}|\boldsymbol{\gamma})$) compares to the Gaussian approximation for the MLEs; see Sect. 2.8.1 and formula (5.14). Also, note that in the case of constant priors, the mode of the posterior and MLE are the same. This is also true if the prior is uniform within a bounded region, provided that the MLE is within this region.

Prior distribution. If one is interested in getting inferences based mainly on observations, then *noninformative* or *vague* priors can be utilised. Informative priors can be used if external data and expert opinions are taken into account, as explained and discussed in Chapter 4.

Predictive distributions. The 0.999 quantile $Q^P_{0.999}$ of the predictive distribution $h(z|\mathbf{y})$ can be calculated using Monte Carlo Algorithm 4.1. Note that this accounts for both the process uncertainty (severity and frequencies are random variables) and the parameter uncertainty (parameters are simulated from their posterior distribution). The parameter uncertainty comes from the fact that we do not know the true values of the parameters.

The predictive distribution of the 0.999 quantile $Q_{0.999}(\boldsymbol{\gamma})$ of the conditional annual loss density $h(z|\boldsymbol{\gamma})$ can be calculated knowing that $\boldsymbol{\gamma}$ is distributed from $\pi(\boldsymbol{\gamma}|\mathbf{y})$. This can be used to form a one-sided or two-sides predictive intervals to contain the true value of the quantile with some probability q. Then one can argue that the conservative estimate of the capital charge should be based on the upper bound of the constructed confidence interval. Calculation of the predictive interval for the quantile $Q_{0.999}(\boldsymbol{\gamma})$ of $h(z|\boldsymbol{\gamma})$ can be accomplished using FFT to calculate $Q_{0.999}(\boldsymbol{\gamma})$ for a given $\boldsymbol{\gamma}$ and samples of $\boldsymbol{\gamma}$ from its posterior; see Monte Carlo Algorithm 4.2.

Posterior distribution. Typically, for models with truncation, direct sampling from the posterior is not possible. In general, it can be accomplished numerically using MCMC; see Sect. 2.11.

Example 5.2 To illustrate, consider the Bayesian inference method for the truncated dataset that was fitted using maximum likelihood method in Example 5.1. That is, the loss amounts X_j and times T_j are simulated from $Posson(10)$-$GPD(0.3, 6)$ over $T = 80$ years and truncated below the reporting level $L = 2$. In calculations below we consider the simulated dataset over three different time periods: $T = 5, 20$, and 80.

A closed-form solution for posterior densities is not available and we utilise MCMC method to get samples from the posterior $\pi(\boldsymbol{\gamma}|\mathbf{y})$. In this example, RW-MH within Gibbs Algorithm 2.4 is adopted. For comparison purposes with MLE results, we are interested in the inferences mainly implied by the data. Thus we choose the independent constant priors bounded as follows: $\lambda \in [5, 20]$, $\xi \in [0.02, 1]$, $\beta \in [1, 13]$. That is, all parameters are independent under the prior $\pi(\boldsymbol{\gamma})$ and distributed uniformly with $\gamma_i \sim \mathcal{U}(a_i, b_i)$ on a wide ranges. Denote by $\boldsymbol{\gamma}^{(k)}$ the state of the chain at iteration k with the initial state $\boldsymbol{\gamma}^{(k=0)}$ taken as MLEs. The algorithm proceeds by proposing a new state γ_i^* sampled from the MCMC proposal transition kernel, that we choose to be the Gaussian distribution truncated below a_i and above b_i, with the density

$$f_N^{(T)}(\gamma_i^*; \gamma_i^{(k)}, \sigma_i) = \frac{f_N(\gamma_i^*; \gamma_i^{(k)}, \sigma_i)}{F_N(b_i; \gamma_i^{(k)}, \sigma_i) - F_N(a_i; \gamma_i^{(k)}, \sigma_i)}. \tag{5.20}$$

Here, $f_N(x;\mu,\sigma)$ and $F_N(x;\mu,\sigma)$ are the normal density and its distribution respectively with the mean μ and standard deviation σ. Then, the proposed move is accepted with the probability

$$p(\boldsymbol{\gamma}^{(k)},\boldsymbol{\gamma}^*) = \min\left\{1, \frac{\pi(\boldsymbol{\gamma}^*|\mathbf{y}) f_N^{(T)}(\gamma_i^*;\gamma_i^{(k)},\sigma_i)}{\pi(\boldsymbol{\gamma}^{(k)}|\mathbf{y}) f_N^{(T)}(\gamma_i^{(k)};\gamma_i^*,\sigma_i)}\right\}, \tag{5.21}$$

where $\mathbf{y}$ is the vector of observations and $\pi(\boldsymbol{\gamma}^*|\mathbf{y})$ is the posterior distribution. Also, here $\boldsymbol{\gamma}^* = (\gamma_1^{(k)},\ldots,\gamma_{i-1}^{(k)},\gamma_i^*,\gamma_{i+1}^{(k-1)},\ldots)$ is a new state, where parameters $1,2,\ldots,i-1$ are already updated while $i+1, i+2,\ldots$ are not updated yet.

Note that the normalisation constant for the posterior distribution is not needed here. If under the rejection rule one accepts the move, then the new state of the i-th parameter at iteration k is given by $\gamma_i^{(k)} = \gamma_i^*$, otherwise the parameter remains in the current state $\gamma_i^{(k)} = \gamma_i^{(k-1)}$ and an attempt to move that parameter is repeated at the next iteration.

Using the chain samples $\boldsymbol{\gamma}^{(k)}$, $k = 1,2,\ldots$ as realisations from the posterior $\pi(\boldsymbol{\gamma}|\mathbf{y})$, we can estimate the expectations such as posterior mean, posterior standard deviation, etc. This has already been discussed in Sect. 2.11. Here, we just mention that in this example σ_i for proposals is chosen to be the MLE standard deviation of the corresponding parameter.

Figure 5.1 presents the chain samples for all parameters λ, ξ and β produced by the described algorithm. The chain is run for 1,010,000 iterations[2]; the initial 10,000 are discarded and 1,000,000 samples are used for estimation of expectations. It is useful to inspect the chain visually. One can see for example that the chain is mixing very well. Formal diagnostics on the stationarity of the chain can also be calculated; see Sect. 2.12. Fig. 5.2 and Table 5.3 present:

- the posterior densities for all parameters (ξ,β,λ) – one can see the change in the posterior as the data sample increases (in particular, the uncertainty decreases);
- the predictive density and distribution for the annual loss (over next year); and
- the density of the 0.999 quantile $Q_{0.999}(\gamma)$, where $\boldsymbol{\gamma}$ is distributed from the posterior $\pi(\boldsymbol{\gamma}|\mathbf{y})$.

One can see that the MLE and Bayesian posterior estimates for the parameters and quantile converge as the data sample increases. Also, the 0.999 quantile of the predictive distribution $Q_{0.999}^P$ converges to the $\mathrm{E}[Q_{0.999}(\gamma)]$.

5.3 Extension to Negative Binomial and Binomial Frequencies

In addition to Poisson, negative binomial and binomial are other distributions often used to model frequencies. A nice property is that binomial has the mean less than

[2] The chain should be long enough so that numerical error in the estimates due to finite number of chain samples is not material.

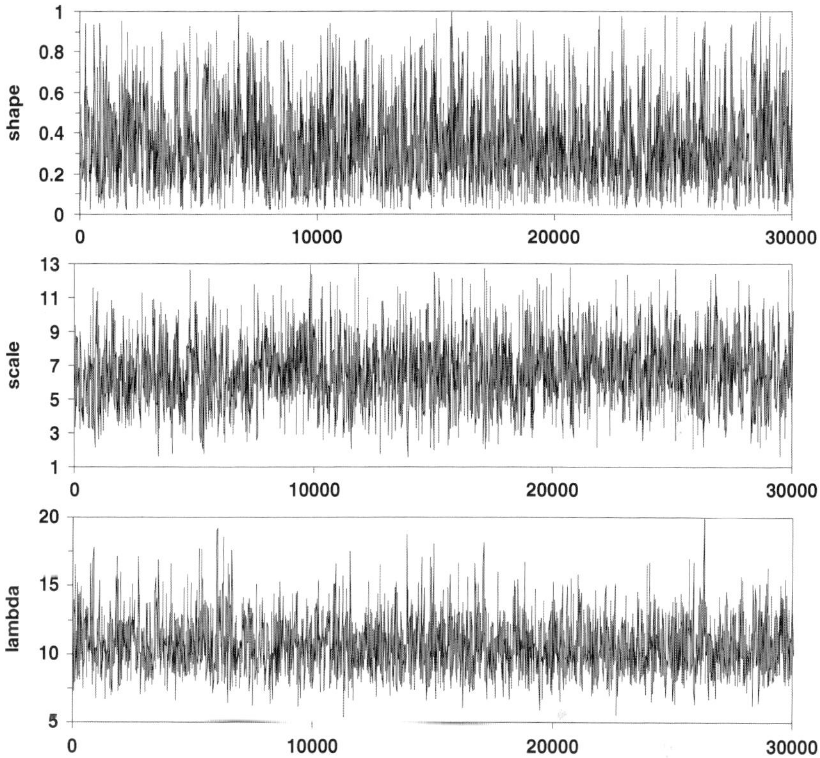

Fig. 5.1 MCMC realisations of the shape ξ, scale β, and Poisson intensity λ parameters in the case of reporting threshold $L = 2$ and data from Table 5.1

the variance; the mean of the negative binomial is larger than its variance; and Poisson mean equals its variance. This is often used as a criterion to choose a frequency distribution.

Another useful property of these distributions is that their type is preserved in the case of loss truncation as given by the following proposition.

Proposition 5.1 (Frequency of truncated losses) *Consider independent losses $X_1, X_2, \ldots, X_N$ with a common distribution $F(x)$ over some time period. Assume that the losses are independent of the loss frequency N. Denote the frequency of the losses above the reporting level L as N_L. Then*

(a) If $N \sim Poisson(\lambda)$, then $N_L \sim Poisson(\lambda(1 - F(L)))$.
(b) If $N \sim NegBin(r, p)$, where the parameter $p = 1/(1 + q)$, then $N_L \sim NegBin(r, \widetilde{p})$ with $\widetilde{p} = 1/(1 + \widetilde{q})$, where $\widetilde{q} = q(1 - F(L))$.
(c) If $N \sim Bin(n, p)$, then $N_L \sim Bin(n, \widetilde{p})$, where $\widetilde{p} = p(1 - F(L))$.

Proof The proof is trivial using a more general result (5.26) derived below. □

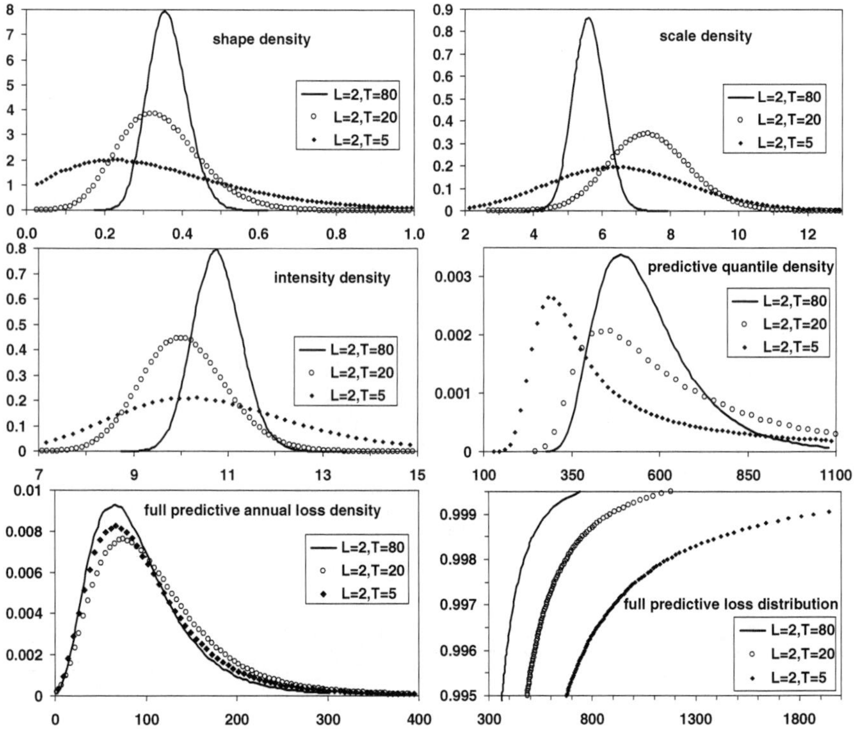

Fig. 5.2 MCMC posterior distributions of the shape ξ, scale β, and Poisson intensity λ parameters in the case of reporting threshold $L = 2$ and data from Table 5.1

Remark 5.2 We have already used this property of Poisson distribution, e.g. in Sect. 5.2.

In general, the relation between the distributions of N and N_L can be calculated as follows. Assume that the probability function for the number of events N is known to be $p_n = \Pr[N = n]$ and its probability generating function is

$$\psi_N(t) = \mathrm{E}[t^N] = \sum_k p_k t^k. \tag{5.22}$$

Consider a compound sum $S = M_1 + \cdots + M_N$, where N is a discrete random variable with probability generating function $\psi_N(t)$, and M_i are independent discrete random variables with probability generating function $\psi_M(t)$. Utilising the fact that the probability generating function of the sum of independent random variables is the product of the individual probability generating functions, the probability generating function of S can be found as

Table 5.3 Characteristics of the posterior distributions for parameters ξ (shape), β (scale), λ (intensity); quantile $Q_{0.999}(\boldsymbol{\gamma})$ and full predictive loss distribution. The data are simulated from $Poisson(\lambda = 10)$-$GPD(\xi = 0.3, \beta = 6)$ with a reporting level $L = 2$ and time horizons $T = 5, 20, 80$. The 5 year data are given in Table 5.1

	$T = 5$	$T = 20$	$T = 80$
	Posterior for model parameters		
E[ξ]	0.343(0.001)	0.346(0.0003)	0.363(0.0002)
stdev[ξ]	0.209(0.0003)	0.106(0.0002)	0.051(0.0001)
E[β]	6.614(0.006)	7.415(0.004)	5.628(0.002)
stdev[β]	2.027(0.003)	1.17 (0.002)	0.465(0.001)
E[λ]	10.716(0.006)	10.093(0.002)	10.777(0.002)
stdev[$\widehat{\lambda}$]	2.048(0.005)	0.903(0.001)	0.509(0.001)
$\rho_{\beta,\xi}$	–0.66	–0.69	–0.69
$\rho_{\lambda,\xi}$	0.34	0.28	0.33
$\widehat{\rho_{\widehat{\lambda},\widehat{\beta}}}$	–0.50	–0.43	–0.52
	Posterior for the 0.999 quantile, $Q_{0.999}(\boldsymbol{\gamma})$		
E[$Q_{0.999}(\boldsymbol{\gamma})$]	1,591(8)	777(4)	571.2(0.6)
stdev[$Q_{0.999}$]	4,037(20)	653(26)	153.1(0.6)
0.25 quantile	318	460	464
median	470	602	540
0.75 quantile	1038	864	642
	Full predictive loss distribution		
$Q^P_{0.999}$	1,864(27)	887(7)	580(5)

$$\begin{aligned}\psi_S(t) &= \sum_k \Pr[S = k]t^k \\ &= \sum_k \sum_n \Pr[M_1 + \cdots + M_n = k|N = n] \Pr[N = n]t^k \\ &= \sum_n \Pr[N = n](\psi_M(t))^n \\ &= \psi_N(\psi_M(t)).\end{aligned} \tag{5.23}$$

The number of events above the threshold can be written as

$$N_L = I_1 + \cdots + I_N,$$

where I_j are independent and identically distributed indicator random variables

$$I_j = \begin{cases} 1, \ \Pr[I_j = 1] = 1 - F(d), & \text{if} \quad X_j > u, \\ 0, \ \Pr[I_j = 0] = F(d) & \text{if} \quad X_j \le u, \end{cases} \tag{5.24}$$

with probability generating function

$$\psi_I(t) = F(L) + t(1 - F(L)) = 1 + (1 - F(L))(t - 1).$$

The probability generating function of the number of events above the threshold L can then be calculated as

$$\psi_{N_L}(t) = \psi_N(\psi_I(t)). \tag{5.25}$$

Moreover, if the distribution of N is parameterised by some θ and its probability generating function has a special form $\psi_N(t;\theta) = g(\theta(t-1))$, i.e. t and θ appear in $\psi_N(t;\theta)$ as $\theta(t-1)$ only, then

$$\psi_{N_L}(t;\theta) = g(\theta(1-F(L))(t-1)) = \psi_N(t;\theta(1-F(L))). \tag{5.26}$$

That is, both N and N_L have the same distribution type with different parameter θ. Specifically, if N is distributed from $P(\cdot|\theta)$ then N_L is distributed from $P(\cdot|\widetilde{\theta})$, where $\widetilde{\theta} = \theta(1-F(L))$. It can be checked directly that this relationship holds for Poisson, binomial and negative binomial. For more details and examples, see Panjer ([181], sections 5.7 and 7.8.2).

5.4 Ignoring Data Truncation

Often, the data below a reported level are simply ignored in the analysis, arguing that the high quantiles are mainly determined by the low-frequency/heavy-tailed severity risks. However, the impact of data truncation for other risks can be significant. Even if the impact is small, often it should be estimated to justify the reporting level. There are several ways to ignore the truncation discussed below.

Assume that the true model is based on the annual number of events N and severities X_j coming from distributions $P(\cdot|\boldsymbol{\lambda})$ and $F(\cdot|\boldsymbol{\beta})$ respectively. Here, $P(\cdot|\boldsymbol{\lambda})$ can be different from Poisson and $\boldsymbol{\lambda}$ denotes all frequency parameters. The density of the distribution $F(\cdot|\boldsymbol{\beta})$ is $f(\cdot|\boldsymbol{\beta})$. If it is further assumed that severities are independent and identically distributed, and independent of frequency. Then the frequency $\widetilde{N}$ and losses $\widetilde{X}_j$ above the threshold L are from $\widetilde{P}(\cdot|\boldsymbol{\theta})$ and

$$F_L(x|\boldsymbol{\beta}) = \frac{F(x|\boldsymbol{\beta}) - F(L|\boldsymbol{\beta})}{1 - F(L|\boldsymbol{\beta})}, \quad x \geq L,$$

respectively. Note that $\boldsymbol{\theta}$ is a function of $\boldsymbol{\lambda}$, $\boldsymbol{\beta}$ and L; see Sect. 5.3. The corresponding truncated severity density is

$$f_L(x|\boldsymbol{\beta}) = \frac{f(x|\boldsymbol{\beta})}{1 - F(L|\boldsymbol{\beta})}, \quad x \geq L.$$

Denote the data above the threshold as $\widetilde{\mathbf{Y}}$. Then fitting of the correct model proceeds as follows.

"True model". Using the frequency $\widetilde{P}(\cdot|\boldsymbol{\theta})$ and severity $F_L(x|\boldsymbol{\beta})$ distributions of the truncated data $\widetilde{\mathbf{Y}}$, fit the model parameters $\boldsymbol{\lambda}$ and $\boldsymbol{\beta}$, using the likelihood of

the observed data $\widetilde{\mathbf{Y}}$ via the MLE or Bayesian inference methods as described in Sect. 5.2. Then calculate the annual loss as

$$Z^{(0)} = \sum_{i=1}^{N} X_i, \quad N \sim P(\cdot|\boldsymbol{\lambda}), \quad X_i \stackrel{iid}{\sim} F(\cdot|\boldsymbol{\beta}). \tag{5.27}$$

Of course it is assumed that data below the threshold are generated from the same process as data above. To simplify the fitting procedure or to avoid making the assumptions about data below the level, several approaches are popular in practice. In particular *"naive model"*, *"shifted model"* and *"truncated model"* are defined below.

"Naive model". Using truncated data $\widetilde{\mathbf{Y}}$, fit frequency distribution $\widetilde{P}(\cdot|\boldsymbol{\theta})$ and severity $F(\cdot|\boldsymbol{\beta}_U)$ assuming that there is no truncation. Then calculate the annual loss as

$$Z^{(U)} = \sum_{i=1}^{N} X_i, \quad N \sim \widetilde{P}(\cdot|\boldsymbol{\theta}), \quad X_i \stackrel{iid}{\sim} F(\cdot|\boldsymbol{\beta}_U). \tag{5.28}$$

"Shifted model". Using truncated data $\widetilde{\mathbf{Y}}$, fit frequency $\widetilde{P}(\cdot|\boldsymbol{\theta})$ and severity $F_L^{(S)}(x) = F(x - L|\boldsymbol{\beta})$. Then calculate the annual loss as

$$Z^{(S)} = \sum_{i=1}^{N} X_i, \quad N \sim \widetilde{P}(\cdot|\boldsymbol{\theta}), \quad X_i \stackrel{iid}{\sim} F_L^{(S)}(\cdot|\boldsymbol{\beta}_S), \tag{5.29}$$

"Truncated model". Using truncated data $\widetilde{\mathbf{Y}}$, fit frequency $\widetilde{P}(\cdot|\boldsymbol{\theta})$ and severity $F_L(x|\boldsymbol{\beta})$. Then calculate the annual loss as

$$Z^{(T)} = \sum_{i=1}^{N} X_i, \quad N \sim \widetilde{P}(\cdot|\boldsymbol{\theta}), \quad X_i \stackrel{iid}{\sim} F_L(\cdot|\boldsymbol{\beta}). \tag{5.30}$$

Denote the 0.999 quantiles of the annual losses under the "true", "naive", "shifted" and "truncated" models as $Q^{(0)}$, $Q^{(U)}$, $Q^{(S)}$ and $Q^{(T)}$ respectively. The bias introduced into the 0.999 quantile of the annual loss distribution from use of the wrong model can be quantified by the relative difference

$$\delta^{(\star)} \equiv \frac{Q^{(\star)} - Q^{(0)}}{Q^{(0)}}, \quad (\star) = \text{"}U\text{"}, \text{"}T\text{"}, \text{"}S\text{"}. \tag{5.31}$$

Calculation of the annual loss quantile using the incorrect model (wrong frequency and severity distributions) will induce a bias. One may think that the bias is not material and take one of the above methods.

Each of the "naive model", "shifted model" and "truncated model" is biased for finite truncation, that is, their quantile estimates will never converge to the true value as the data sample size increases.

The difference (bias) between $Q^{(0)}$ and $Q^{(S)}$, and between $Q^{(0)}$ and $Q^{(U)}$ was studied in Luo, Shevchenko and Donnelly [151]. The difference between $Q^{(T)}$ and $Q^{(0)}$ was studied in Mignola and Ugoccioni [163]. The "naive model" was analysed in Chernobai, Menn, Trück and Rachev [53] and Frachot, Moudoulaud and Roncalli [95].

Example 5.3 (Poisson-lognormal) To demonstrate the impact of ignoring data truncation consider N and X_i modelled by the $Poisson(\lambda)$ and $\mathcal{LN}(\mu,\sigma)$ with the probability mass $p(\cdot|\lambda)$ and the density $f(x|\mu,\sigma)$, $0 < x < \infty$ respectively. The density of a left-truncated lognormal distribution is

$$f_L(x|\mu,\sigma) = \frac{f(x|\mu,\sigma)}{1 - F(L|\mu,\sigma)}; \quad L \leq x < \infty, \tag{5.32}$$

Assuming that losses originating from $f(x|\mu,\sigma)$ and $p(k|\lambda)$ are recorded above known reporting level L, the data above L are counts from $Poisson(\theta)$, $\theta = \lambda(1 - F(L|\mu,\sigma))$ and losses from $f_L(x|\mu,\sigma)$. Then the models are calculated as follows.

- *"True model"* is obtained by using λ, μ and σ in (5.27).
- *"Shifted model"*. Suppose that the shifted lognormal density

$$f_L^{(S)}(x|\mu_s,\sigma_s) = \frac{1}{(x-L)\sqrt{2\pi\sigma_s^2}} \exp\left(-\frac{(\ln(x-L)-\mu_s)^2}{2\sigma_s^2}\right), \tag{5.33}$$

where $L \leq x < \infty$, is fitted to the truncated data using the method of maximum likelihood. In the limit of large sample size, the parameters of this distribution μ_S and σ_S can be determined using first two moment, that is, expressed in terms of the true parameters μ and σ as follows:

$$\mu_S = \int_L^\infty \ln(x-L) f_L^{(T)}(x|\mu,\sigma)dx, \tag{5.34}$$

$$\sigma_S^2 = \int_L^\infty [\ln(x-L)]^2 f_L^{(T)}(x|\mu,\sigma)dx - \mu_S^2. \tag{5.35}$$

The above integrals can be efficiently calculated using Gaussian quadrature (see Sect. 3.5.2) or just using standard adaptive integration routines available from most of software packages (for example, adaptive integration routine QDAGI from IMSL library). In this model the frequency is modelled by $Poisson(\theta)$, that is the losses below L are ignored. Finally, θ, μ_S and σ_S are used in (5.29).

- *"Naive model"*. This model is based on the un-truncated lognormal with density $f(x|\mu_U, \sigma_U)$ defined by (2) and fitted to data above the threshold L using the method of maximum likelihood. Similar to the "shifted model", in the limit of large sample size, parameters μ_U and σ_U can be determined via the true parameters μ and σ as follows; see Chernobai, Menn, Trück and Rachev [53]:

$$\mu_U = \int_L^\infty \ln(x) f_L^{(T)}(x|\mu, \sigma)dx, \tag{5.36}$$

$$\sigma_U^2 = \int_L^\infty (\ln x)^2 f_L^{(T)}(x|\mu, \sigma)dx - \mu_U^2. \tag{5.37}$$

  Unlike the "shifted model" these integrals can be evaluated in closed form. The frequency under the "naive model" is modelled by $Poisson(\theta)$, that is, the losses below the threshold are ignored when the intensity of loss events is estimated. Finally, θ, μ_U and σ_U are used in (5.28).
- *"Truncated model"* is obtained by using θ, μ and σ in (5.30).

Figure 5.3 shows the relative bias in the 0.999 annual loss quantile (5.31) vs a fraction of truncated points $\Psi = F(L|\mu, \sigma) \times 100\%$, for the cases of light and heavy tail severities. In this example, the parameter values are chosen the same as some cases considered in Luo, Shevchenko and Donnelly [151]. In particular, we show the results for $(\theta = 10, \sigma = 1)$ and $(\theta = 10, \sigma = 2)$. The latter corresponds to the heavier tail severity. Here, the calculated bias is due to the model error only, that is, the bias corresponds to the limiting case of a very large data sample. Also note that the actual value of the scale parameter μ is not relevant because only relative quantities are calculated. *"Naive model"* and *"shifted model"* are easy to fit but induced bias can be very large. Typically *"naive model"* leads to a significant underestimation of the capital, even for a heavy tail severity; *"shifted model"* is better than *"naive model"* but worse then *"truncated model"*; the bias from *"truncated model"* is less for heavier tail severities.

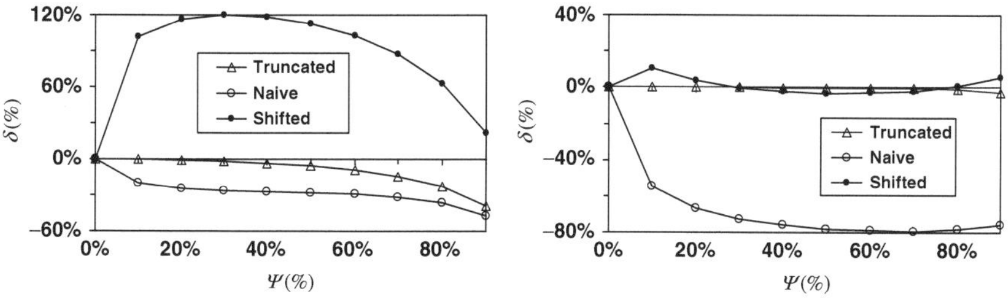

Fig. 5.3 Relative bias in the 0.999 quantile of the annual loss vs % of truncated points for several models ignoring truncation in the case of light tail severities from $\mathcal{LN}(3, 1)$ (*left figure*) and heavy tail severities from $\mathcal{LN}(3, 2)$ (*right figure*). The annual counts above the truncation level are from $Poisson(10)$

Remark 5.3 The biases introduced by the "naive" and "shifted" models, studied in this section, are the biases in the limit of large sample size. The parameters fitted using real data are estimates that have statistical fitting errors due to finite sample size. The true parameters are not known. The impact of parameter uncertainty on quantile estimates can be taken into account using a Bayesian framework. The problem with the use of the simplified models that ignore data truncation, such as "naive" and "shifted" models, is not just the introduced bias but underestimation of extra capital required to cover parameter uncertainty. Typically these simplified models lead to smaller fitting errors. It is not difficult to find a realistic example where the "shifted model" overestimating the quantile leads to under-estimation when the parameter uncertainty is taken into account; for an example, see Luo, Shevchenko and Donnelly ([151], section 6, Table 1). "Naive model" typically underestimates the capital even if the parameter uncertainty is taken into account This is because the "shifted" and "naive" models lead to smaller fitting errors in comparison to the "unbiased model". Of course, as the number of observations increases, the impact of parameter uncertainty diminishes. However, for modest fitting errors 5–10% (often, in modelling operational risk data, the errors are larger) the impact of parameter uncertainty is significant.

5.5 Threshold Varying in Time

Often, in practice, before fitting a specific severity distribution, a modeller scales the losses by some factors (inflation, business factors, etc). The reporting threshold should be scaled correspondingly and thus the losses in the fitted sample will have different threshold levels. In addition, the reporting policy may change affecting the threshold. To model this situation consider the following setup studied in [217].

- In the absence of a threshold, the events follow a homogeneous Poisson process with the intensity λ and the severities X_j are independent with a common distribution $F(\cdot|\boldsymbol{\beta})$; denote $\boldsymbol{\gamma} = (\lambda, \boldsymbol{\beta})$.
- The losses are reported above the known time dependent level $L(t)$. Denote the severities and arrival times of the reported losses as $\widetilde{X}_j$ and $\widetilde{T}_j$, $j = 1, \ldots, J$ respectively and t_0 is the start of the observation period.

Under the above assumptions, the events above $L(t)$ follow a non-homogeneous Poisson process with the intensity

$$\theta(\boldsymbol{\gamma}, L(t)) = \lambda(1 - F(L(t)|\boldsymbol{\beta})). \tag{5.38}$$

Denote

$$\Lambda(t, h) = \int_{t}^{t+h} \theta(\boldsymbol{\gamma}, L(x))dx. \tag{5.39}$$

Then, given that $(j-1)$th event occurred at $\widetilde{t}_{j-1}$, the inter-arrival time for the j-th event $\delta\widetilde{T}_j = \widetilde{T}_j - \widetilde{T}_{j-1}$ is distributed from

$$G_j(\tau|\boldsymbol{\gamma}) = 1 - \exp(-\Lambda(t_{j-1}, \tau)) \tag{5.40}$$

with the density

$$g_j(\tau|\boldsymbol{\gamma}) = \theta(\boldsymbol{\gamma}, L(t_{j-1} + \tau)) \exp(-\Lambda(t_{j-1}, \tau)). \tag{5.41}$$

The implied number of events in year m is $Poisson(\Lambda(s_m, 1))$ distributed, where s_m is the time of the beginning of year m, and the number of events over the non-overlapping periods are independent.

Data for event times and severities. Given $\boldsymbol{\gamma}$, the joint density of the data $\mathbf{Y}$ of the events above $L(t)$ over the time period $[t_0, t_E]$, consisting of the inter-arrival times $\delta\widetilde{T}_j = \widetilde{T}_j - \widetilde{T}_{j-1}$ and severities $\widetilde{X}_j$, $j = 1, \ldots, J$ above $L(t)$, can be written as

$$\begin{aligned} h(\mathbf{y}|\boldsymbol{\gamma}) &= \left(1 - G_J(t_E - \widetilde{t}_J|\boldsymbol{\gamma})\right) \prod_{j=1}^{J} f_{L(\widetilde{t}_j)}(\widetilde{x}_j|\boldsymbol{\beta}) g_j(\widetilde{\tau}_j|\boldsymbol{\gamma}) \\ &= \lambda^J \exp(-\Lambda(t_0, t_E - t_0)) \prod_{j=1}^{J} f(\widetilde{x}_j|\boldsymbol{\beta}). \end{aligned} \tag{5.42}$$

Here, explicitly

$$\Lambda(t_0, t_E - t_0) - \lambda \int_{t_0}^{t_E} [1 - F(L(x)|\boldsymbol{\beta})]dx.$$

Then, the likelihood function for the model is $\ell_{\mathbf{y}}(\boldsymbol{\gamma}) = h(\mathbf{y}|\boldsymbol{\gamma})$ and the maximum likelihood equations are

$$\frac{\partial \ln \ell_{\mathbf{y}}(\boldsymbol{\gamma})}{\partial \lambda} = \frac{J}{\lambda} - \int_{t_0}^{T} [1 - F(L(x)|\boldsymbol{\beta})]dx = 0, \tag{5.43}$$

$$\frac{\partial \ln \ell_{\mathbf{y}}(\boldsymbol{\gamma})}{\partial \boldsymbol{\beta}} = -\frac{\partial}{\partial \boldsymbol{\beta}} \Lambda(t_0, T - t_0) + \sum_{j=1}^{J} \frac{\partial}{\partial \boldsymbol{\beta}} \ln f(\widetilde{x}_j|\boldsymbol{\beta}) = 0. \tag{5.44}$$

The first equation gives

$$\widehat{\lambda} = \frac{J}{\int_{t_0}^{T} [1 - F(L(x)|\widehat{\boldsymbol{\beta}})]dx}, \tag{5.45}$$

that can be substituted into (5.42) and maximisation will be required with respect to $\boldsymbol{\beta}$ only. The likelihood contains integral over the severity distribution. If integration is not possible in closed form then it can be calculated numerically (that can be done efficiently using standard routines available in many numerical packages). For convenience, one can assume that a threshold is constant between the reported events: $L(t) = L(t_j)$, $\widetilde{t}_{j-1} < t \leq \widetilde{t}_j$ and $L(t) = L(t_E)$ for $\widetilde{t}_j < t \leq t_E$, so that

$$\int_{t_0}^{t_E} [1 - F(L(x)|\boldsymbol{\beta})]dx = [1 - F(L(t_E)|\boldsymbol{\beta})](t_E - \widetilde{t}_J) + \sum_{j=1}^{J} [1 - F(L(\widetilde{t}_j)|\boldsymbol{\beta})]\tau_j. \tag{5.46}$$

Of course this assumption is reasonable if the intensity of the events is not small. Typically scaling is done on the annual basis and one can assume a piecewise constant threshold per annum and the integral is replaced by a simple summation.

Data for annual counts and severities. If a data vector $\mathbf{Y}$ of the events above the reporting threshold consists of the annual counts $\widetilde{N}_m$, $m = 1, \ldots, T$ and severities $\widetilde{X}_j$, $j = 1, \ldots, J$ $(J = \widetilde{N}_1 + \ldots + \widetilde{N}_T)$, then the joint density of the data (given $\boldsymbol{\gamma}$) is

$$h(\mathbf{y}|\boldsymbol{\gamma}) = \prod_{j=1}^{J} f_{L(t_j)}(\widetilde{x}_j|\boldsymbol{\beta}) \prod_{m=1}^{T} p(\widetilde{n}_m|\Lambda(s_m, 1)), \tag{5.47}$$

where $p(\cdot|\Lambda(s_m, 1))$ is probability mass function of $Poisson(\Lambda(s_m, 1))$. Usually, in practice, scaling is done on an annual basis. Thus we can consider the case of a piece-wise constant threshold per annum such that for year m:

$$L(t) = L_m, \; \theta(\boldsymbol{\gamma}, L(t)) = \lambda(1 - F(L_m|\boldsymbol{\beta})) = \theta_m, \; s_m \leq t < s_m + 1,$$

where s_m is the time of the beginning of year m. The joint density in this case is

$$h(\mathbf{y}|\boldsymbol{\gamma}) = \prod_{j=1}^{J} f_{L(\widetilde{t}_j)}(\widetilde{x}_j|\boldsymbol{\beta}) \prod_{m=1}^{T} p(\widetilde{n}_m|\theta_m) \tag{5.48}$$

and equations to find MLEs using the likelihood function $\ell_{\mathbf{y}}(\boldsymbol{\gamma}) = h(\mathbf{y}|\boldsymbol{\gamma})$ are

$$\frac{\partial \ln \ell_{\mathbf{y}}(\boldsymbol{\gamma})}{\partial \lambda} = \sum_{m=1}^{T} [1 - F(L_m|\boldsymbol{\beta})] \frac{\partial}{\partial \theta_m} \ln p(\widetilde{n}_m|\theta_m) = 0, \tag{5.49}$$

$$\frac{\partial \ln \ell_{\mathbf{y}}(\boldsymbol{\gamma})}{\partial \boldsymbol{\beta}} = \sum_{j=1}^{J} \frac{\partial}{\partial \boldsymbol{\beta}} \ln f_{L(\widetilde{t}_j)}(\widetilde{x}_j|\boldsymbol{\beta})$$
$$-\lambda \sum_{m=1}^{T} \frac{\partial F(L_m|\boldsymbol{\beta})]}{\partial \boldsymbol{\beta}} \frac{\partial}{\partial \theta_m} \ln p(\widetilde{n}_m|\theta_m) = 0. \tag{5.50}$$

The first equation gives

$$\widehat{\lambda} = \frac{\sum_{m=1}^{T} \widetilde{n}_m}{\sum_{m=1}^{T} (1 - F(L_m|\widehat{\boldsymbol{\beta}}))}. \tag{5.51}$$

The MLEs of the severity parameters should be estimated jointly with the intensity. Given that the intensity MLE can be expressed in terms of the severity parameters MLEs via the above equation, one can substitute (5.51) into the likelihood function (5.48) and find severity parameters MLEs by maximising the obtained likelihood profile.

Often the MLEs for severity parameters calculated marginally (that is by simply maximising $\sum \ln f_{L(t_j)}(\widetilde{x}_j|\boldsymbol{\beta})$) do not differ materially from the results of the joint estimation if the variability of the threshold is not extremely fast. The results of the estimation for the simulated data in the case of exponentially varying threshold, presented in [217], confirm this intuitive observation, although the difference can still be significant if the intensity is small. Also, marginal estimation does not allow for quantification of the covariances between frequency and severity parameters required to account for parameter uncertainty.

Example 5.4 To illustrate the case of time varying threshold, consider data simulated from $Poisson(10) - GPD(0.3, 6)$ over $T = 5$ years (see Example 5.1). Assume that the reporting threshold $L(t)$ is piece-wise constant per annum, that is $L(t) = L_m, m-1 \leq t < m, m = 1, 2, \ldots, T$. Moreover, assume that $L_1 = L_2 = 1$ and $L_3 = L_4 = L_5 = 2$. Then the log-likelihood function for the reported events is

$$\ln \ell_{\mathbf{y}}(\boldsymbol{\gamma}) = J \ln \lambda - \lambda \sum_{m=1}^{T} \left(1 + \frac{\xi L}{\beta}\right)^{-\frac{1}{\xi}} - J \ln \beta$$
$$-\left(1 - \frac{1}{\xi}\right) \sum_{j=1}^{J} \ln \left(1 + \frac{\xi \widetilde{x}_j}{\beta}\right). \tag{5.52}$$

Table 5.4 MLE and MCMC (with 2×10^5 iterations) results in the case of data simulated over 5 years from $Poisson(\lambda = 10) - GPD(0.3, 6)$ and truncated below $L_1 = L_2 = 1$, $L_3 = L_4 = L_5 = 2$. Numerical errors of the MCMC estimates are given in brackets next to the estimates

Maximum likelihood estimates	
$\widehat{\xi} = 0.176$ $\widehat{\text{stdev}[\widehat{\xi}]} = 0.183$	$\widehat{\beta} = 7.593$ $\widehat{\text{stdev}[\widehat{\beta}]} = 2.090$
$\widehat{\lambda} = 9.571$ $\widehat{\text{stdev}[\widehat{\lambda}]} = 1.611$	$\widehat{\rho_{\widehat{\beta},\widehat{\xi}}} = -0.716$ $\widehat{\rho_{\widehat{\lambda},\widehat{\xi}}} = 0.208$ $\widehat{\rho_{\widehat{\lambda},\widehat{\beta}}} = -0.308$
$\widehat{Q}_{0.999} = 305.079$ $\widehat{\text{stdev}[\widehat{Q}_{0.999}]} = 132.504$	
Bayesian MCMC estimates	
$\text{E}[\xi] = 0.295(0.002)$ $\text{stdev}[\xi] = 0.189$	$\text{E}[\beta] = 7.37(0.01)$ $\text{stdev}[\beta] = 1.90$
$\text{E}[\lambda] = 9.99(0.01)$ $\text{stdev}[\lambda] = 1.69$	$\rho_{\beta,\xi} = -0.59$ $\rho_{\lambda,\xi} = 0.20$ $\rho_{\lambda,\beta} = -0.33$
$\text{VaR}_{0.25}[Q_{0.999}(\boldsymbol{\gamma})] = 305.2(0.3)$ $\text{VaR}_{0.75}[Q_{0.999}] = 777(3)$	
$\text{VaR}_{0.5}[Q_{0.999}] = 415.8(0.8)$ $\text{E}[Q_{0.999}] = 1,179(19)$ $Q^P_{0.999} = 1,411(19)$	

Note that the integral is replaced by a simple summation because the threshold is a piece-wise constant function of time. After scaling the threshold L_0 and scale parameter β by USD 10,000 this corresponds to more or less typical values observed with real data. The maximum likelihood and Bayesian MCMC estimations are presented in Table 5.4. The prior distributions and MCMC procedure used to obtain the results are the same as in Example 5.2.

Problems[3]

5.1 ($\star$) Assume that the losses $X_1, X_2, \ldots, X_n$ are independent random variables from a one-parameter Pareto distribution $F(x|\xi) = 1-(x/a)^{-\xi}$, $x \geq a > 0$, $\xi > 0$. Also, assume that the losses below $L > a$ are not reported. Find the likelihood function for the observed losses above L and find the MLE for the parameter ξ.

5.2 ($\star\star$) Suppose that:

- The frequencies $N_1, N_2, \ldots, N_M$ are independent with a common binomial distribution $Bin(n, p)$.
- Corresponding losses $X_1, X_2, \ldots, X_J$, where $J = N_1 + \cdots + N_M$, are independent random variables from a one-parameter Pareto distribution $F(x|\xi) = 1 - (x/a)^{-\xi}$, $x \geq a > 0$, where $\xi > 0$ is unknown and a is known.
- The severities and frequencies are independent.

Suppose that the losses below $L > a$ are not reported. Find the likelihood of the observed frequencies and severities (that is, events with the losses above L). Derive the MLE for parameters p and ξ assuming binomial parameter n is known.

5.3 ($\star\star$) Solve Problem 5.2 in the case of frequencies distributed from the negative binomial distribution $NegBin(n, p)$ instead of binomial distribution $Bin(n, p)$.

5.4 ($\star\star\star$) Suppose that:

[3] Problem difficulty is indicated by asterisks: ($\star$) – low; ($\star\star$) – medium, ($\star\star\star$) – high.

- The annual frequencies $N_1, N_2, \ldots, N_M$ are independent with a common distribution $Poisson(\lambda)$.
- Corresponding losses $X_1, X_2, \ldots, X_J$, where $J = N_1 + \cdots + N_M$, are independent random variables from a one-parameter Pareto distribution $F(x|\xi) = 1 - (x/a)^{-\xi}$, $x \geq a > 0$, where a is known.
- The severities and frequencies are independent.

Simulate the data over 5 years (i.e. $M = 5$) using $\lambda = 5$, $\xi = 3$, $a = 1$. Suppose that the losses below $L = 2$ are not reported. Using truncated data (i.e. the events with losses above L), estimate λ and ξ (assuming that these parameters are unknown) utilising random walk Metropolis-Hastings within Gibbs algorithms. Also, estimate the 0.999 quantile of the predictive distribution for the annual loss. Repeat estimation for the case $M = 20$. Assume constant priors.

5.5 ($\star$) Suppose that:

- The annual frequency N is distributed from the negative binomial with mean 10 and standard deviation 5.
- The independent risk severities are Pareto distributed, $X_i \sim Pareto(\xi, a)$, with $\xi = 4$ and $a = 1$.
- The severities and frequency are independent.

Calculate the true 0.999 quantile of the annual loss $Z = \sum_{i=1}^{N} X_i$ assuming that all model parameters are known. Suppose that the losses below $L > a$ are not reported and ignored using "truncated model"; see Sect. 5.4. Find the 0.999 quantile of the annual loss distribution under "truncated model". Compare it with the true value for different threshold levels corresponding to the fraction of truncated data ranging from 0% to 90%.

5.6 ($\star$) Repeat calculations of Problem 5.5 if the frequency distribution is $Bin(n, p)$ with mean 10 and standard deviation 3.

5.7 ($\star$) Repeat calculations of Problem 5.5 if the frequency is Poisson distributed with mean 10 and the severity distribution is $GPD(\xi, \beta = 1)$, for two cases: $\xi = 0.2$ and $\xi = 0.5$.

5.8 ($\star$) Suppose that risk events follow to Poisson process with time dependent intensity that will change linearly from 5 to 10 over next year. Find the distribution of the number of events over the next year.

Chapter 6
Modelling Large Losses

If there is a possibility of several things going wrong, the one that causes the most damage is the one to go wrong.

Murphy

Abstract Some operational risk events are rare but may have a major impact on a bank. Limited historical data make quantification of such risks difficult. This chapter discusses Extreme Value Theory that allows analysts to rationally extrapolate to losses beyond those historically observed and to estimate their probability. The chapter also discusses several parametric distributions which have been proposed to model the distribution tail of operational risk losses.

6.1 Introduction

Some operational risk events are rare but may have a major impact on a bank. These are often referred to as *low-frequency/high-severity* risks. It is recognised that these operational risks have heavy tailed severity distributions. Due to simple fitting procedure, one of the popular distributions to model severity is the lognormal distribution. It is a heavy-tailed distribution, that is, belongs to the class of so-called sub-exponential distributions where the tail decays slower than any exponential tail. Often it provides a reasonable overall statistical fit, as reported in the literature, and was suggested for operational risk at the beginning of Basel II development; see BCBS ([19], p. 34). However, due to the high quantile level requirement for operational risk capital charge, accurate modelling of extremely high losses (the tail of severity distribution) is critical and other heavy tail distributions are often considered to be more appropriate.

Two studies of operational risk data collected over many institutions are of central importance here: Moscadelli [166] analysing 2002 LDCE (where Extreme Value Theory (EVT) is used for analysis in addition to some standard two-parameter distributions), and Dutta and Perry [77] analysing 2004 LDCE. The latter paper considered the four-parameter g-and-h and GB2 distributions as well as EVT and several two-parameter distributions. There are two types of EVT models: traditional *block maxima* and *threshold exceedances*. *Block maxima* EVT is focused on modelling the largest loss per time period of interest. This is used in the insurance and in many other fields. For example, it is used in the design of dams for flood control where engineers are interested in quantification of the probability of the annual

P. Shevchenko, *Modelling Operational Risk Using Bayesian Inference*, DOI 10.1007/978-3-642-15923-7_6,

maximum water level. It is certainly important to operational risk too. However, for capital calculations, the primary focus is to quantify the impact of all losses per year. Modelling of all large losses exceeding a threshold is dealt by EVT–*threshold exceedances*. The key result of EVT is that the largest losses or losses exceeding a large threshold can be approximated by the limiting distribution – which is the same regardless of the underlying process. This allows for rational extrapolation to losses beyond those historically observed and estimation of their probability. However, as with any extrapolation methods, EVT should be applied with caution.

Typically, to apply EVT (or other extrapolation method) for a dataset, we assume that there is a single physical process responsible for the observed data and any future losses exceeding the observed levels. This is often the case in physical sciences (e.g. hydrology).

However, in assessing operational risk, some people may argue that extreme values are anomalous and are not strongly related to the rest of the data. In addition, multiple processes might be responsible for extreme events within a risk cell and these processes might be different from the processes generating less severe losses. A good discussion on these issues can be found in Cope, Antonini, Mignola and Ugoccioni [62] and Nešlehová, Embrechts and Chavez-Demoulin [174].

Another important issue is that the loss-generating processes in operational risk change in time due to changes in regulations, bank size and policy, political environment, etc. Often, the occurrence of a large operational risk loss event leads to changes in the bank's controls and policies designed to prevent the occurrence of another similar event. One should either discard such data points from a fitting procedure or include them under *'what if'* scenario. Finally, for some risks there might be upper limits on the maximum possible loss (e.g. underwriting risks); extrapolation beyond these limits does not make sense. All these issues should be addressed in practice.

If we assume that a single mechanism is responsible for the losses in dataset and extrapolation can be done, then EVT is a very powerful tool. A detailed presentation of EVT can be found in Embrechts, Klüppelberg and Mikosch [83] or chapter 7 in McNeil, Frey and Embrechts [157]; also see Panjer ([179], chapter 7). In this chapter, we summarise the main results relevant to operational risk. It is important to note that EVT is asymptotic theory. Whether the conditions validating the use of the asymptotic theory are satisfied is often a difficult question to answer. Also, the convergence of some parametric models to the EVT regime is very slow. For example, this is the case for the lognormal and g-and-h distributions studied in Mignola and Ugoccioni [164] and Degen, Embrechts and Lambrigger [71] respectively. In general, EVT should not preclude the use of other parametric distributions. Several severity distributions popular in operational risk practice will be presented in this chapter too.

6.2 EVT – Block Maxima

Consider a sequence of n independent random variables $X_1, \dots, X_n$ from a distribution $F(x)$ representing losses. Denote the maximum loss as

$$M_n = \max(X_1, \ldots, X_n).$$

Because each loss cannot exceed the maximum and due to independence between the losses, the distribution of the maximum is

$$\begin{aligned} F_{M_n}(x) &= \Pr[M_n \le x] = \Pr[X_1 \le x, \ldots, X_n \le x] \\ &= \prod_{i=1}^{n} \Pr[X_i \le x] = (F(x))^n. \end{aligned} \tag{6.1}$$

Remark 6.1 If we were interested to find the distribution of the largest loss per year (*annual maximum*) but the number of observation years is small, then one can study the largest loss per month (*monthly maximum*). This will increase the number of observations by a factor of 12. Suppose the distribution of the monthly maximum is $F_M(x)$, and monthly maxima are independent and identically distributed. Then the distribution of the annual maximum is $(F_M(x))^{12}$.

Given that $F(x) < 1$ or $F(x) = 1$, it is easy to see that if $n \to \infty$, then the distribution of maximum (6.1) converges to the degenerate distribution which is either 0 or 1 (i.e. the density concentrates on a single point) which is not very useful information. That is why the study of the largest losses in the limit $n \to \infty$ requires appropriate normalisation. This is somewhat similar to the central limit theory stating that appropriately normalised sum

$$\widetilde{S}_n = (S_n - a_n)/b_n,$$

where $S_n = X_1 + \cdots + X_n$ and $X_1, \ldots, X_n$ are independent and identically distributed random variables with finite variance, converges to the standard normal distribution as $n \to \infty$. Here, the normalised constants are

$$a_n = n\mathrm{E}[X_1], \quad b_n = \sqrt{n\mathrm{Var}[X_1]}.$$

Similarly, the limiting result for the distribution of the normalised maximum $\widetilde{M}_n = (M_n - d_n)/c_n$ says that for some sequences of $c_n > 0$ and d_n,

$$\lim_{n\to\infty} \Pr[(M_n - d_n)/c_n \le x] = \lim_{n\to\infty} (F(c_n x + d_n))^n = H(x). \tag{6.2}$$

If $H(x)$ is non-degenerate distribution, then F is said to be in the *maximum domain of attraction* of H, which is denoted as $F \in MDA(H_\xi)$. Then the well-known Fisher-Trippet, Gnedenko Theorem essentially says that $H(x)$ must be the generalised extreme value (GEV) distribution $H_\xi((x - \mu)/\sigma)$, $\sigma > 0$, $\mu \in \mathbb{R}$ with the standard form

$$H_\xi(x) = \begin{cases} \exp\left(-(1 + \xi x)^{-1/\xi}\right), & \xi \ne 0, \\ \exp(-\exp(-x)), & \xi = 0, \end{cases} \tag{6.3}$$

where $1 + \xi x > 0$. The standard GEV will often be referred to as $GEV(\xi)$. If convergence takes place, then it is always possible to choose normalising sequences c_n and d_n so that the limit will be in the standard form $H_\xi(x)$. The shape parameter ξ determines a type of distribution: $\xi > 0$ – a Fréchet distribution; $\xi = 0$ – a Gumbel distribution; and $\xi < 0$ – a Weibull distribution. The Weibull distribution ($\xi < 0$) has bounded right tail (i.e. $x \leq -1/\xi$), while Gumbel and Fréchet have unbounded right tail. Also, the decay of the Fréchet tail is much slower than the Gumbel tail.

Remark 6.2

- The GEV is continuous at $\xi = 0$ which is very convenient in statistical modelling because the distribution type is not known a priori and has to be determined by fitting.
- The standard EVT assumes independent and identically distributed data. The maxima of strictly stationary time series (for many processes) has the same limiting distribution, that is GEV; see McNeil, Frey and Embrechts ([157], section 7.1.3).
- There are regularity conditions required from $F(x)$, so that H_ξ is a limiting distribution of the maximum. In particular, the distribution $F(x)$ should satisfy some continuity conditions at the right endpoint; see chapter 3 in Embrechts, Klüppelberg and Mikosch [83]. For the purpose of this book, we just say that essentially all common *continuous* distributions in operational risk are in the $MDA(H_\xi)$. Also note that some discrete distributions do not satisfy the required conditions so that a non-degenerate limit distribution for maxima does not exist. This is the case, for example, for Poisson and negative binomial distributions.

Example 6.1 (Maximum of exponentially distributed losses) If $X_1, \ldots X_n$ are independent with an exponential distribution $F(x) = 1 - \exp(-\beta x)$, $\beta > 0$, $x \geq 0$, then the distribution of their maximum is

$$F_{M_n}(x) = (1 - \exp(-\beta x))^n .$$

Rewriting this as

$$F_{M_n}(x) = (F(x))^n = \left(1 - \frac{1}{n}\exp\left(-\beta\left(x - \beta^{-1}\ln n\right)\right)\right)^n,$$

it is easy to see that, by choosing $c_n = 1/\beta$ and $d_n = (\ln n)/\beta$, we get the following limiting distribution of the normalised maximum $\widetilde{M}_n = (M_n - d_n)/c_n$:

$$\begin{aligned} F_{\widetilde{M}_n}(x) &= (F(c_n x + d_n))^n = \left(1 - \frac{1}{n}\exp(-x)\right)^n \\ &\to \exp(-\exp(-x)), \quad n \to \infty, \end{aligned} \tag{6.4}$$

with the domain $x \geq -d_n/c_n = -\ln n$ (i.e. the domain becomes $(-\infty, \infty)$ as $n \to \infty$). It is easily recognised as GEV, $H_{\xi=0}(x)$.

Example 6.2 (Maximum of Pareto distributed losses) For a two-parameter Pareto distribution

$$F(x) = 1 - \left(1 + \frac{x}{\beta}\right)^{-\alpha}, \quad \alpha > 0, \beta > 0, x \geq 0,$$

choosing

$$c_n = \beta \frac{n^{1/\alpha}}{\alpha}, \quad d_n = \beta n^{1/\alpha} - \beta$$

we get the limiting distribution for the normalised maximum

$$(F(c_n x + d_n))^n = \left(1 - \frac{1}{n}\left(1 + \frac{x}{\alpha}\right)^{-\alpha}\right)^n \to \exp\left(-\left(1 + \frac{x}{\alpha}\right)^{-\alpha}\right), \quad n \to \infty,$$

with the domain

$$x \geq -d_n/c_n = \alpha(n^{-1/\alpha} - 1) \to 1 + x/\alpha > 0 \quad \text{as} \quad n \to \infty.$$

It is easily recognised as GEV, $H_{\xi=1/\alpha}$.

The limiting distribution of the maximum can be applied to any distribution $F(x)$ (satisfying some general conditions) and thus can be used as an approximation to the true distribution of the maximum without a full knowledge of the underlying $F(x)$. The implication of this theory is that the true distribution of the maximum can be approximated by three parameter GEV, $H_{\xi,\mu,\sigma}(x) = H_\xi((x - \mu)/\sigma)$:

$$H_{\xi,\mu,\beta}(x) = \begin{cases} \exp\left(-(1 + \xi(x - \mu)/\sigma)^{-1/\xi}\right), & \xi \neq 0, \\ \exp\left(-\exp(-(x - \mu)/\sigma\right), & \xi = 0. \end{cases} \tag{6.5}$$

The corresponding density function is

$$h_{\xi,\mu,\sigma}(x) = \begin{cases} \frac{1}{\sigma}\exp\left(-\left(1 + \xi\frac{x-\mu}{\sigma}\right)^{-1/\xi}\right)\left(1 + \xi\frac{x-\mu}{\sigma}\right)^{-1-1/\xi}, & \xi \neq 0; \\ \frac{1}{\sigma}\exp\left(-\exp(-\frac{x-\mu}{\sigma})\right)\exp\left(-\frac{x-\mu}{\sigma}\right), & \xi = 0. \end{cases} \tag{6.6}$$

Example 6.3 (Fitting GEV) Consider the weekly maximum losses due to transaction errors $M(1), \ldots, M(m)$ over m weeks. To apply the above described EVT, the losses should be independent and identically distributed, and the number of losses for each week should be large and the same. The latter is not really valid for operational risk where frequency is random; see remark below and Sect. 6.5. Here, we just assume that these conditions are satisfied, then $M(i)$ are independent and their common distribution is approximated by $H_{\xi,\mu,\sigma}(x)$. The log-likelihood function

$$\ln \ell_{\mathbf{M}}(\xi, \mu, \sigma) = \ln \prod_{i=1}^{m} h_{\xi,\mu,\sigma}(M(i))$$

can be calculated explicitly as

$$\begin{aligned}
\ln \ell_{\mathbf{M}}(\xi \neq 0, \mu, \sigma) = & - m \ln \sigma - \left(1 + \frac{1}{\xi}\right) \sum_{i=1}^{m} \ln \left(1 + \xi \frac{M(i) - \mu}{\sigma}\right) \\
& - \sum_{i=1}^{m} \left(1 + \xi \frac{M(i) - \mu}{\sigma}\right)^{-1/\xi}, \qquad (6.7) \\
\ln \ell_{\mathbf{M}}(\xi = 0, \mu, \sigma) = & - m \ln \sigma - \left(1 + \frac{1}{\xi}\right) \sum_{i=1}^{m} \exp \left(\frac{M(i) - \mu}{\sigma}\right) \\
& - \sum_{i=1}^{m} \frac{M(i) - \mu}{\sigma}, \qquad (6.8)
\end{aligned}$$

where $\sigma > 0$ and $1 + \xi(M(i) - \mu)/\sigma > 0$ for all i. Subject to these constraints, parameter estimates can be obtained using the maximum likelihood method by maximising the above log-likelihood or using MCMC. Also, note that the likelihood is continuous at $\xi = 0$. The distribution of the annual maximum loss can be estimated by $\left(H_{\xi,\mu,\sigma}(x)\right)^K$, where K is the number of weeks in a year.

Remark 6.3 The distribution of the maximum (6.1) assumes a deterministic (known) number of events. Moreover, in this framework, fitting GEV using data on maxima over many time periods (blocks), assumes that the number of events n is the same for all blocks. However, in operational risk the number of events is unknown and modelled as a random variable. This does not really affect the application of the block-maxima EVT in practice (i.e. often it can still be done as in the above example); see Sect. 6.5 for more details.

6.3 EVT – Threshold Exceedances

While it is important to understand and measure maximum possible loss over a 1-year time horizon, the primary focus in operational risk capital charge calculations is quantification of overall impact of all losses. For this purpose, the method of EVT threshold exceedances is very useful. Consider a random variable X, whose distribution is $F(x) = \Pr[X \leq x]$. Given a threshold u, the exceedance of X over u is distributed from

$$F_u(y) = \Pr[X - u \leq y | X > u] = \frac{F(y + u) - F(u)}{1 - F(u)}. \qquad (6.9)$$

As the threshold u increases, the limiting distribution of $F_u(\cdot)$ is given by the Pickands-Balkema-de Haan theorem; see McNeil, Frey and Embrechts ([157], section 7.2). The theorem essentially states that *if and only if* $F(x)$ is the distribution for which the distribution of the maximum (6.2) is $GEV(\xi)$ given by (6.3), then, as u increases, the excess distribution $F_u(\cdot)$ converges to a generalised Pareto distribution (GPD), $GPD(\xi, \beta)$:

$$G_{\xi,\beta}(y) = \begin{cases} 1 - (1 + \xi y/\beta)^{-1/\xi}, & \xi \neq 0, \\ 1 - \exp(-y/\beta), & \xi = 0. \end{cases} \tag{6.10}$$

Here, the shape parameter ξ is the same as the shape parameter of GEV distribution H_ξ. More strictly, we can find a function $\beta(u)$ such that

$$\lim_{u \to a} \sup_{0 \leq y \leq a-u} |F_u(y) - G_{\xi,\beta(u)}(y)| = 0, \tag{6.11}$$

where $a \leq \infty$ is the right endpoint of $F(x)$, ξ is the GPD shape parameter and $\beta > 0$ is the GPD scale parameter. The domain of GPD is

$$y \in \begin{cases} [0, \infty), & \text{if} \quad \xi \geq 0, \\ \left[0, -\beta/\xi\right], & \text{if} \quad \xi < 0. \end{cases} \tag{6.12}$$

The properties of GPD depend on the value of the shape parameter ξ:

- The case $\xi = 0$ corresponds to an exponential distribution with the right tail unbounded.
- If $\xi > 0$, the GPD right tail is unbounded and the distribution is heavy-tailed, so that some moments do not exist. In particular, if $\xi \geq 1/m$ then the m-th and higher moments do not exist. For example, for $\xi \geq 1/2$ the variance and higher moments do not exist. The analysis of operational risk data in Moscadelli [166] reported even the cases of $\xi \geq 1$ for some business lines, that is, infinite mean distributions; also see discussions in Nešlehová, Embrechts and Chavez-Demoulin [174].
- $\xi < 0$ leads to a bounded right tail, that is, $x \in \left[0, -\beta/\xi\right]$. It seems that this case is not relevant to modelling operational risk as all reported results indicate a non-negative shape parameter. One could think though of a risk control mechanism restricting the losses by an upper level and then the case of $\xi < 0$ might be relevant.
- The density of GPD is

$$h(x|\xi, \beta) = \begin{cases} \frac{1}{\beta}(1 + \xi x/\beta)^{-\frac{1}{\xi}-1}, & \xi \neq 0, \\ \frac{1}{\beta}\exp(-x/\beta), & \xi = 0, \end{cases} \tag{6.13}$$

  where, $h(x = 0) = 1/\beta$. Note some special cases of negative shape parameter: if $\xi = -1/2$ then $h(x) = \frac{1}{\beta}(1 - \frac{1}{2}x/\beta)$ is linear function; if $\xi = -1$ then

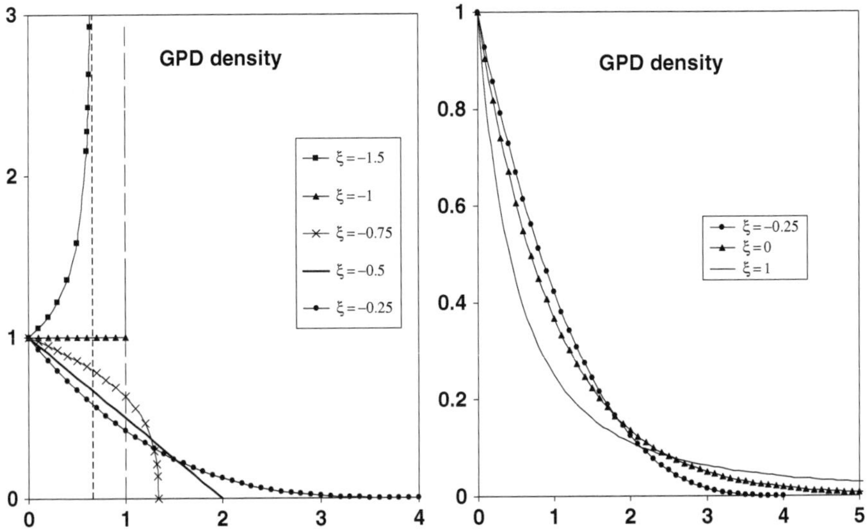

Fig. 6.1 The density of $GPD(\xi, \beta = 1)$ for several values of the shape parameter ξ

$h(x) = 1/\beta$ is constant; if $\xi < -1$ then $h(x)$ is infinity at the boundary of the domain $-\beta/\xi$. The latter case is certainly not relevant to operational risk in practice and can be excluded during fitting procedures. Figure 6.1 shows the density of GPD for different values of the shape parameter ξ.

- The GPD has a special stability property with respect to excesses. Specifically, if $X \sim G_{\xi,\beta}(x)$, $x > 0$, then the distribution of the conditional excesses $X - L|X > L$ over the threshold L is also the GPD with the same shape parameter ξ and changed scale parameter from β to $\beta + \xi L$:

$$\Pr[X - L \leq y|X > L] = G_{\xi,\beta+\xi L}(y), \quad y > 0. \tag{6.14}$$

The proof is simple and left to the reader as Problem 6.8. This stability property implies that if $\xi < 1$, then the mean excess function is

$$e(L) = \mathrm{E}[X - L|X > L] = \frac{\beta + \xi L}{1 - \xi}. \tag{6.15}$$

That is, the mean excess function is linear in L. This is often used as a diagnostic to check that the data follow the GPD model. In particular, it is used in a graphical method (plotting the mean excess of the data versus the threshold) to choose a threshold when the plot becomes approximately "linear".

GPD maxima. It is easy to verify that the distribution of the normalised maximum of the $GPD(\xi, \beta)$ exceedances is $GEV(\xi)$. In particular, calculate the distribution of the maximum of $GPD(\xi, \beta)$ independent and identically distributed exceedances $Y_1, \ldots, Y_n$

$$F_{M_n}(x) = \left(1 - (1 + \xi x/\beta)^{-1/\xi}\right)^n .$$

Rewrite this as

$$F_{M_n}(x) = (F(x))^n = \left(1 - \frac{1}{n}\left(1 + \frac{\xi}{n^\xi \beta}(x - (n^\xi - 1)\beta/\xi)\right)^{-1/\xi}\right)^n .$$

Then, it is easy to see that by choosing $c_n = \beta n^\xi$ and $d_n = (\beta/\xi)(n^\xi - 1)$, the limiting distribution of the normalised maximum $\widetilde{M}_n = (M_n - d_n)/c_n$ is

$$F_{\widetilde{M}_n}(x) = \left(1 - \frac{1}{n}(1 + \xi x)^{-1/\xi}\right)^n \to \exp\left(-(1 + \xi x)^{-1/\xi}\right), \quad n \to \infty,$$

with the domain $1 + \xi x > 0$, i.e. $GEV(\xi)$. The case of $\xi = 0$ can be obtained by simply taking a limit $\xi \to 0$ in the above.

GPD likelihood function. In practice, the implication of the limiting result for the distribution of threshold exceedances is that we can approximate F_u by $GPD(\xi, \beta)$ when u is large. Given independent and identically distributed exceedances Y_i, $i = 1, \ldots, K$, the log-likelihood function is

$$\begin{aligned}\ln \ell_{\mathbf{Y}}(\xi, \beta) &= \sum_{i=1}^{K} \ln h(Y_i|\beta, \xi) \\ &= -K \ln \beta - \left(1 + \frac{1}{\xi}\right) \sum_{i=1}^{K} \ln\left(1 + \xi \frac{Y_i}{\beta}\right) \qquad (6.16)\end{aligned}$$

which is continuous at $\xi = 0$, where

$$\ln \ell_{\mathbf{Y}}(\xi = 0, \beta) = -K \ln \beta - \frac{1}{\beta}\sum_{i=1}^{K} Y_i. \qquad (6.17)$$

This can be maximised (subject to $\beta > 0$ and $1 + \xi Y_i/\beta > 0$ for all i) to get maximum likelihood estimators for the parameters (see Sect. 6.4) or used in MCMC procedure to obtain Bayesian inferences (see Sect. 6.6).

Modelling the whole severity distribution. Often we have to model the whole severity distribution, rather than the tail only. Assume that losses X_k, $k = 1, 2, \ldots, K$ are independent and identically distributed. Then we can try to model the losses above a selected threshold u using $G_{\xi,\beta}$ and the losses below using an empirical distribution. That is,

$$F(x) \approx \begin{cases} G_{\xi,\beta}(x - u)(1 - F_n(u)) + F_n(u), & x \geq u, \\ F_n(x), & x < u. \end{cases} \qquad (6.18)$$

Here,

$$F_n(x) = \frac{1}{K}\sum\nolimits_{k=1}^{K} 1_{\{X_k \le x\}}$$

is an empirical distribution. This approach is a so-called *'splicing' method* when the density is modelled as

$$f(x) = w_1 f_1(x) + w_2 f_2(x), \quad w_1 + w_2 = 1, \tag{6.19}$$

where $f_1(x)$ and $f_2(x)$ are proper density functions defined on $x < u$ and $x \ge u$ respectively. In (6.18), $f_1(x)$ is modelled by the empirical distribution but one may choose a parametric distribution instead. Splicing can be viewed as a mixture of distributions defined on non-overlapping regions, while a standard mixture distribution is a combining of distributions defined on the same range. More than two components can be considered in the mixtures but typically only two components are used in the operational risk context. The choice of the threshold u is critical; for details of the methods to choose a threshold the reader is referred to McNeil, Frey and Embrechts [157].

Threshold exceedances. Finally, it is important to note that frequency of the exceedances changes when the threshold changes. Consider n independent losses $X_1, \dots, X_n$ with a common distribution $F(x)$. If there is a threshold u, then the probability of loss exceeding u is $\Pr[X > u] = 1 - F(u)$. If we define the indicator random variables

$$I_j = \begin{cases} 1, & \text{if} \quad X_j > u, \\ 0, & \text{if} \quad X_j \le u, \end{cases} \tag{6.20}$$

then $\Pr[I_j = 1] = 1 - F(u)$ and the number of losses above u

$$N_u = I_1 + \cdots + I_n$$

is a random variable from a binomial distribution, $Bin(n, 1 - F(u))$. As the threshold increases, the probability of exceedance becomes smaller and it is argued that for a fixed time period, if n is large, then N_u follows a Poisson distribution with mean $n(1-F(u))$. This follows from the well-known convergence of the $Bin(n, p)$ to $Poisson(\lambda = np)$ when $p \to 0$ for a fixed $\lambda = np$. However, in operational risk the number of losses per time period, n, is not known and is modelled as a random variable N. This case will be considered in Sect. 6.5.

6.4 A Note on GPD Maximum Likelihood Estimation

Given the likelihood, the estimation of the GPD can be done by the maximum likelihood or Bayesian MCMC methods. For the latter, the knowledge of MLEs is also very useful. In particular, the MLEs are often used as starting point for the Markov chain. Moreover, the standard deviations of the MLEs are often used as the standard deviations for the proposal Gaussian densities; see Example 6.4. Below, we consider the maximum likelihood method.

Maximisation of the likelihood (6.16) with respect to ξ and β provides their estimators $\widehat{\xi}$ and $\widehat{\beta}$. Formally, maximisation of the likelihood is subject to

$$\begin{cases} \beta > 0, \\ 1 + \xi y_{\max}/\beta > 0, \end{cases} \tag{6.21}$$

where $y_{\max} = \max(y_1, \ldots, y_K)$. It is important to note that if $\xi < -1$, then the likelihood (6.16) is infinity when $-\beta/\xi \to y_{\max}$. This is because the GPD density is infinity at the upper bound if $\xi < -1$. Thus, to get maximum likelihood estimates, one has to maximise likelihood (6.16) subject to conditions (6.21) and $\xi \geq -1$. It is convenient to introduce a new variable $\tau = -\xi/\beta$ and then maximise the log-likelihood function

$$\ln \ell_{\mathbf{y}}(\xi, \tau) = -K \ln(-\xi/\tau) - \left(1 + \frac{1}{\xi}\right) \sum_{i=1}^{K} \ln(1 - \tau \times y_i) \tag{6.22}$$

subject to $\tau < 1/y_{\max}$ and $\xi \geq -1$. Also, the extremum condition $\partial \ln \ell(\xi, \tau)/\partial\xi = 0$, gives

$$\xi(\tau) = \frac{1}{K} \sum_{i=1}^{K} \ln(1 - \tau \times y_i). \tag{6.23}$$

Then, the estimator $\widehat{\tau}$ can be obtained by maximisation of $\ln \ell(\xi(\tau), \tau)$ with respect to one parameter τ only, subject to $\tau < 1/y_{\max}$. Finally

$$\widehat{\xi} = \frac{1}{K} \sum_{i=1}^{K} \ln(1 - \widehat{\tau} \times y_i), \quad \widehat{\beta} = -\widehat{\xi}/\widehat{\tau}.$$

Note that $\ln \ell(\xi(\tau), \tau)$ is continuous at $\tau = 0$: if $\widehat{\tau} = 0$, then

$$\widehat{\xi} = 0, \quad \widehat{\beta} = \sum_{i=1}^{K} y_i/K.$$

To ensure that $\xi \geq -1$, the condition $\tau < 1/y_{\max}$ should be modified to $\tau \leq (1 - \epsilon)/y_{\max}$, where ϵ can be found from the condition $\xi(\tau) \geq -1$.

Another approach to avoid explicit condition $\xi \geq -1$ while avoiding the problem with infinite likelihood is to discretise the GPD using a precision δ of the data;

see section 13 in Schmock [210]. For example, if losses are measured in USD 1000, then take $\delta = 1{,}000$. Then the likelihood function

$$\ell_{\mathbf{y}}(\xi,\beta) = \prod_{i=1}^{K} \left(G_{\xi,\beta}\left(y_i+\delta\right) - G_{\xi,\beta}\left(y_i\right)\right) \tag{6.24}$$

has no singularity if $\xi < -1$.

The covariances of the MLEs can be estimated using the inverse of the observed information matrix (2.36). The latter can be estimated using second derivatives of the log-likelihood at the maximum. In the case of likelihood (6.16) these are:

$$\frac{\partial^2}{\partial\beta^2}\ln \ell_{\mathbf{y}}(\xi,\beta) = -\frac{1}{\beta}(\xi+1)\sum_{i=1}^{K}\frac{y_i}{(\beta+\xi y_i)^2}; \tag{6.25}$$

$$\frac{\partial^2}{\partial\xi^2}\ln \ell_{\mathbf{y}}(\xi,\beta) = -\frac{2}{\xi}\sum_{i=1}^{K}\frac{y_i}{\beta+\xi y_i} + \left(1+\frac{1}{\xi}\right)\sum_{i=1}^{K}\frac{y_i^2}{(\beta+\xi y_i)^2}; \tag{6.26}$$

$$\frac{\partial^2}{\partial\xi\partial\beta}\ln \ell_{\mathbf{y}}(\xi,\beta) = \sum_{i=1}^{K}\frac{y_i}{(\beta+\xi y_i)^2} - \frac{1}{\beta}\sum_{i=1}^{K}\frac{y_i^2}{(\beta+\xi y_i)^2}. \tag{6.27}$$

In the case of $\xi > -1/2$, it was shown that the MLE vector $(\widehat{\xi},\widehat{\beta})$ is asymptotically consistent and distributed from the bivariate normal distribution; see Smith [221] and sect. 6.5.1 in Embrechts, Klüppelberg and Mikosch [83]. The asymptotic covariance matrix for $(\widehat{\xi},\widehat{\beta})$, calculated using the inverse of the Fisher information matrix (2.35), can be found in closed form

$$\frac{1}{K}\begin{pmatrix} (1+\xi)^2 & -(1+\xi)\beta \\ -(1+\xi)\beta & 2(1+\xi)\beta^2 \end{pmatrix}. \tag{6.28}$$

This matrix (with ξ and β replaced by $\widehat{\xi}$ and $\widehat{\beta}$ respectively) is often used to estimate the precision of the MLEs.

6.5 EVT – Random Number of Losses

As has been mentioned above, in operational risk, the number of losses per time period is not fixed and is modelled as a random variable N with $p_n = \Pr[N = n]$. This has some implications for the use of the above described EVT.

Distribution of maximum. If the frequency N is random then, instead of (6.1), the distribution of a maximum M_N is calculated as

$$F_{M_N}(x) = \sum_{n=0}^{\infty} \Pr[M_N \leq x | N = n] \Pr[N = n]$$
$$= \sum_{n=0}^{\infty} (F(x))^n \Pr[N = n] = \psi_N(F(x)), \tag{6.29}$$

where

$$\psi_N(t) = \mathrm{E}[t^N] = \sum_k p_k t^k$$

is the probability generating function of the frequency distribution; also see (3.6). Note that there is a finite probability for zero maximum, that is, $\Pr[M_N = 0] = \psi_N(F(0))$. Typically, severity distribution has $F(0) = 0$ and frequency distribution has a finite probability at zero, $\Pr[M_N = 0] = \Pr[N = 0]$.

For example, if the annual number of losses $N \sim Poisson(\lambda)$ then $\psi(t) = \exp(-\lambda(1 - t))$ and thus the distribution of the maximum loss (per annum) is $F_{M_N}(x) = \exp(-\lambda(1-F(x))$. The distribution of the maximum loss over m years is $(F_{M_N}(x))^m = \exp(-m\lambda(1 - F(x))$. If the severities are from GPD $G_{\xi,\beta}(x)$, $x \geq 0$, defined by (6.10), then in the case of $\xi \neq 0$:

$$F_{M_N}(x) = \exp\left(-\lambda(1 + \xi x/\beta)^{-1/\xi}\right)$$
$$= \exp\left(-(1 + \xi(x - \mu)/\sigma)^{-1/\xi}\right), \quad x \geq 0, \tag{6.30}$$

where $\sigma = \beta\lambda^{\xi}$ and $\mu = (\beta\lambda^{\xi} - \beta)/\xi$. In the case of $\xi = 0$,

$$F_{M_N}(x) = \exp\left(-\lambda \exp(-x/\beta)\right)$$
$$= \exp\left(-\exp(-(x - \mu)/\sigma)\right), \quad x \geq 0, \tag{6.31}$$

where $\mu = \beta \ln \lambda$ and $\sigma = \beta$. Thus the distribution of M_N is a three parameter GEV $H_{\xi,\mu,\sigma}(x)$ for $x \geq 0$ and zero for $x < 0$. Note that $F_{M_N}(x)$ is continuous at $\xi = 0$. It is important to note that $F_{M_N}(x)$ is not GEV for all x but for $x \geq 0$ and there is a finite probability at zero $\Pr[M_N = 0] = \exp(-\lambda)$. If λ increases, then $\Pr[M_N = 0] \to 0$ and $F_{M_N}(x)$ will converge to GEV on the whole domain. In the case of other severity distributions, one can consider the limit of large λ when the distribution of $F_{M_N}(x)$ converges to a continuous distribution function, then it can be argued that the limiting distribution of maxima over many time periods is GEV.

Frequency of exceedances. Consider N independent losses $X_1, \ldots, X_N$ with a common distribution $F(x)$, where N is a discrete random variable with $p_n = \Pr[N = n]$. Then the number of losses above a threshold u is

$$N_u = I_1 + \cdots + I_N,$$

where I_j are independent indicator random variables. Then, the probability generating function of N_u can be calculated using probability generating function of N, $\psi_N(t)$, and probability generating function of I_j, $\psi_I(t)$, as

$$\psi_{N_u} = \psi_N(\psi(I));$$

see (5.25). Moreover, in the case when N is from Poisson, binomial or negative binomial, the distribution of N_u is the same as distribution of N with only one parameter changed; see Sect. 5.3. As a reminder:

- **Poisson.** If the frequency of losses is from $Poisson(\lambda)$ then the frequency of losses above u is $Poisson(\widetilde{\lambda})$ with $\widetilde{\lambda} = \lambda(1 - F(u))$.
- **Negative Binomial.** If the frequency of losses is from $NegBin(r, p)$, where the parameter $p = 1/(1+q)$, then the frequency of losses above u is $NegBin(r, \widetilde{p})$ with $\widetilde{p} = 1/(1+\widetilde{q})$, where $\widetilde{q} = q(1 - F(u))$.
- **Binomial.** If the frequency of losses is from $Bin(n, p)$, then the frequency of losses above L is $Bin(n, \widetilde{p})$, where $\widetilde{p} = p(1 - F(u))$.

It can be argued that binomial and negative binomial distributions will converge to Poisson when u increases, that is, when $1-F(u) \to 0$. This is intuitively expected because in this limit, the variance and mean of these distributions converge to the mean of Poisson. There is an extensive literature on point processes of extremes; see Sect. 7.4 in McNeil, Frey and Embrechts [157] or chapter 5 in Embrechts, Klüppelberg and Mikosch [83]. For the purposes of this book, we just say that in general it is argued that the distribution of the exceedances over a high threshold is well approximated by Poisson distribution. Moreover, the theory suggests to model the high threshold exceedances by the Poisson process.

Remark 6.4 The EVT *threshold exceedances* says that for a large threshold, the distribution of loss exceedances $F(x)$ can be approximated by GPD. Thus one can always argue that the distribution of the maximum of independent and identically distributed excesses over a high threshold can be approximated by GEV in the case of Poisson frequency; see (6.30). The assumption of Poisson distribution for the frequency of exceedances is also typical for high threshold exceedances.

6.6 EVT – Bayesian Approach

The above described EVT provides theoretical results for limiting distributions assuming that the model parameters are known. Of course in real life the parameters should be estimated using data. The uncertainty in parameter estimates will have implications for our inferences about maximum possible losses or loss sizes over the time period of prediction. Here we take Bayesian approach, which is a convenient way to quantify the uncertainty; see Sect. 2.9. Under this approach unknown parameters are modelled as random variables. Denote the vector of all model parameters by $\boldsymbol{\gamma}$, then the Bayesian approach for EVT proceeds as follows:

Model Assumptions 6.1

- *Suppose that, given* $\boldsymbol{\gamma}$, *we consider independent and identically distributed loss exceedances*

$$Y_1(1), \ldots, Y_{N_1}(1), \ldots, Y_1(m+1), \ldots, Y_{N_{m+1}}(m+1)$$

above a high threshold over $m+1$ *time periods with the independent frequencies*

$$N_1, \ldots, N_{m+1}$$

which are independent of the exceedances, i.e. N_i *is the number of events in the* i*-th period.*

- *Assume that the threshold is high enough, so that EVT is applicable. That is the distribution of the exceedances and frequencies can be modelled by* $GPD(\xi, \beta)$ *and* $Poisson(\lambda)$ *respectively, i.e. the vector of all model parameters is* $\boldsymbol{\gamma} = (\lambda, \xi, \beta)$.
- *Denote the data vector by* $\mathbf{X} = (\mathbf{Y}, \mathbf{N})$, *where:*

$$\mathbf{Y} = (Y_1(1), \ldots, Y_{N_1}(1), \ldots, Y_1(m), \ldots, Y_{N_m}(m));$$
$$\mathbf{N} = (N_1, \ldots, N_m).$$

In the context of operational risk, the objective is to predict loss events over the next time period given available data. Under the above assumptions, we consider the data $\mathbf{X}$ over m time periods and we try to make predictions for period $m+1$. The Bayesian approach models unknown parameters $\boldsymbol{\gamma}$ as random variables with a conditional, given data, density (*posterior*):

$$\pi(\boldsymbol{\gamma}|\mathbf{x}) \propto \ell_{\mathbf{x}}(\boldsymbol{\gamma})\pi(\boldsymbol{\gamma}),$$

where $\ell_{\mathbf{x}}(\boldsymbol{\gamma})$ is the likelihood of the data and $\pi(\boldsymbol{\gamma})$ is the prior density (specified before the data are available). The frequencies and severities are assumed independent. Thus

$$\ell_{\mathbf{x}}(\boldsymbol{\gamma}) = \ell_{\mathbf{n}}(\lambda)\ell_{\mathbf{y}}(\xi, \beta),$$

where

$$\ell_{\mathbf{n}}(\lambda) = \prod_{i=1}^{m} \exp(-\lambda)\lambda^{n_i}/n_i!$$

is the likelihood of counts and $\ell_{\mathbf{y}}(\xi, \beta)$ is the likelihood of GPD severities. The latter is given by (6.16). We can also write the posterior for frequency and severity parameters

$$\pi(\lambda|\mathbf{n}) \propto \ell_{\mathbf{n}}(\lambda)\pi(\lambda),$$
$$\pi(\xi,\beta|\mathbf{y}) \propto \ell_{\mathbf{y}}(\xi,\beta)\pi(\xi,\beta),$$

where $\pi(\lambda)$ and $\pi(\xi,\beta)$ are the prior densities. If λ and (ξ,β) are independent in prior, then $\pi(\boldsymbol{\gamma}) = \pi(\lambda)\pi(\xi,\beta)$ and thus $\pi(\boldsymbol{\gamma}|\mathbf{x}) = \pi(\lambda|\mathbf{n})\pi(\xi,\beta|\mathbf{y})$.

Frequency prediction. Given data, $\mathbf{N} = \mathbf{n}$, the density of number of events over next period N_{m+1} is

$$p(n|\mathbf{n}) = \int p(n|\lambda)\pi(\lambda|\mathbf{n})d\lambda,$$

where $p(n|\lambda)$ is $Poisson(\lambda)$. This is a *full predictive distribution* for the frequency. Note that the model assumptions imply that the number of events over next time period N_{m+1} is independent of $\mathbf{N}$. As it is shown in Sect. 4.3.3, if there is no prior information then the prior can be assumed to be a noninformative improper constant and the posterior $\pi(\lambda|\mathbf{n})$ is the density of $Gamma(\alpha_m,\beta_m)$, with

$$\alpha_m = 1 + \sum_{j=1}^{m} n_j, \quad \beta_m = 1/m.$$

Also, if the prior $\pi(\boldsymbol{\gamma})$ is the density of $Gamma(\alpha,\beta)$ then the posterior is $Gamma(\alpha_m,\beta_m)$ with

$$\alpha_m = \alpha + \sum_{j=1}^{m} n_j, \quad \beta_m = \beta/(1+\beta m).$$

It is easy to show that if the posterior $\pi(\lambda|\mathbf{n})$ is $Gamma(\alpha_m,\beta_m)$, then the predictive density $p(n|\mathbf{n})$ of N_{m+1} corresponds to a negative binomial distribution, $NegBin(\alpha_m, 1/(1+\beta_m))$; see (4.19).

Severity prediction. Given data for severities, $\mathbf{Y} = \mathbf{y}$, the predictive density of possible losses for the next and subsequent periods is

$$f(y|\mathbf{y}) = \int f(y|\boldsymbol{\theta})\pi(\boldsymbol{\theta}|\mathbf{y})d\boldsymbol{\theta}.$$

Here we assumed that, for a given $\boldsymbol{\theta} = (\xi,\beta)$, all subsequent losses are independent of the data $\mathbf{Y}$. There is no closed form for the posterior in this case, but it can easily be estimated numerically using MCMC. In the following example we will use RW-MH within Gibbs described in Sect. 2.11.3. The complication here is that $\xi > -\beta/Y_{\max}$ and thus the domain for ξ parameter is formally dependent on the data. Using this domain for the prior $\pi(\xi)$ contradicts to the concept that the prior is specified before the data. The way to overcome this problem is not to impose any specific restriction on the prior but to modify the acceptance probability of MCMC so that parameter samples outside the domain are rejected.

Remark 6.5 It is important to note that if the prior $\pi(\xi)$ has finite probability for $\xi \geq 1$, that is, $\Pr[\xi \geq 1] > 0$, then the prediction distributions of loss has infinite mean and higher moments, that is, $\mathrm{E}[Y_i(m+1)|\mathbf{Y}] = \infty$. If we do not want to allow for infinite mean behaviour, then the prior for ξ should be restricted to the domain $\xi < 1$.

Aggregate loss prediction. Overall, we are interested in the aggregated possible loss over $m+1$,

$$Z_{m+1} = \sum_{i=1}^{N_{m+1}} (L + Y_i(m+1)).$$

Its predictive density is simply

$$h(z|\mathbf{x}) = \int h(z|\boldsymbol{\gamma})\pi(\boldsymbol{\gamma}|\mathbf{x})d\boldsymbol{\gamma}.$$

Here, $h(z|\boldsymbol{\gamma})$ is the density of aggregate loss $Z = \sum_{i=1}^{N}(L + Y_i)$, where N is from $Poisson(\lambda)$ and Y_i are independent $GPD(\xi, \beta)$ exceedances above threshold L.

Remark 6.6 If the prior $\pi(\xi)$ has finite probability for $\xi \geq 1$, that is, $\Pr[\xi \geq 1] > 0$, then not only predicted severity (see Remark 6.5) but also predicted aggregate loss has infinite mean, i.e. $\mathrm{E}[Z_{m+1}|\mathbf{X}] = \infty$. So, if we would like to avoid infinite mean distribution for predicted aggregate loss over next time period, then the prior should be defined on $\xi < 1$.

Maximum exceedance prediction. The predictive density of maximum exceedance over $m+1$ is

$$f_{M_N}(w|\mathbf{x}) = \int f_{M_N}(w|\boldsymbol{\gamma})\pi(\boldsymbol{\gamma}|\mathbf{x})d\boldsymbol{\gamma},$$

where $f_{M_N}(w|\boldsymbol{\gamma})$ is given by (6.30) which is essentially GEV apart of finite probability at zero. Again, the closed-form solution is not available for the cases of practical interest but it is trivial to calculate using MCMC.

Example 6.4 As an illustrative example of Bayesian inference calculations, we simulate the loss exceedances Y_j above USD 1 million and event times T_j from $Poisson(10)$-$GPD(\xi_0, 6)$ over 4 years in the case of $\xi_0 = -0.1$ (i.e. right tail is bounded) and $\xi_0 = 0.25$ (i.e. right tail is unbounded). The simulated loss amounts and times are given in Table 6.1.

For comparison purposes with maximum likelihood method, assume constant independent priors bounded as follows: $\lambda \in [5, 20]$, $\xi \in [-1, 1]$, $\beta \in [1, 13]$. That is, all parameters are independent under the prior distribution $\pi(\boldsymbol{\gamma})$ and distributed uniformly with $\gamma_i \sim U(a_i, b_i)$ on a wide ranges, so that the inference is mainly implied by the data only. The full posterior $\pi(\boldsymbol{\gamma}|\mathbf{x})$ is not available in closed form and to simulate from the posterior we adopt RW-MH within Gibbs Algorithm 2.4.

Denote by $\boldsymbol{\gamma}^{(k)}$ the state of the chain at iteration k with the initial state $\boldsymbol{\gamma}^{(k=0)}$ taken as MLEs. The algorithm proceeds by proposing a new state γ_i^* sampled from

Table 6.1 Event times t_i and loss exceedances x_i (in USD 1 million, over the threshold USD 1 million simulated from $Poisson(10)$-$GPD(\xi_0, 6)$ over 4 years in the case of $\xi_0 = -0.1$ (i.e. right tail is bounded) and $\xi_0 = 0.25$ (i.e. right tail is unbounded)

		$\xi_0 = -0.1$	$\xi_0 = 0.25$			$\xi_0 = -0.1$	$\xi_0 = 0.25$
Index, i	t_i	x_i	x_i	Index, i	t_i	x_i	x_i
1	0.0257	6.063	7.323	20	1.9538	7.297	9.190
2	0.2540	2.881	3.141	21	1.9897	3.912	4.407
3	0.4662	2.019	2.145	22	2.1843	14.851	24.861
4	0.5784	1.936	2.051	23	2.2377	17.847	34.011
5	0.7248	5.719	6.830	24	2.3737	13.915	22.419
6	0.7399	10.266	14.366	25	2.5410	0.474	0.480
7	0.7803	0.472	0.478	26	2.6488	3.940	4.442
8	0.8533	1.789	1.887	27	2.7531	14.790	24.698
9	0.9065	7.385	9.328	28	2.9669	5.421	6.410
10	1.2136	5.426	6.418	29	3.1671	1.018	1.049
11	1.2265	2.704	2.933	30	3.2638	8.471	11.112
12	1.3274	0.219	0.221	31	3.2988	1.306	1.357
13	1.5192	9.696	13.289	32	3.3984	11.309	16.454
14	1.5728	6.570	8.072	33	3.6000	5.147	6.032
15	1.8030	22.662	54.563	34	3.7285	1.990	2.112
16	1.8641	3.554	3.958	35	3.7799	6.264	7.617
17	1.8648	3.999	4.517	36	3.9074	9.693	13.284
18	1.8755	3.256	3.592	37	3.9117	3.634	4.058
19	1.9202	0.630	0.642				

the MCMC proposal transition kernel, chosen to be the Gaussian distribution truncated below a_i and above b_i, with the density

$$f_N^{(T)}\left(\gamma_i^*; \gamma_i^{(k)}, \sigma_i\right) = \frac{f_N\left(\gamma_i^*; \gamma_i^{(k)}, \sigma_i\right)}{F_N\left(b_i; \gamma_i^{(k)}, \sigma_i\right) - F_N\left(a_i; \gamma_i^{(k)}, \sigma_i\right)} \tag{6.32}$$

where $f_N(x; \mu, \sigma)$ and $F_N(x; \mu, \sigma)$ are the normal density and its distribution respectively with the mean μ and standard deviation σ. For the proposal standard deviations σ_i we take the MLE standard deviation of corresponding parameters. Then the proposed move is accepted with the probability

$$p\left(\boldsymbol{\gamma}^{(k)}, \boldsymbol{\gamma}^*\right) = \min\left\{1, \frac{\pi(\boldsymbol{\gamma}^*|\mathbf{y}) f_N^{(T)}\left(\gamma_i^*; \gamma_i^{(k)}, \sigma_i\right)}{\pi\left(\boldsymbol{\gamma}^{(k)}|\mathbf{y}\right) f_N^{(T)}\left(\gamma_i^{(k)}; \gamma_i^*, \sigma_i\right)} 1_{\{\xi^* > -\beta^*/y_{\max}\}}\right\}, \tag{6.33}$$

where $\mathbf{y}$ is the vector of observations and $\pi(\boldsymbol{\gamma}^*|\mathbf{y})$ is the posterior density. Also, here $\boldsymbol{\gamma}^* = (\gamma_1^{(k)}, \ldots, \gamma_{i-1}^{(k)}, \gamma_i^*, \gamma_{i+1}^{(k-1)}, \ldots)$; that is, $\boldsymbol{\gamma}^*$ is a new state, where parameters $1, 2, \ldots, i-1$ are already updated while $i+1, i+2, \ldots$ are not updated yet; finally, ξ^* and β^* are components of γ^* corresponding to parameters ξ and β respectively.

The procedure is the same as in Example 5.2, except an indicator function

$$1_{\{\xi^* > -\beta^*/y_{\max}\}}$$

Table 6.2 MLE and MCMC estimates of the $Poisson(\lambda)$-$GPD(\xi, \beta)$ model. The data for datasets 1 and 2 were simulated from $Poisson(10)$-$GPD(-0.1, 6)$ and $Poisson(10)$-$GPD(0.25, 6)$ respectively; see Table 6.1

Maximum likelihood estimates, dataset 1	
$\widehat{\xi} = -0.210$ $\widehat{\text{stdev}[\widehat{\xi}]} = 0.156$	$\widehat{\beta} = 7.485$ $\widehat{\text{stdev}[\widehat{\beta}]} = 1.675$
$\widehat{\lambda} = 9.250$ $\widehat{\text{stdev}[\widehat{\lambda}]} = 1.521$	$\widehat{\rho_{\widehat{\beta},\widehat{\xi}}} = -0.824$ $\widehat{\rho_{\widehat{\lambda},\widehat{\xi}}} = 0.0$ $\widehat{\rho_{\widehat{\lambda},\widehat{\beta}}} = 0.0$
$\widehat{Q}_{0.999} = 168.715$	$\widehat{\text{stdev}[\widehat{Q}_{0.999}]} = 28.369$
Bayesian MCMC estimates, dataset 1	
$\text{E}[\xi] = -0.12$ $\text{stdev}[\xi] = 0.19$	$\text{E}[\beta] = 7.57$ $\text{stdev}[\beta] = 1.70$
$\text{E}[\lambda] = 9.51$ $\text{stdev}[\lambda] = 1.55$	$\rho_{\beta,\xi} = -0.73$ $\rho_{\lambda,\xi} = 0.0$ $\rho_{\lambda,\beta} = 0.0$
$\text{VaR}_{0.25}[Q_{0.999}(\boldsymbol{\gamma})] = 165.9$	$\text{VaR}_{0.75}[Q_{0.999}(\boldsymbol{\gamma})] = 213.1$
$\text{VaR}_{0.5}[Q_{0.999}(\boldsymbol{\gamma})] = 186.8$	$\text{E}[Q_{0.999}(\boldsymbol{\gamma})] = 228$ $Q^P_{0.999} = 292$
Maximum likelihood estimates, dataset 2	
$\widehat{\xi} = 0.177$ $\widehat{\text{stdev}[\widehat{\xi}]} = 0.197$	$\widehat{\beta} = 7.578$ $\widehat{\text{stdev}[\widehat{\beta}]} = 1.934$
$\widehat{\lambda} = 9.250$ $\widehat{\text{stdev}[\widehat{\lambda}]} = 1.521$	$\widehat{\rho_{\widehat{\beta},\widehat{\xi}}} = -0.662$ $\widehat{\rho_{\widehat{\lambda},\widehat{\xi}}} = 0.0$ $\widehat{\rho_{\widehat{\lambda},\widehat{\beta}}} = 0.0$
$\widehat{Q}_{0.999} = 314.419$	$\widehat{\text{stdev}[\widehat{Q}_{0.999}]} = 145.757$
Bayesian MCMC estimates, dataset 2	
$\text{E}[\xi] = 0.26$ $\text{stdev}[\xi] = 0.21$	$\text{E}[\beta] = 7.86$ $\text{stdev}[\beta] = 1.87$
$\text{E}[\lambda] = 9.50$ $\text{stdev}[\lambda] = 1.54$	$\rho_{\beta,\xi} = -0.56$ $\rho_{\lambda,\xi} = 0.0$ $\rho_{\lambda,\beta} = 0.0$
$\text{VaR}_{0.25}[Q_{0.999}(\boldsymbol{\gamma})] = 297$	$\text{VaR}_{0.75}[Q_{0.999}(\boldsymbol{\gamma})] = 766$
$\text{VaR}_{0.5}[Q_{0.999}(\boldsymbol{\gamma})] = 399$	$\text{E}[Q_{0.999}(\boldsymbol{\gamma})] = 1,293(24)$ $Q^P_{0.999} = 1,614(21)$

in acceptance probability (6.33). This indicator is to ensure that the proposed move is rejected if $\boldsymbol{\gamma}^*$ is outside of the parameter domain $1 + \xi^* y_{\max}/\beta^* > 0$.

Note that the normalisation constant for posterior distribution is not needed here. If under the rejection rule one accepts the move then the new state of the i-th parameter at iteration k is given by $\gamma_i^{(k)} = \gamma_i^*$, otherwise the parameter remains in the current state $\gamma_i^{(k)} = \gamma_i^{(k-1)}$ and an attempt to move that parameter is repeated at the next iteration.

Using the chain samples $\boldsymbol{\gamma}^{(k)}, k = 1, 2, \ldots$ from the posterior $\pi(\boldsymbol{\gamma}|\mathbf{x})$, we estimate the characteristics of the posterior distributions. The results are presented in Table 6.2 and Fig. 6.2.

6.7 Subexponential Severity

Operational risk losses are typically modelled by *heavy-tailed* or the so-called *subexponential* distributions, for example lognormal, Weibull, Pareto. These are formally defined as follows.

Definition 6.1 (Subexponential distribution) A distribution $F(x)$, $x \in (0, \infty)$, is subexponential if

$$\lim_{x \to \infty} \frac{\overline{F^{(n)*}(x)}}{\overline{F}(x)} = n \quad \text{for all} \quad n \geq 2. \tag{6.34}$$

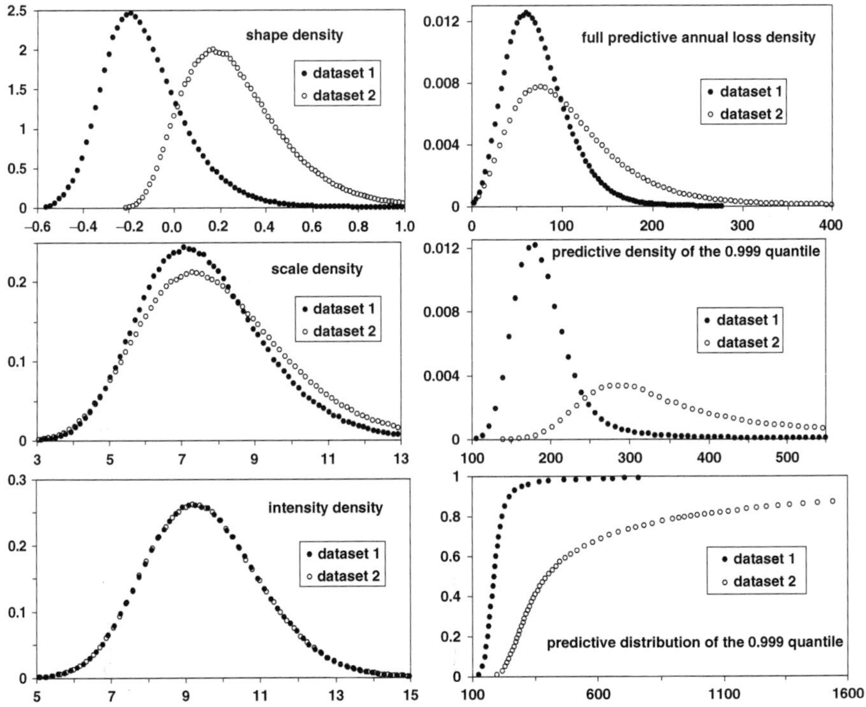

Fig. 6.2 MCMC posterior distributions of the $Poisson(\lambda)$-$GPD(\xi, \beta)$ model parameters: ξ (*shape*); β (*scale*); λ (*intensity*); full predictive annual loss distribution and the distribution of the 0.999 annual loss quantile. The data for datasets 1 and 2 were simulated from $Poisson(10)$-$GPD(-0.1, 6)$ and $Poisson(10)$-$GPD(0.25, 6)$ respectively; see Table 6.1

Here, $\overline{F}(x) = 1 - F(x)$ and $\overline{F^{(n)*}} = 1 - F^{(n)*}(x)$, where $F^{(n)*}(x)$ is the n-fold convolution of $F(x)$, i.e. $F^{(n)*}(x) = \Pr[X_1 + \cdots + X_n \leq x]$ with $X_1, \ldots, X_n$ independent with a common distribution $F(x)$; also see Sect. 3.1.1 for definition of a convolution.

The following results for subexponential distributions are of central importance for operational risk.

- If $F(x)$ is a subexponential distribution, then it can be shown that for all $\epsilon > 0$,

$$\exp(\epsilon x)\overline{F}(x) \to \infty, \quad x \to \infty. \tag{6.35}$$

  This property justifies the name *subexponential* because the severity distribution tail decays to 0 slower than any exponential $\exp(-\epsilon x)$, $\epsilon > 0$. For a proof, see Lemma 1.3.5 in Embrechts, Klüppelberg and Mikosch [83].
- If $X_1, \ldots, X_n$ are independent subexponential random variables with a common distribution $F(x)$, then

$$\lim_{x\to\infty}\frac{\Pr[X_1+\cdots+X_n>x]}{\Pr[\max(X_1,\ldots,X_n)>x]}=1,\quad n\geq 1. \tag{6.36}$$

That is, the tail of the sum of subexponentially distributed random variables has the same order of magnitude as the tail of the maximum of these random variables. This means that the tail (high quantiles) of the aggregated loss is mainly determined by the tail (high quantiles) of the maximum loss. A popular interpretation of this property is to say that the severe overall loss is due to a single large loss rather than due to accumulated small losses. Another useful result closely related to (6.36) is

$$\Pr[\max(X_1,\ldots,X_n)>x]\to n\overline{F}(x),\quad x\to\infty. \tag{6.37}$$

- If $X_1,\ldots,X_N$ are independent subexponential random variables with a common distribution $F(x)$, where N is random with a probability mass function $p_n = \Pr[N=n]$ satisfying

$$\sum_{n=0}^{\infty}(1+\epsilon)^n p_n<\infty \tag{6.38}$$

for some $\epsilon>0$, then

$$\Pr[X_1+\cdots+X_N>x]\to \mathrm{E}[N](1-F(x)),\quad x\to\infty; \tag{6.39}$$

see Theorem 1.3.9 in Embrechts, Klüppelberg and Mikosch [83]. Examples 1.3.10 and 1.3.11 in Embrechts, Klüppelberg and Mikosch [83] demonstrate that the conditions are satisfied when N is distributed from Poisson and negative binomial respectively. As is shown in Böcker and Klüppelberg [29], approximation (6.39) can be used to get a closed-form approximation for the high quantiles of the aggregate loss distribution $F_Z(x)=\Pr[X_1+\cdots+X_N\leq x]$, also see (3.72):

$$F_Z^{-1}(q)\to F^{-1}\left(1-\frac{1-q}{\mathrm{E}[N]}\right),\quad q\to 1. \tag{6.40}$$

It is often referred to as the *single-loss approximation* because compound distribution is expressed in terms of the single loss distribution (severity distribution).

- The tail of the sum of subexponential random variables with different tails will typically follow the heaviest tail. This will be discussed more in Chap. 7. That is, the risk measure over many risk cells will be mainly determined by the risk cell with the heaviest tail.
- As shown in Embrechts, Goldie and Veraverbeke [82], the subexponential class includes lognormal distribution $\mathcal{LN}(\mu,\sigma)$ whose tail satisfies

$$1 - F(x) \to \frac{\sigma}{\sqrt{2\pi}(\ln x - \mu)} \exp\left(-\frac{(\ln x - \mu)^2}{2\sigma^2}\right), \quad x \to \infty. \tag{6.41}$$

- Subexponential class includes the distributions with *regularly varying tail*, also referred to as Pareto type distributions

$$\overline{F}(x) = x^{-\alpha} C(x), \quad x \to \infty, \ \alpha \ge 0, \tag{6.42}$$

where α is the so-called *power tail index* and $C(x)$ is *slowly varying function*. The latter is defined as the function that satisfies

$$\lim_{x\to\infty} \frac{C(tx)}{C(x)} = 1 \quad \text{for all} \quad t > 0. \tag{6.43}$$

Examples of slowly varying functions include positive functions converging to the constant and logarithm function $\ln(x)$. A mathematical theory of heavy-tailed functions is *Karamata's theory* of regular variation; for an excellent summary, see Embrechts, Klüppelberg and Mikosch [83].
- It is convenient to characterise distributions by their tail behaviour. Two distributions $F_1(x)$ and $F_2(x)$ are said to be *tail equivalent* if

$$\lim_{x\to\infty} \frac{\overline{F}_1(x)}{\overline{F}_2(x)} = C(x), \tag{6.44}$$

where $C(x)$ is slowly varying function. If the right hand endpoint of the distributions is finite then $x \to \infty$ means x increasing to the right endpoint.

Of course the above definition and results for the subexponential distributions are asymptotic properties that can be difficult to prove with limited data and thus should be used with caution. In practice, *heavy-tailed* distribution for losses means that the observed losses are ranging over several orders of magnitude. For example, the largest losses are greater than the median by several orders of the magnitude; see Table 6.1 for the case $\xi_0 = 0.25$. This is typically observed in practice for many operational risks. Usually, subexponential distributions provide a good in-sample fit to the real datasets. Estimation of these distributions using limited data is a difficult task because individual large losses can dominate over the impact of other losses. As a result, the uncertainty of the estimates is very large. Results in Table 5.4 show that the uncertainty in the estimate of the 0.999 quantile is even comparable with the actual estimate. Often, the estimates based on the observed losses are not stable. The internal datasets for inference on the 0.999 quantile are nearly always limited because we attempt to quantify a 1 in 1000 year event. The following simple example demonstrates the instability of the sample averages for the heavy-tailed distribution.

Example 6.5 Consider the losses simulated from $Poisson(10)$-$GPD(0.25, 6)$ over 4 years and presented in Table 6.1. The mean of the losses is approximately USD 9.2

million. Assume that the next loss observed is the 0.9999 quantile of $GPD(0.25, 6)$ which is approximately USD 216 million. Note that the 0.9999 quantile of the severity in the case of $Poisson(10)$ corresponds to the 0.999 quantile of the annual loss approximately as implied by asymptotic formula (6.40). The new mean of the observed losses is $\approx$ USD 14.6 million which is a 60% increase in the mean as a result of a single loss.

6.8 Flexible Severity Distributions

It is important to remember that EVT is an asymptotic theory. Whether the conditions validating the use of the asymptotic theory are satisfied is often a difficult question to answer in operational risk practice. The convergence of some parametric models to the EVT regime is very slow; see the lognormal and g-and-h distributions studied in Mignola and Ugoccioni [164] and Degen, Embrechts and Lambrigger [71] respectively. In general, EVT should not preclude the use of other parametric distributions. Often other severity distributions are fitted to the datasets and compared to the EVT. For example, g-and-h and GB2 four-parameter distributions were used in Dutta and Perry [77] as a benchmark model alternative to EVT. Many parametric distributions are standard and can be found in textbooks; for example, see Panjer [181]. Below, we present several less known severity distributions suggested for use in operational risk.

6.8.1 g-and-h Distribution

A random variable X is said to have g-and-h distribution if

$$X = a + b\frac{\exp(gZ) - 1}{g}\exp(hZ^2/2), \tag{6.45}$$

where Z is a random variable from the standard normal distribution, $\mathcal{N}(0, 1)$, and $(a, b, g, h) \in \mathbb{R}$ are the parameters. In the case of $g = 0$, it is interpreted as $X = a + bZ\exp(hZ^2/2)$, i.e. the limit of (6.45) as $g \to 0$.

The g-and-h random variable is a strictly increasing transformation of the standard normal random variable, introduced by Tukey [235]. Thus it is trivial to simulate from the g-and-h distribution. The parameters a and b are the location and scale parameters respectively. The g and h parameters are responsible for skewness and kurtosis of the distribution. The advantage of this distribution is its ability to approximate a variety of data and distributions as shown in Martinez and Iglewicz [155]. Often g and h are constant but in general, if data do not fit well, one can generalise g and h parameters to be polynomials of Z^2. For example:

$$g = \gamma_0 + \gamma_1 Z^2 + \gamma_2 Z^4 + \cdots \quad \text{and} \quad h = \eta_0 + \eta_1 Z^2 + \eta_2 Z^4 + \cdots .$$

In particular, Dutta and Perry [77] had to use $h = \eta_0 + \eta_1 Z^2 + \eta_2 Z^4 + \eta_3 Z^4$ and constant g to fit operational risk losses for some banks/business lines. One of the main empirical findings in Dutta and Perry [77], analysing operational risk data of the 2004 LDCE, is that g-and-h distribution is a good choice to model operational risk losses. Typical estimates for g and h parameters obtained in Dutta and Perry [77] are in the ranges $g \in [1.79, 2.3]$ and $h \in [0.1, 0.35]$. It seems that only the case $g > 0$ and $h > 0$ is relevant for operational risk.

If $h > 0$, the transformation (6.45) is strictly increasing. Thus the quantiles of g-and-h distribution can easily be calculated as

$$F_X^{-1}(\alpha) = y\left(F_N^{-1}(\alpha)\right), \tag{6.46}$$

where

$$y(z) = a + b\frac{\exp(gz) - 1}{g}\exp(hz^2/2)$$

is just a transformation (6.45), and $F_N^{-1}(\alpha)$ is the inverse of the standard normal distribution at the quantile level α. Also, the distribution of X can be found as

$$F_X(x) = F_N\left(y^{-1}(x)\right), \tag{6.47}$$

where $y^{-1}(\cdot)$ is the inverse of the function $y(\cdot)$.

Because calculation of the quantile is simple, typically a quantile-based method, such as that presented in Hoaglin ([123], [124]), is used to fit g-and-h distribution. This is the approach used in Dutta and Perry [77] too.

Differentiating (6.47) with respect to x, the density of the g-and-h distribution can be calculated analytically as

$$f_X(x) = \frac{f_N\left(y^{-1}(x)\right)}{y'\left(y^{-1}(x)\right)}, \tag{6.48}$$

where $f_N(\cdot)$ is the density of the standard normal distribution and $y'(u) = dy(u)/du$. However, the inverse $y^{-1}(\cdot)$ cannot be found in closed form and thus numerical search procedure is required to calculate the density. If the density is calculated numerically, then one can use maximum likelihood method to estimate the g-and-h parameters. However, there are some shortcomings in this case, as discussed in Rayner and MacGillivray [198], and quantile-based method is usually preferred. Moments and other standard characteristics of the g-and-h distribution can be found in the above referenced literature. For the case of $h > 0$, the g-and-h distribution is heavy-tailed. It was shown in Degen, Embrechts and Lambrigger [71] that

$$1 - F_X(x) = x^{-1/h}C(x), \tag{6.49}$$

where $C(x)$ is slowly varying function (6.43). For $h = 0$ and $g > 0$, the g-and-h distribution is simply a scaled lognormal which is subexponential (6.35) but not regularly varying. Degen, Embrechts and Lambrigger [71] demonstrated that for the g-and-h distribution, convergence of the excess distribution to the GPD is extremely slow. Therefore, *if the data are well modelled by the g-and-h distribution*, the quantile estimation using EVT may lead to inaccurate results. This is consistent with empirical findings in Dutta and Perry [77].

Conceptually, there is no problem with estimating the g-and-h distribution using Bayesian inference. This can be done in the same way as for GPD in Sect. 6.6. The posterior is not available in closed form and thus MCMC should be used to get samples from the posterior distribution of the parameters. One can use RW-MH algorithm as in Example 6.4; also see Sect. 2.11.3. However, the density of g-and-h distribution (6.48) should be calculated numerically. This can make MCMC very slow. Another approach is to use the MCMC-ABC method described in Sect. 2.11.4; see Peters and Sisson [188]. This is because it is easy to simulate from the g-and-h distribution by simple transformation of the standard normal random variable.

6.8.2 GB2 Distribution

The GB2 (the generalised beta distribution of the second kind) is another four-parameter distribution that nests many important one- and two-parameter distributions. Its density is defined as

$$h(x) = \frac{|a|x^{ap-1}}{b^{ap}B(p,q)(1+(x/b)^a)^{p+q}}, \quad x > 0, \tag{6.50}$$

where $B(p,q) = \Gamma(p)\Gamma(q)/\Gamma(p+q)$ is the beta function, $\Gamma(\cdot)$ is the gamma function and (a, b, p, q) are parameters. It is shown in Bookstabber and McDonald [34] that GB2 nests many standard distributions for certain limits of parameters, examples being the lognormal ($a \to 0, q \to \infty$) and Weibull/Gamma ($q \to \infty$). GB2 can accommodate a variety of values for skewness and kurtosis. The properties and applications of GB2 can be found in McDonald and Xu [156]. In short: b is a scale parameter; a is a location parameter that also determines the tail decay; $a \times q$ drives the kurtosis; moments greater than $a \times q$ do not exist; p and q affect the skewness.

The GB2 distribution has been used in Dutta and Perry [77] to analyse operational risk data. They found a good fit in many cases but also noted that all GB2 distributions that fitted the data well could be approximated to a very high degree by the g-and-h distribution. Since the density is available in closed form (while the quantile calculation requires numerical procedure), typically the maximum likelihood method is used to fit GB2.

Note that simulation from GB2 can be achieved through the following representation of the GB2 random variable

$$X = b\left(\frac{Y_1}{Y_2}\right)^{1/a}, \tag{6.51}$$

where $Y_1 \sim Gamma(p, 1)$ and $Y_2 \sim Gamma(q, 1)$ are independent; see Devroye [76].

Estimation of the parameters using Bayesian inference requires calculation of the posterior distribution that cannot be found in closed form in the case of GB2 distribution. Thus MCMC should be utilised to estimate the posterior. Given that the GB2 density has closed form, the RW-MH or Metropolis-Hastings algorithms described in Sect. 2.11.1 can successfully be used; for example, see Peters and Sisson [188]. Given a simple simulation from GB2, MCMC ABC algorithms (see Sect. 2.11.4) can also be used but these might not be as efficient as Metropolis-Hastings algorithms.

6.8.3 Lognormal-Gamma Distribution

A random variable X is said to have the lognormal-gamma distribution if

$$Y \equiv \ln X = \mu + \sigma\sqrt{W}Z, \tag{6.52}$$

where $\mu \in \mathbb{R}$, $\sigma > 0$, Z is a random variable from the standard normal distribution, W is a random variable from the gamma distribution $Gamma(1/\alpha, \alpha)$, and Z and W are independent. It is another well-known heavy-tailed distribution. It is used for modelling insurance losses and pricing of catastrophe bonds where heavy-tailed distributions are used for modelling large losses; see, for example, Ibragimov and Walden [125], and Burnecki, Kukla and Taylor [47]. Recently, Ergashev [90] suggested using this distribution for operational risk. Fitting this distribution using MLE or quantile methods is difficult because the density (and quantile) does not have closed form. In particular, the density of random variable Y in (6.52) is the integral

$$f(y|\boldsymbol{\theta}) = \frac{\alpha^{-1/\alpha}}{\Gamma(1/\alpha)\sqrt{2\pi}\sigma}\int_0^\infty \frac{1}{\sqrt{w}}\exp\left(-\frac{(y-\mu)^2}{2\sigma^2 w}\right) w^{-1+1/\alpha}\exp(-w/\alpha)dw$$

that should be evaluated numerically. Here, $\boldsymbol{\theta} = (\mu, \sigma, \alpha)$ are model parameters. This difficulty can be avoided by using MCMC under the Bayesian inference approach. Consider the independent and identically distributed log-losses $\mathbf{Y} = (Y_1, \ldots, Y_n)'$ from the model (6.52). Denote the corresponding independent gamma shocks as $\mathbf{W} = (W_1, \ldots, W_n)'$. The gamma shocks are latent (not observable) variables. The posterior density of the parameters is

$$\pi(\boldsymbol{\theta}|\mathbf{y}) \propto \pi(\mathbf{y}|\boldsymbol{\theta})\pi(\boldsymbol{\theta}) = \pi(\boldsymbol{\theta})\prod_{i=1}^{n} f(y_i|\boldsymbol{\theta}),$$

where evaluation of the density $\pi(\mathbf{y}|\boldsymbol{\theta})$ requires numerical integration. However, this integration can be avoided by sampling $(\boldsymbol{\Theta}, \mathbf{W})$ from the joint density

$$\pi(\boldsymbol{\theta}, \mathbf{w}|\mathbf{y}) \propto \pi(\mathbf{y}|\mathbf{w}, \boldsymbol{\theta})\pi(\mathbf{w}|\boldsymbol{\theta})\pi(\boldsymbol{\theta}),$$

where $\pi(\mathbf{y}|\mathbf{w}, \boldsymbol{\theta})$ and $\pi(\mathbf{w}|\boldsymbol{\theta})$ are in closed form. This is because $f(y|w, \boldsymbol{\theta})$ is just the density of $\mathcal{N}(\mu, \sigma\sqrt{w})$. Then, marginally taken samples of $\boldsymbol{\Theta}$ are the samples from $\pi(\boldsymbol{\theta}|\mathbf{y})$. Of course, as usual, specification of the prior $\pi(\boldsymbol{\theta})$ is required. Sampling from $\pi(\boldsymbol{\theta}, \mathbf{w}|\mathbf{y})$ can be accomplished using different MCMC procedures. For example, Ergashev [90] derives the conditional densities for all parameters and gamma shocks:

$$\begin{aligned}
&\pi(w_i|\mathbf{y}, \mu, \sigma, \alpha) \propto \pi(y_i|w_i, \mu, \sigma, \alpha)\pi(w_i|\alpha),\\
&\pi(\mu|\mathbf{y}, \mathbf{w}, \sigma, \alpha) \propto \pi(\mathbf{y}|\mathbf{w}, \mu, \sigma)\pi(\mu),\\
&\pi(\sigma^2|\mathbf{y}, \mathbf{w}, \mu, \alpha) \propto \pi(\mathbf{y}|\mathbf{w}, \mu, \sigma^2),\\
&\pi(\alpha|\mathbf{y}, \mathbf{w}, \mu, \sigma) \propto \pi(w|\alpha)\pi(\alpha).
\end{aligned}$$

Then, the Gibbs sampler algorithm is used (simulating from the above conditional distributions iteratively) to get the samples of $(\boldsymbol{\Theta}, \mathbf{W})$, where the samples from $\pi(w_i|\mathbf{y}, \mu, \sigma, \alpha)$ and $\pi(\alpha|\mathbf{y}, \mathbf{w}, \mu, \sigma)$ are obtained using the Metropolis-Hastings algorithm; see also Sect. 2.11.

6.8.4 Generalised Champernowne Distribution

The density function of the generalised Champernowne distribution (GCD) is

$$f(x|\alpha, M, c) = \frac{\alpha(x+c)^{\alpha-1}((M+c)^\alpha - c^\alpha)}{((x+c)^\alpha + (M+c)^\alpha - 2c^\alpha)^2}, \quad x \geq 0, \tag{6.53}$$

with three parameters $\alpha > 0, M > 0, c \geq 0$. This distribution was suggested in Buch-Larsen, Nielsen, Guillen and Bolance [41] for semi-parametric fitting of heavy-tailed distributions. In the case of $c = 0$, it is a distribution introduced by Champernowne [51] and used in Clements, Hurn and Lindsay [57]. For recent use of GCD in operational risk, see Gustafsson and Thuring [115] and Buch-Kromann [40].

The GCD behaves as lognormal in the middle and as Pareto in the tail – an appealing feature for modelling operational risk. In particular:

$$f(x|\alpha, M, c) \to \text{const} \times x^{-\alpha-1}, \quad x \to \infty, \tag{6.54}$$

where const $= \alpha\left((M+c)^\alpha - c^\alpha\right)$. The distribution of GCD is available in closed form

$$F(x|\alpha, M, c) = \frac{(x+c)^\alpha - c^\alpha}{(x+c)^\alpha + (M+c)^\alpha - 2c^\alpha}, \quad x \geq 0. \tag{6.55}$$

The properties and estimation using the maximum likelihood method can be found in the above referenced literature. To estimate the model parameters using the Bayesian inference approach, the standard MCMC (e.g. RW-MH Algorithm 2.4) can be used.

6.8.5 *α*-Stable Distribution

A random variable X is said to have an α-stable distribution, denoted as $X \sim \alpha\text{Stable}(\alpha, \beta, \sigma, \mu)$, if its characteristic function is

$$\mathrm{E}[e^{itX}] = \begin{cases} \exp\left(-\sigma^\alpha |t|^\alpha \left(1 - i\beta \operatorname{sign}(t) \tan\left(\frac{1}{2}\pi\alpha\right)\right) + i\mu t\right), & \alpha \neq 1, \\ \exp\left(-\sigma |t| \left(1 + i\beta \frac{2}{\pi} \operatorname{sign}(t) \ln |t|\right) + i\mu t\right), & \alpha = 1, \end{cases} \tag{6.56}$$

where $\alpha \in (0, 2]$ is the shape parameter (also called stability index), $\beta \in [-1, 1]$ is skewness parameter, $\sigma > 0$ is the scale parameter, $\mu \in (-\infty, \infty)$ is the location parameter and $\operatorname{sign}(t) = t/|t|$.

An attractive property of this distribution class is that if $X_1, \ldots, X_n$ are independent random variables from the alpha stable distribution, then

$$X_1 + \cdots + X_n = c_n X + d_n,$$

where X is a random variable from the same α-stable distribution for some constants $c_n > 0$ and d_n. Equality in the above formula means the equality in distribution.

The case $\alpha = 2$ corresponds to the normal distribution, $\mathcal{N}(\mu, \sigma)$. Also, there are few other cases when α-stable distribution has a closed-form density. However, in general, the closed-form density is not available that makes the use of this distribution difficult in practice. Still, their flexibility in modelling heavy-tailed distributions is very attractive and their use in modelling operational risk has been receiving increasing recognition. Early applications of the α-stable distributions to financial data can be traced back to 1960s. The detailed analysis of α-stable distributions can be found in Rachev and Mittnik [197]. For application to modelling heavy-tailed distributions in insurance, see Embrechts and Klüppelberg and Mikosch [83]. The use of α-stable distributions in operational risk is relatively new; here we refer to Chernobai, Rachev and Fabozzi ([55], chapter 7), Chernobai and Rachev [54], and Giacometti, Rachev, Chernobai and Bertocchi [104] and references therein.

The distribution has four parameters and is very flexible in modelling nonsymmetric and heavy-tailed data. For $0 < \alpha < 2$, it has a power tail decay property

$$\Pr(|X| > x) \to \text{const} \times x^{-\alpha}, \quad x \to \infty.$$

That is, it belongs to the subexponential family of distributions (see Sect. 6.7) and thus allows capture of extreme events in the tails. The k-th moment is infinite when $k \geq \alpha$, i.e. $\mathrm{E}[X] = \mu$ for $\alpha > 1$ and $\mathrm{E}[X] = \infty$ for $0 < \alpha \leq 1$. For $0 < \alpha < 2$, the variance and higher moments are infinite.

The density has no closed form and thus the estimation of the parameters is non-trivial task. Several approaches can be used here:

- Estimate the parameters by minimising the distance between the sample and theoretical characteristic functions; for example, see Kogon and William [137].
- Use a traditional maximum likelihood method where the density is evaluated numerically by the Fourier inversion of the characteristic function, using for example FFT method; see Menn and Rachev [160] and Nolan [177].
- Though the density has no closed form, simulation from the model is relatively simple; see Appendix B.2. Thus one can use ABC-MCMC methods, described in Sect. 2.11.4; see also Peters, Sisson and Fan [189].

There are different parameterisations for the α-stable distribution. The representation (6.56) is often not convenient for estimation purposes because it is not jointly continuous with respect to parameters. The parameterisation typically used for numerical purposes, denoted as $\alpha\mathrm{Stable}_0(\alpha, \beta, \sigma, \mu_0)$, is

$$\mathrm{E}[e^{itX}] = \begin{cases} \exp\left(-|\sigma t|^{\alpha}(1 + i\beta \mathrm{sign}(t)\tan\left(\frac{\pi\alpha}{2}\right)(|\sigma t|^{1-\alpha} - 1)) + i\mu_0 t\right), & \alpha \neq 1, \\ \exp\left(-\sigma|t|\left(1 + i\beta\frac{2}{\pi}\mathrm{sign}(t)\ln(\sigma|t|)\right) + i\mu_0 t\right), & \alpha = 1; \end{cases}$$

see Nolan [177]. This representation is continuous in all four parameters and related to the parameterisation (6.56) through the change of location parameter

$$\mu = \begin{cases} \mu_0 - \beta\sigma\tan\left(\frac{1}{2}\pi\alpha\right), & \alpha \neq 1, \\ \mu_0 - \beta\sigma\frac{2}{\pi}\ln\sigma, & \alpha = 1. \end{cases} \tag{6.57}$$

Under this representation $X = \mu_0 + \sigma Z$, where $Z \sim \alpha\mathrm{Stable}_0(\alpha, \beta, 1, 0)$. Note that under parameterisation (6.56): $X = \mu + \sigma Z$ if $\alpha \neq 1$; and $X = \mu + \sigma Z + \frac{2}{\pi}\beta\sigma\ln\sigma$ if $\alpha = 1$, where $Z \sim \alpha\mathrm{Stable}(\alpha, \beta, 1, 0)$.

Since the operational losses take positive values only, α-stable distribution is typically transformed before applying to the data. The following transformations are popular:

- Symmetric α-stable distributions: the original data set $\mathbf{X}$ is symmetrised to get the dataset $\mathbf{Y} = \{\mathbf{X}, -\mathbf{X}\}$. Then $\alpha\mathrm{Stable}(\alpha, 0, \sigma, 0)$ is fitted to the dataset $\mathbf{Y}$. Only two parameters α and σ should be fitted in this case that simplifies the procedure. Analysis of real operation loss data in Giacometti, Rachev, Chernobai and Bertocchi [104] reported a good fit for symmetric alpha stable distributions.

- Log-α-stable distribution: logarithm transform is applied to the data $\mathbf{X}$ and then $\alpha\text{Stable}(\alpha, \beta, \sigma, \mu)$ is fitted to the transformed data $\ln X_1, \ldots, \ln X_n$.
- Truncated α-stable distribution: fitting the original dataset $\mathbf{X}$ using α-stable distribution truncated below zero, that is, fitting the density

$$f_T(x) = \frac{f(x)}{1 - F(0)} 1_{\{x>0\}},$$

where $f(x)$ is the density of $\alpha\text{Stable}(\alpha, \beta, \sigma, \mu)$ and $F(0)$ is its distribution at zero.

Problems[1]

6.1 ($\star$) Assume that the annual number of losses follows to a negative binomial distribution $NegBin(r, p)$ and the loss severities are independent random variables from the exponential distribution $F(x) = 1 - \exp(-x/\beta)$. Find the distribution of the maximum loss for a 1-year period. Assuming that the annual number of losses over m years are independent, find the distribution of the maximum loss over m-year period.

6.2 ($\star\star\star$) Simulate $K = 100$ independent realisations $x_k, k = 1, \ldots, K$ from GB2 distribution with parameters $(a = 1, b = 1, p = 3, q = 4)$; see Sect. 6.8.2. Assume now that all distribution parameters are unknown. Using simulated samples $x_k, k = 1, \ldots, K$ as the observed data, estimate parameters (a, b, p, q) utilising the Metropolis-Hastings within Gibbs algorithm; see Sect. 2.11.3. Assume vague priors.

6.3 ($\star\star\star$) Simulate $K = 100$ independent realisations $x_k, k = 1, \ldots, K$ from g-and-h distribution with parameters $(a = 0, b = 1, g = 2, h = 0.3)$; see Sect. 6.8.1. Assume now that g and h distribution parameters are unknown. Using simulated samples $x_k, k = 1, \ldots, K$ as the observed data, estimate parameters g and h utilising ABC algorithm; see Sect. 2.11.4. Assume vague priors.

6.4 ($\star$) Simulate $K = 10{,}000$ independent realisations $x_k, k = 1, \ldots, K$ from the generalised Champernowne distribution with parameters $\alpha = 3$, $M = 5$ and $c = 4$; see Sect. 6.8.4. Using the simulated sample, estimate the 0.9 and 0.99 quantiles. Compare with the true values. Plot the histogram of the simulated sample and compare with the true density function.

6.5 ($\star\star$) Simulate $K = 10{,}000$ independent realisations $x_k, k = 1, \ldots, K$ from the α-stable distribution $\alpha\text{Stable}(\alpha, \beta, \sigma, \mu)$, defined by (6.56). Assume that $\alpha = 1.5$, $\beta = 0.5$, $\sigma = 1$ and $\mu = 0$. Using the simulated sample, plot the empirical distribution and histogram for the density. Compare with the distribution and density calculated from the characteristic function (6.56) using FFT; see Sect. 3.4.

[1] Problem difficulty is indicated by asterisks: ($\star$) – low; ($\star\star$) – medium, ($\star\star\star$) – high.

6.6 (⋆⋆) Suppose that the risk annual frequency is $Poisson(\lambda)$ distributed and the severities are $GPD(\xi, \beta)$ distributed, where $\xi = 0.5$ and $\beta = 1$ are known. Assume that past data imply that the posterior distribution for λ is $Gamma(\alpha, \beta)$ with mean 20 and standard deviation 10. Find the predictive distribution for the frequency. Using Monte Carlo, calculate the 0.999 quantile of the predictive distribution for the annual loss and compare with the result obtained from a closed-form approximation (6.40).

6.7 (⋆⋆) Suppose that the risk annual frequency is $Poisson(\lambda)$ distributed and the severities are $GPD(\xi, \beta)$ distributed, where all parameters are unknown. Assume that past data imply that the posterior distribution for λ is the gamma distribution with mean 20 and standard deviation 10; the posterior for ξ is the gamma distribution with mean 0.5 and standard deviation 0.2; and the posterior for β is the gamma distribution with mean 1 and standard deviation 0.5. Using Monte Carlo, calculate the 0.999 quantile of the annual loss predictive distribution and compare with the result obtained in Problem 6.6.

6.8 (⋆) Prove a stability property of the GPD, that if $X \sim G_{\xi,\beta}(x)$, $x > 0$, then

$$\Pr[X - L \leq y | X > L] = G_{\xi,\beta+\xi L}(y), \quad y > 0.$$

Here, $G_{\xi,\beta}(x)$ is $GPD(\xi, \beta)$. That is, the distribution of the conditional excesses $X - L | X > L$ over the threshold L is also the GPD with the same shape parameter ξ and changed scale parameter from β to $\beta + \xi L$.

6.9 (⋆) Simulate $K = 10{,}000$ realisations $x_i > 0$, $i = 1, \ldots, K$ from $GPD(\xi, \beta)$ with $\xi = 0.2$ and $\beta = 1$. Calculate the mean of the exceedances over the threshold L and plot it vs the threshold L. Find when the plot displays linear behaviour. Repeat the calculation and graphical analysis for realisations sampled from $\mathcal{LN}(0, 2)$.

Chapter 7
Modelling Dependence

Remember that all models are wrong; the practical question is how wrong do they have to be to not be useful.
George Box and Norman Draper

Abstract Aggregation of operational risks in order to estimate a bank's capital is a challenging problem. This chapter considers how dependence between operational risks can be modelled. It presents different approaches and issues debated in the literature. It also discusses conceptual problems with the dominance of the heavy-tailed risks in the capital charge and possible failed diversification.

7.1 Introduction

The aim of this chapter is to address the issue of aggregation across many operational risks. The LDA model discussed throughout this book so far has focused on the case of a single risk. This chapter considers modelling of dependence between the risks. The LDA for a bank's total loss in year t is calculated as

$$Z_t = \sum_{j=1}^{J} Z_t^{(j)}, \tag{7.1}$$

where $Z_t^{(j)}$ is the annual loss in the j-th risk cell (business line/event type) modelled as a compound random variable,

$$Z_t^{(j)} = \sum_{s=1}^{N_t^{(j)}} X_s^{(j)}(t). \tag{7.2}$$

Here:

- $t = 1, 2, \ldots, T, T+1$ is discrete time (in annual units) with $T+1$ corresponding to the next year. For simplicity of notation in this chapter, this subscript is often dropped.

P. Shevchenko, *Modelling Operational Risk Using Bayesian Inference*,
DOI 10.1007/978-3-642-15923-7_7,

- The upper script j is used to identify the risk cell. Formally for operational risk $J = 56$ (eight business lines times seven event types), but this may differ depending on the financial institution and type of problem.
- The annual number of events $N_t^{(j)}$ is a random variable distributed according to a frequency distribution $P_j(\cdot|\lambda_t^{(j)})$, typically Poisson, which also depends on parameter(s) $\lambda_t^{(j)}$ that can be time-dependent.
- The severities, in year t, are represented by random variables $X_s^{(j)}(t)$, $s \geq 1$, distributed according to a severity distribution $F_j(\cdot|\psi_t^{(j)})$ with parameter(s) $\psi_t^{(j)}$.
- The index j on the distributions $P_j(\cdot)$ and $F_j(\cdot)$ reflects the fact that distribution type can be different for different risks. For simplicity of notation, often we shall omit this j if the parameter index is presented, that is, using $P(\cdot|\lambda_t^{(j)})$ and $F(\cdot|\psi_t^{(j)})$.
- The variables $\lambda_t^{(j)}$ and $\psi_t^{(j)}$ generically represent distribution (model) parameters of the jth risk that we refer to hereafter as the risk profiles.
- Typically, it is assumed that given $\lambda_t^{(j)}$ and $\psi_t^{(j)}$, the frequency and severities of the jth risk are independent, and the severities within the jth risk are also independent.

Modelling dependence between different risk cells and factors is an important challenge in operational risk management. The difficulties of correlation modelling are well known and, hence, regulators typically take a conservative approach when considering correlation in risk models. For example, the Basel II operational risk regulatory requirement for the Advanced Measurement Approach, BCBS ([17], p. 152), states as follows[1]:

> Risk measures for different operational risk estimates must be added for purposes of calculating the regulatory minimum capital requirement. However, the bank may be permitted to use internally determined correlations in operational risk losses across individual operational risk estimates, provided it can demonstrate to the satisfaction of the national supervisor that its systems for determining correlations are sound, implemented with integrity, and take into account the uncertainty surrounding any such correlation estimates (particularly in periods of stress). The bank must validate its correlation assumptions using appropriate quantitative and qualitative techniques.

The current risk measure specified by regulatory authorities is Value-at-Risk (VaR) at the 0.999 level for a 1-year holding period. In this case simple summation over VaRs corresponds to an assumption of perfect dependence between risks. This can be very conservative as it ignores any diversification effects. If the latter are allowed in the model, it is expected that the capital may reduce, providing a strong incentive to model dependence in the banking industry. At the same time, limited data do not allow for reliable estimates of correlations and there are attempts to

---

[1] The original text is available free of charge on the BIS website www.BIS.org/bcbs/publ.htm

estimate these using expert opinions. In such a setting a transparent dependence model is very important from the perspective of model interpretation, understanding of model sensitivity and with the aim of minimising possible model risk.

However, it is important to note that VaR is not a coherent risk measure; see Definition 2.12 in Sect. 2.6. This means that, in principle, dependence modelling could also increase VaR; see Embrechts, Nešlehová and Wüthrich [86] and Embrechts, Lambrigger and Wüthrich [84]. This issue will be discussed in Sect. 7.3.

Another potential problem debated in the literature is that the capital is mainly determined by the risk with the heaviest tail severity. This will be discussed in Sect. 7.2.

The pitfalls with the use of linear correlation as a measure of dependence are now widely known and copula functions to model dependence structures are now widely used in financial risk management. This was not the case until the publication of the highly influential paper by Embrechts, McNeil and Straumann [85], that was first available as a RiskLab (ETH Zurich) report in early 1999. These will be discussed throughout this chapter. A textbook reference for modelling dependence between financial risks is McNeil, Frey and Embrechts [157] that also contains an extensive bibliography on this subject.

Conceptually, under model (7.2), the dependence between the annual losses $Z_t^{(j)}$ and $Z_t^{(i)}$, $i \neq j$, can be introduced in several ways:

- Modelling dependence between frequencies $N_t^{(j)}$ and $N_t^{(i)}$ directly through copula methods; see Frachot, Roncalli and Salomon [97], Bee [25] and Aue and Klakbrener [12]. Here, we note that the use of copula methods, in the case of discrete random variables, needs to be done with care.
- Common shocks; see Lindskog and McNeil [145] and Powojowski, Reynolds and Tuenter [194]. The approach of common shocks is proposed as a method to model events affecting many cells at the same time. Formally, this leads to a dependence between frequencies of the risks if superimposed with cell internal events. Dependence between severities occurring at the same time is considered in Lindskog and McNeil [145].
- Modelling dependence between the k-th severities or between k-th event times of different risks; see Chavez-Demoulin, Embrechts and Nešlehová [52] (e.g. 1st, 2nd, etc losses/event times of the j-th risk are correlated to the 1st, 2nd, etc losses/event times of the i-th risk respectively). This can be difficult to interpret especially when one considers high-frequency versus low-frequency risks.
- Modelling dependence between annual losses directly via copula methods; see Giacometti, Rachev, Chernobai and Bertocchi [104], and Embrechts and Puccetti [88]. However, this may create irreconcilable problems with modelling insurance for operational risk that directly involves event times. Additionally, it will be problematic to quantify these correlations using historical data, and the LDA model (7.2) will lose its structure. One can, however, consider dependence between losses aggregated over shorter periods.
- Using the multivariate compound Poisson model based on Lévy copulas as suggested in Böcker and Klüppelberg [30] and Böcker and Klüppelberg [31].

- Using structural models with common (systematic) factors that can lead to the dependence between severities and frequencies of different risks and within risk; see Sect. 7.11 below.
- Modelling dependence between severities and frequencies from different risks and within risk using dependence between risk profiles, as considered in Peters, Shevchenko and Wüthrich [187].
- In the general case, when no information about the dependence structure is available, Embrechts and Puccetti [87] work out bounds for aggregated operational risk capital; see also Embrechts, Nešlehová and Wüthrich [86].

Below, we describe the main concepts and issues behind some of these approaches. The choice of appropriate dependence structures is crucial and determines the amount of diversification – it is still an open challenging problem.

Remark 7.1 (Dependence on macroeconomic factors) It is important to note that there is empirical evidence, as reported in Allen and Bali [8], that some operational risks are dependent on macroeconomic variables such as GDP, unemployment, equity indices, interest rates, foreign exchange rates, regulatory environment variables and others. For example, some operational risks typically increase during economic downturns, high unemployment and low interest rates. This will be discussed more in Sect. 7.11.

7.2 Dominance of the Heaviest Tail Risks

It is a well-known phenomenon in operational risk practice that most of the capital estimate and its uncertainty are due to a few low-frequency/high-severity risks. For a methodological insight, consider J independent risks, where each risk is modelled by a compound Poisson. Then, the sum of risks is a compound Poisson with the intensity and severity distribution given by the following proposition.

Proposition 7.1 *Consider J independent compound Poisson random variables*

$$Z^{(j)} = \sum_{s=1}^{N^{(j)}} X_s^{(j)}, \quad j = 1, \ldots, J, \tag{7.3}$$

where the frequencies $N^{(j)} \sim Poisson(\lambda_j)$ and the severities $X_s^{(j)} \sim F_j(x)$, $j = 1, \ldots, J$ and $s = 1, 2, \ldots$ are all independent. Then, the sum $Z = \sum_{j=1}^{J} Z^{(j)}$ is a compound Poisson random variable with the frequency distribution $Poisson(\lambda)$ and severity distribution

$$F(x) = \sum_{j=1}^{J} \frac{\lambda_j}{\lambda} F_j(x),$$

where $\lambda = \lambda_1 + \cdots + \lambda_J$.

Proof The characteristic function of the compound Poisson with the intensity λ_j and severity distribution function F_j is $\chi_j(t) = \exp(-\lambda_j + \lambda_j\phi_j(t))$, where $\phi_j(t)$ is the characteristic function of the severity. Then the characteristic function of the sum Z is

$$\chi(t) = \prod_{j=1}^{J} \chi_j(t) = \exp\left(-\sum_j \lambda_j + \sum_j \lambda_j\phi_j(t)\right)$$
$$= \exp\left(-\lambda\left(1 - \sum_j \frac{\lambda_j}{\lambda}\phi_j(t)\right)\right).$$

This is easily recognised as a characteristic function of the compound Poisson random variable with the intensity $\lambda = \lambda_1 + \cdots + \lambda_J$ and severity characteristic function $\phi(t) = \sum_j \lambda_j\phi_j(t)/\lambda$. The latter corresponds to the severity distribution function $F(x) = \sum_{j=1}^{J} \frac{\lambda_j}{\lambda} F_j(x)$, completing the proof. Note that $F(x)$ is simply a mixture distribution.

Suppose that all severity distributions $F_j(x)$ are heavy-tailed, that is,

$$\overline{F}_j(x) = x^{-\alpha_j} C_j(x),$$

where $\alpha_1 < \cdots < \alpha_J$ and $C_j(x)$ are slowly varying functions; see Sect. 6.7. Then, $F(x) = \sum_{j=1}^{J} \frac{\lambda_j}{\lambda} F_j(x)$ is a heavy-tailed distribution too, with the tail index α_1 for $x \to \infty$. Thus, using the result (6.39) for heavy-tailed distributions, we obtain that

$$\lim_{x\to\infty} \frac{\Pr[Z > x]}{1 - F_1(x)} = \lambda_1. \tag{7.4}$$

This means that high quantiles of the total loss are due to the high losses of the risk with the heaviest tail.

Example 7.1 ***Real data example***. For illustration of this phenomenon with the real data from ORX database, see Cope, Antonini, Mignola and Ugoccioni [62]. In their example, $\mathcal{LN}(8, 2.24)$ gave a good fit for 10 business lines with average 100 losses per year in each line using 10,000 observations. The estimated capital across these 10 business lines was Euro 634 million with 95% confidence interval (uncertainty in the capital estimate due to finite data size) of width Euro 98 million. Then, extra risk cell (corresponding to the "Clients, Products and Business Practices" event type in the "Corporate Finance" business line) was added with one loss per year on average and the $\mathcal{LN}(9.67, 3.83)$ severity estimated using 300 data points. The obtained estimate for the capital over the ten business units plus the additional one was Euro 5,260 million with 95% confidence interval of the width Euro 19 billion. This shows that one high-severity risk cell contributes 88% to the capital estimate and 99.5% to the uncertainty range. In this example, the high-severity unit accounts for 0.1% of the bank's losses.

7.3 A Note on Negative Diversification

As has already been discussed in Sect. 2.6, VaR is not a coherent risk measure; see Artzner, Delbaen, Eber and Heath [10]. In particular, under some circumstances VaR measure may fail a sub-additivity property

$$\mathrm{VaR}_q[Z] \leq \sum_{j=1}^{J} \mathrm{VaR}_q[Z^{(j)}]; \tag{7.5}$$

see Embrechts, Nešlehová and Wüthrich [86] and Embrechts, Lambrigger and Wüthrich [84]. That is, dependence modelling could also increase VaR. Note that if there is a perfect positive dependence between risks, that is, $Z^{(j)} = H_j^{-1}(U)$, $j = 1, \ldots, J$, where $U \sim \mathcal{U}(0, 1)$ and $H_j(\cdot)$ is a distribution of $Z^{(j)}$, then

$$\mathrm{VaR}_q[Z] = \sum_{j=1}^{J} \mathrm{VaR}_q[Z^{(j)}]. \tag{7.6}$$

That is, the failure of the subadditivity means that the VaR for the sum of risks is larger than the VaR in the case of perfectly dependent risks. This is very counterintuitive given a typical expectation of diversification benefits. In particular, the diversification

$$D_q = 1 - \frac{\mathrm{VaR}_q[\sum_j Z^{(j)}]}{\sum_j \mathrm{VaR}_q[Z^{(j)}]} \tag{7.7}$$

is expected to be positive while the subadditivity failure corresponds to the negative diversification. The latter may occur even for independent risks when the risks are heavy-tailed. It was shown and discussed in Nešlehová, Embrechts and Chavez-Demoulin [174] that if independent risks are Pareto type, $Z^{(j)} \sim F_j(x) = 1 - x^{-\alpha_j} C_j(x)$, with the tail indexes $0 < \alpha_j < 1$, then

$$\mathrm{VaR}_q[Z] > \sum_{j=1}^{J} \mathrm{VaR}_q[Z^{(j)}], \tag{7.8}$$

at least for sufficiently large q. The case of $0 < \alpha_j \leq 1$ corresponds to infinite mean distribution, that is, $\mathrm{E}[Z^{(j)}] = \infty$.

Remark 7.2 To simplify notation, the index of discrete time (year) is dropped. Implicitly, in the above discussion of diversification issues, we refer to the next year.

Example 7.2 Assume that we have two independent risks, $X \sim Pareto(\beta, 1)$ and $Y \sim Pareto(\beta, 1)$, where $Pareto(\beta, a) = 1 - (x/a)^{-\beta}$. Calculating the $\mathrm{VaR}_{0.999}[X + Y]$ using for example FFT, we can easily find the diversification D_q as defined in (7.7). Figure 7.1 shows the results for $D_{0.999}$ vs β that demonstrate negative diversification for $\beta < 1$.

Example 7.3 In the previous example, we found that the diversification is positive for $\beta > 1$. In particular, $D_{0.999} \approx 0.27$ when $\beta = 4$, that is mean, variance and skewness are finite. It is important to realise that diversification depends on the quantile level. Figure 7.1 shows the results for D_q vs q in the case of $\beta = 4$. One can see that diversification is positive for high level quantiles but may become zero and negative for lower quantiles.

7.4 Copula Models

Copula functions have become popular and flexible tools in modelling multivariate dependence among risks. In general, a copula is a d-dimensional multivariate distribution on $[0, 1]^d$ with uniform marginal distributions. Given a copula function $C(u_1, \ldots, u_d)$, the joint distribution of random variables $Y_1, \ldots, Y_d$ with marginal distributions $F_1(y_1), \ldots, F_d(y_d)$ can be constructed as

$$F(y_1, \ldots, y_d) = C(F_1(y_1), \ldots, F_d(y_d)). \tag{7.9}$$

The well-known theorem due to Sklar, published in 1959, says that one can always find a unique copula $C(\cdot)$ for a joint distribution with given continuous marginals. Note that in the case of discrete distributions this copula may not be unique. Given (7.9), the joint density can be written as

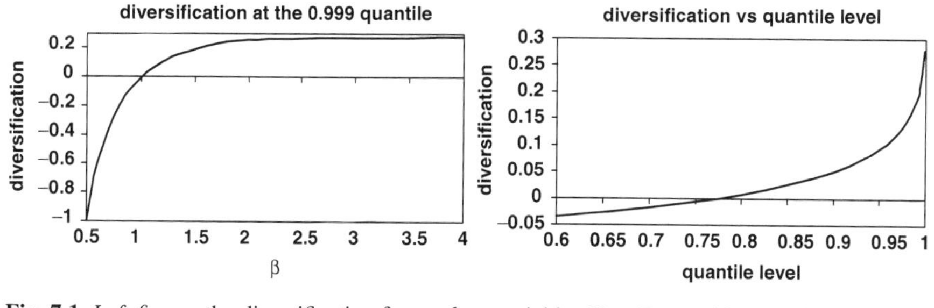

Fig. 7.1 *Left figure:* the diversification for random variables $X \sim Pareto(\beta, 1)$ and $Y \sim Pareto(\beta, 1)$ vs β. *Right figure:* the diversification for random variables $X \sim Pareto(4, 1)$ and $Y \sim Pareto(4, 1)$ vs quantile level q

$$f(y_1, \ldots, y_d) = c(F_1(y_1), \ldots, F_d(y_d)) \prod_{i=1}^{d} f_i(y_i). \tag{7.10}$$

where $c(\cdot)$ is a copula density and $f_1(y_1), \ldots, f_d(y_d)$ are marginal densities. There are many different copulas discussed in the literature and these can be found in many textbooks; for example, see McNeil, Frey and Embrechts ([157], section 5). Below, for illustration of the concept and notation, we give definitions for the Gaussian, Clayton, Gumbel and t copulas (Clayton and Gumbel copulas belong to a so-called family of the Archimedean copulas). An important difference between these three copulas is that they each display different tail dependence properties. The Gaussian copula has no upper and lower tail dependence, the Clayton copula will produce greater lower tail dependence as ρ increases whereas the Gumbel copula will produce greater upper tail dependence as ρ increases.

For a general description of copulas and their properties in the context of financial risk modelling, see McNeil, Frey and Embrechts ([157], chapter 5) and Panjer ([181], chapter 8); multivariate extreme value copulas are described in McNeil, Frey and Embrechts ([157], sections 7.5 and 7.6).

For a model choice of copula using frequentist goodness-of-fit testing, see Klugman and Parsa [135] and Panjer ([181], section 14.5). One can also use Akaike information criterion (AIC) to choose a copula. However, formally it does not hold for copulas fitted using data marginally transformed into $[0, 1]^d$ – a proper correction, referred to as copula information criterion, is derived in Grønneberg and Hjort [113]. Under the Bayesian approach, model choice can be done using Bayesian criteria presented in Sect. 2.13; for a case study of t-copula choice, see Luo and Shevchenko [150].

7.4.1 Gaussian Copula

The d-dimensional Gaussian copula is obtained by transformation of the multivariate normal distribution:

$$C(u_1, \ldots, u_d) = F_N^{\boldsymbol{\Sigma}}\left(F_N^{-1}(u_1), \ldots, F_N^{-1}(u_d)\right) \tag{7.11}$$

and its density is

$$c(u_1, \ldots, u_d) = \frac{f_N^{\boldsymbol{\Sigma}}\left(F_N^{-1}(u_1), \ldots, F_N^{-1}(u_d)\right)}{\prod_{i=1}^{d} f_N\left(F_N^{-1}(u_i)\right)}. \tag{7.12}$$

Here, $F_N(\cdot)$ and $f_N(\cdot)$ are the standard normal distribution and its density respectively; $f_N^{\boldsymbol{\Sigma}}(\cdot)$ and $F_N^{\boldsymbol{\Sigma}}(\cdot)$ are the standard multivariate normal density and distribution respectively with zero means, unit variances and correlation matrix $\boldsymbol{\Sigma}$.

Simulation of the random variates from a Gaussian copula is very simple and can be done as follows.

Algorithm 7.1 (Simulation from Gaussian copula)

1. Simulate d-variate $(x_1, \ldots, x_d)'$ from the standard multivariate normal distribution $\mathcal{N}_d(\mathbf{0}, \mathbf{\Sigma})$ with zero means, unit variances and correlation matrix $\mathbf{\Sigma}$.
2. Calculate $u_1 = F_N(x_1), \ldots, u_d = F_N(x_d)$. Obtained $(u_1, \ldots, u_d)'$ is a d-variate from a Gaussian copula.

7.4.2 Archimedean Copulas

The d-dimensional Archimedean copulas can be written as

$$C(u_1, \ldots, u_d) = \phi^{-1}(\phi(u_1) + \cdots + \phi(u_d)), \tag{7.13}$$

where ϕ is a decreasing function known as the generator for the given copula; see Frees and Valdez [98]. Important members of this family are Clayton and Gumbel copulas defined as follows:

Clayton copula. The Clayton copula is given by

$$C(u_1, \ldots, u_d) = \left(1 - d + \sum_{i=1}^{d} (u_i)^{-\rho}\right)^{-\frac{1}{\rho}} \tag{7.14}$$

and its density is

$$c(u_1, \ldots, u_d) = \left(1 - d + \sum_{i=1}^{d} (u_i)^{-\rho}\right)^{-d-\frac{1}{\rho}} \prod_{i=1}^{d} \left((u_i)^{-\rho-1} \{(i-1)\rho + 1\}\right), \tag{7.15}$$

where $\rho > 0$ is a dependence parameter. The generator and inverse generator for the Clayton copula are given by

$$\phi_C(t) = \left(t^{-\rho} - 1\right); \quad \phi_C^{-1}(s) = (1 + s)^{-\frac{1}{\rho}}. \tag{7.16}$$

Gumbel copula. Tee Gumbel copula and its density are

$$C(u_1, \ldots, u_d) = \exp\left\{-\left(\sum\nolimits_{i=1}^{d} (-\ln u_i)^{\rho}\right)^{\frac{1}{\rho}}\right\}, \tag{7.17}$$

$$c(u_1, \ldots, u_d) = \frac{\partial^d}{\partial u_1 \ldots \partial u_d} C(u_1, \ldots, u_d), \tag{7.18}$$

where $\rho \geq 1$ is a dependence parameter. In the bivariate case the explicit expression for the Gumbel copula is given by

$$\begin{aligned} c(u_1, u_2) &= \frac{\partial^2}{\partial u_1 \partial u_2} C(u_1, u_2) \\ &= C(u_1, u_2)\, u_1^{-1} u_2^{-1} \left[\sum_{i=1}^{2} (-\ln u_i)^{\rho}\right]^{2\left(\frac{1}{\rho}-1\right)} (\ln u_1 \ln u_2)^{\rho-1} \\ &\quad \times \left[1 + (\rho - 1)\left[\sum_{i=1}^{2} (-\ln u_i)^{\rho}\right]^{-\frac{1}{\rho}}\right]. \end{aligned}$$

The generator and inverse generator for the Gumbel copula are given by

$$\phi_G(t) = (-\ln t)^{\rho}; \qquad \phi_G^{-1}(s) = \exp\left(-s^{\frac{1}{\rho}}\right), \tag{7.19}$$

where ρ is a copula parameter.

Simulation from Archimedean copulas can be accomplished using the algorithm provided in Melchiori [158]:

Algorithm 7.2 (Simulation from Archimedean copula)

1. Sample d independent random variates $v_1, \ldots, v_d$ from a uniform distribution $\mathcal{U}(0, 1)$.
2. Simulate y from distribution $D(\cdot)$ such that $D(0) = 0$ and Laplace transform of $D(\cdot)$ is $\mathcal{L}(D) = \phi^{-1}$.
3. Find $s_i = -(\ln v_i)/y$ for $i = 1, \ldots, d$.
4. Calculate $u_i = \phi^{-1}(s_i)$ for $i = 1, \ldots, d$.

The obtained $(u_1, \ldots, u_d)'$ is a d-variate from the d-dimensional Archimedean copula. What remains is to define the relevant distribution $D(\cdot)$ for the Clayton and Gumbel copulas. For the Clayton copula, $D(\cdot)$ is a gamma distribution with the shape parameter given by ρ^{-1} and unit scale. For the Gumbel copula, $D(\cdot)$ is from the α-stable family $\alpha\text{Stable}(\alpha, \beta, \gamma, \delta)$ with the following parameters: shape

$\alpha = \rho^{-1}$, skewness $\beta = 1$, scale $\gamma = (\cos(\frac{1}{2}\pi/\rho))^{\rho}$, and location $\delta = 0$. In the Gumbel case, the density for $D(\cdot)$ has no analytic form and the simulation from this distribution can be achieved using the algorithm from Nolan [176] to efficiently generate the required samples from the univariate stable distribution; also, see Appendix B.2.

7.4.3 t-Copula

In practice, one of the most popular copula in modelling multivariate financial data is perhaps the t-copula, implied by the multivariate t-distribution; see Embrechts, McNeil and Straumann [85], Fang, Fang and Kotz [91] and Demarta and McNeil [72]. This is due to its simplicity in terms of simulation and calibration, combined with its ability to model tail dependence, which is often observed in financial returns data. The t-copulas are most easily described and understood by a stochastic representation, as discussed below. We introduce notation and definitions as follows:

- $\mathbf{Z} = (Z_1, \ldots, Z_n)'$ is a random vector from the standard n-variate normal distribution $F_N^{\boldsymbol{\Sigma}}(\mathbf{z})$ with zero mean vector, unit variances and correlation matrix $\boldsymbol{\Sigma}$;
- $\mathbf{U} = (U_1, U_2, \ldots, U_n)'$ is defined on $[0, 1]^n$ domain;
- V is a random variable from the uniform (0,1) distribution independent of $\mathbf{Z}$;
- $W = G_\nu^{-1}(V)$, where $G_\nu(\cdot)$ is the distribution function of $\sqrt{\nu/S}$ with S distributed from the chi-square distribution with ν degrees of freedom, that is, random variables W and $\mathbf{Z}$ are independent; and
- $t_\nu(\cdot)$ is the standard univariate t-distribution and $t_\nu^{-1}(\cdot)$ is its inverse.

Then we have the following representations:

Standard t-copula. The random vector

$$\mathbf{X} = W \times \mathbf{Z} \tag{7.20}$$

is distributed from a multivariate t-distribution and random vector

$$\mathbf{U} = (t_\nu(X_1), \ldots, t_\nu(X_n))' \tag{7.21}$$

is distributed from the standard t-copula.

Grouped t-copula. The standard t-copula is sometimes criticised due to the restriction of having only one parameter for the degrees of freedom ν, which may limit its ability to model tail dependence in multivariate cases. To overcome this problem, Daul, De Giorgi, Lindskog and McNeil [69] proposed the use of the grouped t-copula, where risks are grouped into classes and each class has its own t-copula with a specific degrees-of-freedom parameter. Specifically, partition $\{1, 2, \ldots, n\}$ into m non-overlapping sub-groups of sizes $n_1, \ldots, n_m$. Then the copula of the distribution of the random vector

$$\mathbf{X} = (W_1 Z_1, \ldots, W_1 Z_{n_1}, W_2 Z_{n_1+1}, \ldots, W_2 Z_{n_1+n_2}, \ldots, W_m Z_n)', \tag{7.22}$$

where $W_k = G^{-1}_{\nu_k}(V)$, $k = 1, \ldots, m$, is the grouped t-copula. That is,

$$\mathbf{U} = (t_{\nu_1}(X_1), \ldots, t_{\nu_1}(X_{n_1}), t_{\nu_2}(X_{n_1+1}), \ldots, t_{\nu_2}(X_{n_1+n_2}), \ldots, t_{\nu_m}(X_n))'$$

is a random vector from the grouped t-copula. Here, the copula for each group is a standard t-copula with its own degrees-of-freedom parameter.

Generalised t-copula with multiple degrees-of-freedom parameters. It is not always obvious how the risk factors should be divided into sub-groups. An adequate choice of grouping configurations requires substantial additional effort if there is no natural grouping, for example by sector or class of asset. The above described grouped t-copula can be generalised, so that each group will have only one member; see Luo and Shevchenko [148]. The generalised t-copula has the advantages of a grouped t-copula with flexible modelling of multivariate dependencies. At the same time, it overcomes the difficulties with a priori choice of groups. Specifically, the copula of the random vector

$$\mathbf{X} = (W_1 Z_1,\ W_2 Z_2, \ldots, W_n Z_n)' \tag{7.23}$$

is said to have a t-copula with multiple degrees-of-freedom parameters, which we denote as $\widetilde{t}_{\boldsymbol{\nu}}$-copula, that is,

$$\mathbf{U} = (t_{\nu_1}(X_1), t_{\nu_2}(X_2), \ldots, t_{\nu_n}(X_n))' \tag{7.24}$$

is a random vector distributed according to this copula. Note, all W_i are perfectly dependent.

Given the above stochastic representation, simulation of the $\widetilde{t}_{\boldsymbol{\nu}}$-copula is straightforward. In the case of a standard t-copula $\nu_1 = \cdots = \nu_n = \nu$; and in the case of grouped t-copula the corresponding subsets have the same degrees-of-freedom parameter. Note that the standard t-copula and grouped t-copula are special cases of $\widetilde{t}_{\boldsymbol{\nu}}$-copula.

From the stochastic representation (7.23), it is easy to show that the $\widetilde{t}_{\boldsymbol{\nu}}$-copula distribution has the following explicit integral expression

$$C^{\boldsymbol{\Sigma}}_{\boldsymbol{\nu}}(\mathbf{u}) = \int_0^1 F^{\boldsymbol{\Sigma}}_N(z_1(u_1, s), \ldots, z_n(u_n, s))ds \tag{7.25}$$

and its density is

$$\begin{aligned} c^{\boldsymbol{\Sigma}}_{\boldsymbol{\nu}}(\mathbf{u}) &= \frac{\partial^n C^{\boldsymbol{\Sigma}}_{\boldsymbol{\nu}}(\mathbf{u})}{\partial u_1 \ldots \partial u_n} \\ &= \frac{1}{\prod_{k=1}^{n} f_{\nu_k}(x_k)} \int_0^1 f^{\boldsymbol{\Sigma}}_N(z_1(u_1, s), \ldots, z_n(u_n, s)) \prod_{k=1}^{n} (w_k(s))^{-1} ds. \end{aligned} \tag{7.26}$$

Here

- $z_k(u_k, s) = t_{\nu_k}^{-1}(u_k)/w_k(s), \ k = 1, 2, \ldots, n$;
- $w_k(s) = G_{\nu_k}^{-1}(s)$;
- $f_N^{\boldsymbol{\Sigma}}(z_1, \ldots, z_n) = \exp(-\frac{1}{2}\mathbf{z}'\boldsymbol{\Sigma}^{-1}\mathbf{z})/\left((2\pi)^{n/2}(\det\boldsymbol{\Sigma})^{1/2}\right)$ is the standard multivariate normal density;
- $x_k = t_{\nu_k}^{-1}(u_k), \ k = 1, 2, \ldots, n$; and
- $f_\nu(x) = \left(1 + x^2/\nu\right)^{-(\nu+1)/2} \frac{\Gamma((\nu+1)/2)}{\Gamma(\nu/2)\sqrt{\nu\pi}}$ is the univariate t-density.

The multivariate density (7.26) involves a one-dimensional integration that should be done numerically. This makes the calculation of the copula density more demanding computationally in comparison with the standard t-copula. However, it is still practical, because fast and accurate algorithms are available for the one-dimensional numerical integration; see Sect. 3.5.2. If all degrees-of-freedom parameters are equal (i.e. $\nu_1 = \cdots = \nu_n = \nu$) then it is easy to show that the copula defined by (7.25) becomes the standard t-copula; see Luo and Shevchenko [148] for a proof.

7.5 Dependence Measures

Measuring dependence between risks is of critical importance for capital calculations. Several popular scalar measures of dependence are discussed below.

7.5.1 Linear Correlation

Linear correlation is a measure of *linear dependence* between random variables

$$\rho[X_i, X_j] = \frac{\text{Cov}[X_i, X_j]}{\sqrt{\text{Var}[X_i]\text{Var}[X_j]}}. \tag{7.27}$$

It is invariant under strictly increasing linear transformations

$$\rho[\alpha_i + \beta_i X_i, \alpha_j + \beta_j x_j] = \rho[X_i, X_j], \ \beta_i, \beta_j > 0.$$

The problems with using the linear correlation coefficient as a measure of dependence between operational risks can be summarised as follows.

- It is defined if variances of X_i and X_j are finite. As has already been discussed, some operational risks are modelled by heavy-tailed distributions with infinite variance and even the cases of infinite mean are reported.
- It is not invariant under strictly increasing nonlinear transformations $T(\cdot)$ and $\widetilde{T}(\cdot)$. In general, $\rho[T(X_i), \widetilde{T}(X_j)] \neq \rho[X_i, X_j]$.

- Independence between random variables implies that linear correlation is zero. However, in general, zero linear correlation does not imply independence. For example if $X \sim \mathcal{N}(0,1)$ and $Y = X^2$, then $\rho[X,Y] = 0$ while it is obvious that there is as strong dependence between X and Y. Zero linear correlation and independence are equivalent only in the case of a multivariate normal distribution as a joint distribution for random variables.
- The linear correlation is bounded to the region $[\rho_{\min}, \rho_{\max}]$, where $-1 \le \rho_{\min} \le \rho_{\max} \le 1$. For example, if $X \sim \mathcal{LN}(0,1)$ and $Y \sim \mathcal{LN}(0,\sigma)$, then the minimum and maximum bounds for correlation are plotted in Fig. 7.2a as functions of σ; for more details, see McNeil, Frey and Embrechts ([157], Example 5.26). Figure 7.2b presents the correlation bounds for the case of $X \sim Pareto(2.1, 1)$ and $Y \sim Pareto(\beta, 1)$, where $Pareto(\beta, a) = 1 - (x/a)^{-\beta}$; for more details see Nešlehová, Embrechts and Chavez-Demoulin ([174], Example 3.1).

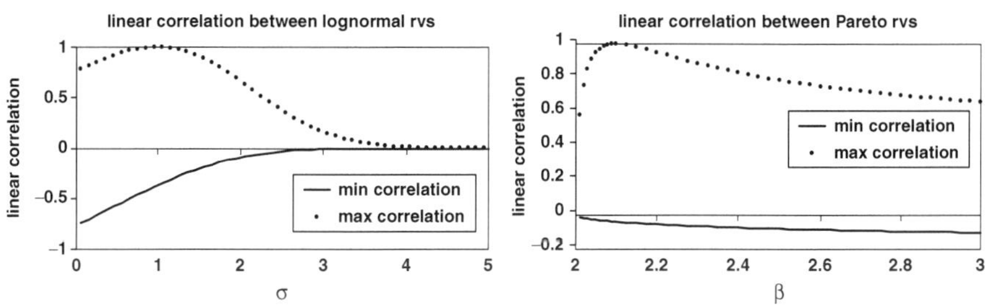

Fig. 7.2 *Left figure*: The minimum and maximum possible linear correlation between the random variables $X \sim \mathcal{LN}(0,1)$ and $Y \sim \mathcal{LN}(0,\sigma)$. *Right figure*: The minimum and maximum possible linear correlation between the random variables $X \sim Pareto(2.1, 1)$ and $Y \sim Pareto(\beta, 1)$

7.5.2 Spearman's Rank Correlation

Spearman's rank correlation (often referred to as *Spearman's rho*) is a simple scalar measure of dependence that depends on the copula of two random variables but not on their marginal distributions. More precisely, Spearman's rank correlation for two random variables X_1 and X_2 with marginal distributions $F_1(x_1)$ and $F_2(x_2)$ is given by

$$\rho_S[X_1, X_2] = \rho[F_1(X_1), F_2(X_2)]. \quad (7.28)$$

That is, Spearman's rank correlation is simply the linear correlation of the probability transformed random variables. For multivariate case $(X_1, \ldots, X_d)$, Spearman's rho matrix is defined by the matrix coefficients $\rho_S[X_i, X_j] = \rho[F_i(X_i), F_j(X_j)]$. The main properties can be summarised as follows.

- The range for possible values of $\rho_S[X_1, X_2]$ is $[-1, 1]$.
- For independent random variables $\rho_S[X_1, X_2] = 0$. However, zero Spearman's rank correlation does not necessarily imply independence.

- $\rho_S[X_1, X_2] = 1$ if X_1 and X_2 are comonotonic (perfect positive dependence); and $\rho_S[X_1, X_2] = -1$ if X_1 and X_2 are countermonotonic (perfect negative dependence). Note that this is not the case for the linear correlation coefficient $\rho[X_1, X_2]$.
- In the case of bivariate Gaussian copula with correlation parameter ρ, the following relation is true:

$$\rho_S[X_1, X_2] = \frac{6}{\pi} \arcsin\left(\frac{1}{2}\rho\right) \approx \rho; \tag{7.29}$$

see McNeil, Frey and Embrechts ([157], Theorem 5.36). This relationship between Spearman's rank correlation and the Gaussian copula correlation parameter is often used to calibrate the Gaussian copula. The error in approximating the right-hand side of the above equation by ρ itself is very small:

$$\left|\frac{6}{\pi} \arcsin\left(\frac{1}{2}\rho\right) - \rho\right| \leq (\pi - 3)|\rho|/\pi \leq 0.0181.$$

7.5.3 Kendall's tau Rank Correlation

Kendall's tau rank correlation for random variables X_1 and X_2 is

$$\begin{aligned}\rho_\tau[X_1, X_2] &= \Pr[(X_1 - \widetilde{X}_1)(X_2 - \widetilde{X}_2) > 0] - \Pr[(X_1 - \widetilde{X}_1)(X_2 - \widetilde{X}_2) < 0] \\ &- \mathrm{E}[\mathrm{sign}((X_1 - \widetilde{X}_1)(X_2 - \widetilde{X}_2))],\end{aligned} \tag{7.30}$$

where $(\widetilde{X}_1, \widetilde{X}_2)$ and (X_1, X_2) are independent random vectors from the same distribution. It can also be written as

$$\rho_\tau[X_i, X_j] = \mathrm{Cov}[\mathrm{sign}(X_i - \widetilde{X}_i)\mathrm{sign}(X_j - \widetilde{X}_j)]. \tag{7.31}$$

Similar to Spearman's rank correlation, Kendall's tau rank correlation is a simple scalar measure of dependence that depends on the copula of two random variables but not on their marginal distributions.

- The range for possible values of $\rho_\tau[X_1, X_2]$ is $[-1, 1]$.
- For independent random variables $\rho_\tau[X_1, X_2] = 0$, although zero Kendall's tau does not necessarily imply independence.
- $\rho_\tau[X_1, X_2] = 1$ if X_1 and X_2 are comonotonic (perfect positive dependence); and $\rho_\tau[X_1, X_2] = -1$ if X_1 and X_2 are countermonotonic (perfect negative dependence).
- In the case of the bivariate Gaussian copula with correlation parameter ρ, the following relation is true:

$$\rho_\tau[X_1, X_2] = \frac{2}{\pi} \arcsin(\rho) \approx \rho; \tag{7.32}$$

see McNeil, Frey and Embrechts ([157], Theorem 5.36). This relationship is also true for a general class of normal variance mixture distributions such as t-copula (it is often used to calibrate t-copula). Strictly speaking, it is true for the bivariate case only. That is, for the multivariate case $(X_1, \ldots, X_d)$, if Kendall's tau rank correlation is found for all pairs $\rho_\tau[X_i, X_j]$, then the correlation matrix coefficients ρ_{ij} calculated using (7.32) may not form a positive definite matrix. If this is the case, then eigenvalue method can be used to adjust the correlation coefficients so that the matrix is well defined; see McNeil, Frey and Embrechts ([157], Example 5.54 and Algorithm 5.5).

7.5.4 Tail Dependence

Similar to rank correlations, the tail dependence coefficient is a simple scalar measure of dependence that depends on the copula of two random variables but not on their marginal distributions. Formally, the coefficient of the upper tail dependence between random variables $X_1 \sim F_1(x_1)$ and $X_2 \sim F_2(x_2)$ is defined as

$$\lambda_u = \lim_{q \to 1} \Pr[X_2 > F_2^{-1}(q) | X_1 > F_1^{(-1)}(q)]. \tag{7.33}$$

The lower tail dependence coefficient is defined similarly as

$$\lambda_l = \lim_{q \to 0} \Pr[X_2 \le F_2^{-1}(q) | X_1 \le F_1^{(-1)}(q)]. \tag{7.34}$$

Both λ_u and λ_l belong to the range [0, 1], provided that the above limits exist. Essentially, these coefficients are measures of the dependence in the tails of bivariate distribution. For operational risk purposes, the upper tail dependence (a chance that X_1 is very large if X_2 is very large) is of primary importance.

If the marginal distributions $F_1(\cdot)$ and $F_2(\cdot)$ are continuous, then the tail dependence coefficients can be expressed in terms of the unique copula $C(u_1, u_2)$ between X_1 and X_2:

$$\lambda_u = \lim_{q \to 1} \frac{1 - 2q + C(q, q)}{1 - q}, \tag{7.35}$$

$$\lambda_l = \lim_{q \to 0} \frac{C(q, q)}{q}. \tag{7.36}$$

Detailed discussion of tail dependence can be found in McNeil, Frey and Embrechts ([157], section 5.2.3). Here, we just mention that the tail dependence coefficient can be very useful for comparing different copulas. In particular:

- For the bivariate Gaussian copula, defined by (7.11): $\lambda_l = \lambda_u = 0$, if the correlation coefficient of the copula $\rho < 1$;

- For the bivariate t-copula, defined by stochastic representation (7.20) and (7.21):

$$\lambda_l = \lambda_u = 2t_{\nu+1}\left(-\sqrt{\frac{(\nu+1)(1-\rho)}{1+\rho}}\right), \tag{7.37}$$

which is positive if $\rho > -1$. Here, ρ is a correlation coefficient parameter of the t-copula and ν is a copula degrees-of-freedom parameter.
- For the bivariate Clayton copula defined by (7.15): $\lambda_u = 0$ and $\lambda_l = 2^{-1/\rho}$, for $\rho > 0$;
- For the bivariate Gumbel copula defined by (7.18): $\lambda_l = 0$ and $\lambda_u = 2 - 2^{1/\rho}$ for $\rho > 1$.

7.6 Dependence Between Frequencies via Copula

The most popular approach in operational risk practice is to consider a dependence between the annual counts of different risks via a copula. Assuming a J-dimensional copula $C(\cdot)$ and the marginal distributions $P_j(\cdot)$ for the annual counts $N_t^{(1)}, \ldots, N_t^{(J)}$ leads to a model

$$N_t^{(1)} = P_1^{-1}(U_t^{(1)}), \ldots, N_t^{(J)} = P_J^{-1}(U_t^{(J)}), \tag{7.38}$$

where $U_t^{(1)}, \ldots, U_t^{(J)}$ are the uniform $\mathcal{U}(0,1)$ random variables from a copula $C(\cdot)$ and $P_j^{-1}(\cdot)$ is the inverse marginal distribution of the counts in the j-th risk. Here, t is discrete time (typically in annual units but shorter steps might be needed to calibrate the model). Usually, the counts are assumed to be independent between different t steps.

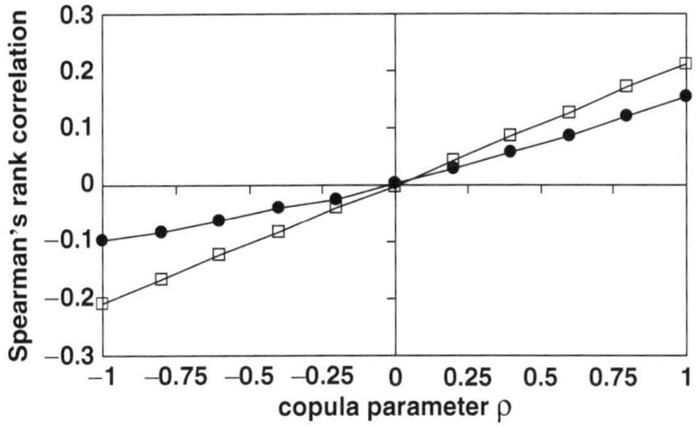

Fig. 7.3 Spearman's rank correlation between the annual losses $\rho_S[Z^{(1)}, Z^{(2)}]$ vs the Gaussian copula parameter ρ: (□) – copula between counts $N^{(1)}$ and $N^{(2)}$; (•) – copula between inter-arrival times of two Poisson processes. Marginally, the frequencies are from $Poisson(5)$ and $Poisson(10)$ respectively and the severities are from $\mathcal{LN}(1, 2)$ for both risks

The approach allows us to model both positive and negative dependence between counts. As reported in the literature, the implied dependence between annual losses even for a perfect dependence between counts is relatively small and as a result the impact on capital is small too. Some theoretical reasons for the observation that frequency dependence has only little impact on the operational risk capital charge are given in Böcker and Klüppelberg [30].

As an example, in Fig. 7.3 we plot Spearman's rank correlation between the annual losses of two risks, $Z^{(1)}$ and $Z^{(2)}$, induced by the Gaussian copula dependence between frequencies. Marginally, the frequencies $N^{(1)}$ and $N^{(2)}$ are from the $Poisson(\lambda = 5)$ and $Poisson(\lambda = 10)$ distributions respectively and the severities are from $\mathcal{LN}(\mu = 1, \sigma = 2)$ distributions for both risks.

7.7 Common Shock Processes

Modelling operational risk events affecting many risk cells can be done using common shock process models; see Johnson, Kotz and Balakrishnan ([128], section 37). In particular, consider J risks with the event counts

$$N_t^{(j)} = N_t^{(C)} + \widetilde{N}_t^{(j)},$$

where $\widetilde{N}_t^{(j)}$, $j = 1, \ldots, J$ and $N_t^{(C)}$ are generated by independent Poisson processes with the intensities $\widetilde{\lambda}_j$ and λ_C respectively. Then, $N_t^{(j)}$, $j = 1, \ldots, J$ are Poisson distributed marginally with the intensities

$$\lambda_j = \widetilde{\lambda}_j + \lambda_C$$

and are dependent via the common events $N_t^{(C)}$. The linear correlation and covariance between risk counts are

$$\rho[N_t^{(i)}, N_t^{(j)}] = \lambda_C / \sqrt{\lambda_i \lambda_j}$$

and

$$\mathrm{Cov}[N_t^{(i)}, N_t^{(j)}] = \lambda_C$$

respectively.

Only a positive dependence between counts can be modelled using this approach. Note that the covariance for any pair of risks is the same though the correlations are different. More flexible dependence can be achieved by allowing a common shock process to contribute to the k-th risk process with some probability p_k; then

$$\mathrm{Cov}[N_t^{(i)}, N_t^{(j)}] = \lambda_C p_i p_j.$$

This method can be generalised to many common shock processes; see Lindskog and McNeil [145] and Powojowski, Reynolds and Tuenter [194]. It is also reasonable to consider the dependence between the severities in different risk cells that occurred due to the same common shock event.

7.8 Dependence Between Aggregated Losses via Copula

Dependence between aggregated losses can be introduced in a manner similar to (7.38). In this approach, one can model the aggregated losses as

$$Z_t^{(1)} = F_1^{-1}(U_t^{(1)}), \ldots, Z_t^{(J)} = F_J^{-1}(U_t^{(J)}), \tag{7.39}$$

where $U_t^{(1)}, \ldots, U_t^{(J)}$ are uniform $\mathcal{U}(0, 1)$ random variables from a copula $C(\cdot)$ and $F_j^{-1}(\cdot)$ is the inverse marginal distribution of the aggregated loss of the j-th risk.

Note that the marginal distribution $F_j(\cdot)$ should be calculated using the frequency and severity distributions. Typically, the data are available over several years only and a short time step t (e.g. quarterly) is needed to calibrate the model.

This approach is probably the most flexible in terms of the range of achievable dependencies between risks, for example, perfect positive dependence between the annual losses is achievable. However, it may create difficulties with incorporation of insurance into the overall model. This is because an insurance policy may apply to several risks with the cover limit applied to the aggregated loss recovery; see Sect. 2.4.

7.9 Dependence Between the k-th Event Times/Losses

Theoretically, one can introduce dependence between the k-th severities or between the k-th event inter-arrival times or between the k-th event times of different risks. For example, 1st, 2nd, etc losses of the j-th risk are correlated to the 1st, 2nd, etc losses of the i-th risk respectively while the severities within each risk are independent. The actual dependence can be done via a copula similar to (7.38); for an accurate description we refer to Chavez-Demoulin, Embrechts and Nešlehová [52]. Here, we would like to note that a physical interpretation of such models can be difficult. Also, an example of dependence between annual losses induced by dependence between the k-th inter-arrival times is presented in Fig. 7.3.

7.10 Modelling Dependence via Lévy Copulas

An interesting approach was suggested in Böcker and Klüppelberg [30, 31] to model dependence in frequency and severity between different risks at the same time using a new concept of Lévy copulas; see Sects. 5.4, 5.5, 5.6, and 5.7 in Cont and Tankov [61]. It is assumed that each risk follows to a univariate compound Poisson process (that belongs to a class of Lévy processes). Then, the idea is to introduce the dependence between risks in such a way that any conjunction of different risks constitutes a univariate compound Poisson process. It is achieved using the multivariate compound Poisson processes based on Lévy copulas. Note that if dependence between frequencies or annual losses is introduced via copula as in (7.38) or (7.39), then the conjunction of risks does not follow to a univariate compound Poisson.

The precise definitions of Lévy measure and Lévy copula are beyond the scope of this book and can be found in the above-mentioned literature. Here, we would like to mention that in the case of a compound Poisson process, Lévy measure is the expected number of losses per unit of time with a loss amount in a pre-specified interval,

$$\overline{\Pi}_j(x) = \lambda_j \Pr(X_j > x).$$

Then the multivariate Lévy measure can be constructed from the marginal measures and a Lévy copula $\widetilde{C}$ as

$$\overline{\Pi}(x_1, \ldots, x_d) = \widetilde{C}(\overline{\Pi}_1(x_1), \ldots, \overline{\Pi}_d(x_d)). \tag{7.40}$$

This is somewhat similar to (7.9) in a sense that the dependence structure between different risks can be separated from the marginal processes. However, it is quite a different concept. In particular, a Lévy copula for processes with positive jumps is $[0, \infty)^d \to [0, \infty)$ mapping while a standard copula (7.9) is $[0, 1]^d \to [0, 1]$ mapping. Also, a Lévy copula controls dependence between frequencies and dependence between severities (from different risks) at the same time.

The interpretation of this model is that dependence between different risks is due to the loss events occurring at the same time. An important implication of this approach is that a bank's total loss can be modelled as a compound Poisson process with some intensity and independent severities. If this common severity distribution is sub-exponential then a closed-form approximation (3.72) can be used to estimate the VaR of the total annual loss.

7.11 Structural Model with Common Factors

Common (systematic) factors are useful for identifying dependent risks and for reducing the number of required correlation coefficients that must be estimated; for example, see McNeil, Frey and Embrechts ([157], section 3.4). Structural models with common factors to model dependence are widely used in credit risk; see industry examples in McNeil, Frey and Embrechts ([157], section 8.3.3). For operational risk, these models are qualitatively discussed in Marshall ([154], sections 5.3 and 7.4) and there are unpublished examples of practical implementation. As an example, assume a Gaussian copula for the annual counts of different risks and consider one common (systematic) factor Ω_t affecting the counts as follows:

$$\begin{aligned} Y_t^{(j)} &= \rho_j \Omega_t + \sqrt{1-\rho_j^2} W_t^{(j)}, \quad j = 1, \ldots, J; \\ N_t^{(1)} &= P_1^{-1}\left(F_N(Y_t^{(j)})\right), \ldots, N_t^{(J)} = P_J^{-1}\left(F_N(Y_t^{(J)})\right). \end{aligned} \tag{7.41}$$

Here, $W_t^{(1)}, \ldots, W_t^{(J)}$ and Ω_t are independent random variables from the standard normal distribution. All random variables are independent between different time steps t. Given Ω_t, the counts are independent; unconditionally, the risk profiles are dependent if the corresponding ρ_j are nonzero. In this example, one should identify J correlation parameters ρ_j only instead of $J(J-1)/2$ parameters of the full correlation matrix.

Extension of this approach to many factors $\Omega_{t,k}$, $k = 1, \ldots, K$ is easy:

$$Y_t^{(j)} = \sum_{k=1}^{K} \rho_{jk} \Omega_{t,k} + \sqrt{1 - \sum_{k=1}^{K} \sum_{m=1}^{K} \rho_{jk} \rho_{jm} \mathrm{Cov}[\Omega_{t,k} \Omega_{t,m}]} W_t^{(j)}, \tag{7.42}$$

where $(\Omega_{t,1}, \ldots, \Omega_{t,K})'$ is from the standard multivariate normal distribution with zero means, unit variances and some correlation matrix.

This approach can also be extended to introduce a dependence between both severities and frequencies. For example, in the case of one factor, one can structure the model as follows:

$$\begin{aligned}
Y_t^{(j)} &= \rho_j \Omega_t + \sqrt{1 - \rho_j^2} W_t^{(j)}, \quad j = 1, \ldots, J; \\
N_t^{(j)} &= P_j^{-1}\left(F_N(Y_t^{(j)})\right), \quad j = 1, \ldots, J; \\
R_s^{(j)}(t) &= \widetilde{\rho}_j \Omega_t + \sqrt{1 - \widetilde{\rho}_j^2} V_s^{(j)}(t), \quad s = 1, \ldots, N_t^{(j)}, \quad j = 1, \ldots, J; \\
X_s^{(j)}(t) &= F_j^{-1}\left(F_N(R_s^{(j)}(t))\right), \quad s - 1, \ldots, N_t^{(j)}, \quad j = 1, \ldots, J.
\end{aligned}$$

Here $W_t^{(j)}$, $V_s^{(j)}(t)$, $s = 1, \ldots, N_t^{(j)}$, $j = 1, \ldots, J$ and Ω_t are independent random variables from the standard normal distribution. Again, the logic is that there is a factor affecting severities and frequencies within a year such that conditional on this factor, severities and frequencies are independent. The factor is changing stochastically from year to year, so that unconditionally there is dependence between frequencies and severities. Also note that in such setup, there is a dependence between severities within a risk category.

Often, common factors are unobservable and practitioners use generic intuitive definitions such as changes in political, legal and regulatory environments, economy, technology, system security, system automation, etc. Several external and internal factors are typically considered, so that some of the factors affect frequencies only (e.g. system automation), some factors affect severities only (e.g. changes in legal environment) and some factors affect both the frequencies and severities (e.g. system security).

It is possible to derive a full joint distribution for all data (frequencies and severities) given model parameters; however, in general it will not have a closed form because the latent variables (factors) should be integrated out. Thus standard methods cannot be used to maximise corresponding likelihood function and one should use more technically involved methods, for example a slice sampler used in Peters, Shevchenko and Wüthrich [187].

The common factor models are supported by empirical evidence, reported in Allen and Bali [8], that some operational risks are dependent on macroeconomic variables such as GDP, unemployment, equity indices, interest rates, foreign exchange rates and regulatory environment variables.

7.12 Stochastic and Dependent Risk Profiles

Consider the LDA for risk cells $j = 1, \dots, J$:

$$Z_j(t) = \sum_{s=1}^{N_j(t)} X_j^{(s)}(t), \quad t = 1, 2, \dots \quad , \tag{7.43}$$

where $N_j(t) \sim P(\cdot|\lambda_t^{(j)})$ and $X_j^{(s)}(t) \sim F(\cdot|\psi_t^{(j)})$. It is realistic to consider that the risk profiles $\boldsymbol{\lambda}_t = (\lambda_t^{(1)}, \dots, \lambda_t^{(J)})$ and $\boldsymbol{\psi}_t = (\psi_t^{(1)}, \dots, \psi_t^{(J)})$ are not constant but changing in time stochastically due to changing risk factors (e.g. changes in business environment, politics, regulations). That is, we may model risk profiles $\boldsymbol{\lambda}_t = \left(\lambda_t^{(1)}, \dots, \lambda_t^{(J)}\right)$ and $\boldsymbol{\psi}_t = \left(\psi_t^{(1)}, \dots, \psi_t^{(J)}\right)$ by random variables $\boldsymbol{\Lambda}_t = \left(\Lambda_t^{(1)}, \dots, \Lambda_t^{(J)}\right)$ and $\boldsymbol{\Psi}_t = \left(\Psi_t^{(1)}, \dots, \Psi_t^{(J)}\right)$, respectively.

Now consider a sequence $(\boldsymbol{\Lambda}_1, \boldsymbol{\Psi}_1), \dots, (\boldsymbol{\Lambda}_{T+1}, \boldsymbol{\Psi}_{T+1})$. It is naive to assume that risk profiles of all risks are independent. Intuitively these are dependent, for example, due to changes in politics, regulations, law, economy, technology (sometimes called drivers or external risk factors) that jointly impact on many risk cells. One can model this by assuming some copula $C(\cdot)$ and marginal distributions for the risk profiles $\boldsymbol{\Lambda}_t$ and $\boldsymbol{\Psi}_t$ (as developed in Peters, Shevchenko and Wüthrich [187]), that gives the following joint distribution of the risk profiles

$$F\left(\boldsymbol{\lambda}_t, \boldsymbol{\psi}_t\right) = C\left(G_1(\lambda_t^{(1)}), \dots, G_J(\lambda_t^{(J)}), H_1(\psi_t^{(1)}), \dots, H_J(\psi_t^{(J)})\right),$$

where $G_j(\cdot)$ and $H_j(\cdot)$ are the marginal distributions of $\lambda_t^{(j)}$ and $\psi_t^{(j)}$ respectively.

Dependence between the risk profiles will induce a dependence between the annual losses. This general model can be used to model the dependencies between the annual counts; between the severities of different risks; between the severities within a risk; and between the frequencies and severities. The likelihood of data (counts and severities) can be derived but involves a multidimensional integral with respect to latent variables (risk profiles). Advanced MCMC methods (such as the slice sampler method described in Sect. 2.11.5 and used in Peters, Shevchenko and Wüthrich [187]) can be used to fit the model.

Stochastic modelling of risk profiles may appeal to intuition. For example, consider the annual number of events for the jth risk modelled as random variables from the Poisson distribution $Poisson\left(\Lambda_t^{(j)} = \lambda_t^{(j)}\right)$. Conditional on $\Lambda_t^{(j)}$, the expected number of events per year is $\Lambda_t^{(j)}$. The latter is not only different for different banks and different risks but also changes from year to year for a risk in the same bank. In general, the evolution of $\Lambda_t^{(j)}$, can be modelled as having deterministic (trend, seasonality) and stochastic components. In actuarial mathematics this is called a mixed Poisson model.

Remark 7.3 The use of common (systematic) factors is useful to identify dependent risks and to reduce the number of required correlation coefficients that must be estimated. For example, assuming a Gaussian copula between risk profiles, consider one common factor Ω_t affecting all risk profiles as follows:

$$
\begin{aligned}
Y_t^{(i)} &= \rho_i \Omega_t + \sqrt{1-\rho_i^2}\, W_t^{(i)}, \quad i = 1, \ldots, 2J; \\
\Lambda_t^{(j)} &= G^{-1}(F_N(Y_t^{(j)})), \quad \Psi_t^{(j)} = H^{-1}(F_N(Y_t^{(j+J)})), \quad j = 1, \ldots, J,
\end{aligned}
$$

where $W_t^{(1)}, \ldots, W_t^{(2J)}$ and Ω_t are independent random variables from the standard normal distribution and all random variables are independent between different time steps t. Given Ω_t, all risk profiles are independent but unconditionally the risk profiles are dependent if the corresponding ρ_i are nonzero. One can consider many factors: some factors affect frequency risk profiles, some factors affect severity risk profiles, and some factors affect both frequency and severity risk profiles.

As an example, consider the following possible model setup for stochastic and dependent risk profiles, proposed in Peters, Shevchenko and Wüthrich [187].

Model Assumptions 7.1 *Consider J risks each with a general model (7.2) for the annual loss in year t, $Z_t^{(j)}$, and each modelled by severity $X_s^{(j)}(t)$ and frequency $N_t^{(j)}$. The frequency and severity risk profiles are modelled by random vectors*

$$
\mathbf{\Lambda}_t = (\Lambda_t^{(1)}, \ldots, \Lambda_t^{(J)})' \quad \text{and} \quad \mathbf{\Psi}_t = (\Psi_t^{(1)}, \ldots, \Psi_t^{(J)})'
$$

respectively and parameterised by risk characteristics

$$
\boldsymbol{\theta}_\Lambda = (\theta_\Lambda^{(1)}, \ldots, \theta_\Lambda^{(J)})' \quad \text{and} \quad \boldsymbol{\theta}_\Psi = (\theta_\Psi^{(1)}, \ldots, \theta_\Psi^{(J)})'
$$

correspondingly. Additionally, the dependence between risk profiles is parameterised by $\boldsymbol{\theta}_\rho$. Assume that, given $\boldsymbol{\theta} = (\boldsymbol{\theta}_\Lambda, \boldsymbol{\theta}_\Psi, \boldsymbol{\theta}_\rho)$:

1. *The random vectors*

$$\left(\boldsymbol{\Psi}_1, \boldsymbol{\Lambda}_1, N_1^{(j)}, X_s^{(j)}(1)\,; j = 1, \ldots, J, s \geq 1\right)'$$

$$\vdots$$

$$\left(\boldsymbol{\Psi}_{T+1}, \boldsymbol{\Lambda}_{T+1}, N_{T+1}^{(j)}, X_s^{(j)}(T+1)\,; j = 1, \ldots, J, s \geq 1\right)'$$

are independent. That is, between different years the risk profiles for frequencies and severities as well as the number of losses and actual losses are independent.

2. *The vectors* $(\Psi_1, \Lambda_1)', \ldots, (\Psi_{T+1}, \Lambda_{T+1})'$ *are independent and identically distributed from a joint distribution with marginal distributions* $\Lambda_t^{(j)} \sim G\left(\cdot|\theta_\Lambda^{(j)}\right)$, $\Psi_t^{(j)} \sim H\left(\cdot|\theta_\Psi^{(j)}\right)$ *and* $2J$*-dimensional copula* $C(\cdot|\boldsymbol{\theta}_\rho)$.
3. *Given* $\boldsymbol{\Lambda}_t = \boldsymbol{\lambda}_t$ *and* $\boldsymbol{\Psi}_t = \boldsymbol{\psi}_t$*, the compound random variables* $Z_t^{(1)}, \ldots, Z_t^{(J)}$ *are independent with* $N_t^{(j)}$ *and* $X_1^{(j)}(t), X_2^{(j)}(t), \ldots$ *independent; frequencies* $N_t^{(j)} \sim P\left(\cdot|\lambda_t^{(j)}\right)$*; and independent severities* $X_s^{(j)}(t) \sim F\left(\cdot|\psi_t^{(j)}\right)$, $s \geq 1$.

Calibration of the above model requires estimation of $\boldsymbol{\theta}$. It can be treated within a Bayesian framework as a random variable $\boldsymbol{\Theta}$ to incorporate expert opinions and external data into the estimation procedure (in Sect. 7.13, we describe the estimation procedure for frequencies). Also note that for simplicity of notation, we assumed one severity risk profile $\Psi_t^{(j)}$ and one frequency risk profile $\Lambda_t^{(j)}$ per risk – extension is trivial if more risk profiles are required.

In general, a copula can be introduced between all risk profiles. For illustration, consider the bivariate case ($J = 2$). That is, we assume that the above Model Assumptions 7.1 are fulfilled for the aggregated losses

$$Z_t^{(1)} = \sum_{s=1}^{N_t^{(1)}} X_s^{(1)}(t) \text{ and } Z_t^{(2)} = \sum_{s=1}^{N_t^{(2)}} X_s^{(2)}(t)\,. \tag{7.44}$$

As marginals, for $j = 1, 2$ we choose:

- $N_t^{(j)} \sim Poisson(\lambda_t^{(j)})$ and $X_s^{(j)}(t) \sim \mathcal{LN}(\mu_j(t), \sigma_j(t))$;
- $\lambda_t^{(1)} \sim Gamma(2.5, 2)$, $\lambda_t^{(2)} \sim Gamma(5, 2)$, $\mu_j(t) \sim \mathcal{N}(1, 1)$, $\sigma_j(t) = 2$;
- The dependence between $\lambda_t^{(1)}$, $\lambda_t^{(2)}$, $\mu_1(t)$ and $\mu_2(t)$ is a Gaussian copula.

The parameters in the above marginal distributions correspond to $\boldsymbol{\theta}_\Lambda$ and $\boldsymbol{\theta}_\Psi$ in Model Assumptions 7.1. Here, we assume the parameters are known a priori. In Sect. 7.13 we will demonstrate the Bayesian inference model and associated methodology to perform an estimation of the model parameters.

Given marginal and copula parameters $(\boldsymbol{\theta}_\Lambda, \boldsymbol{\theta}_\Psi, \boldsymbol{\theta}_\rho)$, the simulation of the annual losses for year $t = T + 1$, when risk profiles are dependent via a copula, can be accomplished using the following procedure.

Algorithm 7.3

1. Simulate $2J$-variate $u_1, \ldots, u_J, v_1, \ldots, v_J$ from a $2J$ dimensional copula $C(\cdot|\boldsymbol{\theta}_\rho)$.
2. Calculate $\lambda_t^{(j)} = G^{-1}\left(u_j|\theta_\Lambda^{(j)}\right)$ and $\psi_t^{(j)} = H^{-1}\left(v_j|\theta_\Psi^{(j)}\right)$, $j = 1, \ldots, J$.
3. Sample $n_t^{(j)}$ from $P\left(\cdot|\lambda_t^{(j)}\right)$, $j = 1, \ldots, J$.
4. Sample independent $x_s^{(j)}(t), s = 1, \ldots, n_t^{(j)}, j = 1, \ldots, J$ from $F\left(\cdot|\psi_t^{(j)}\right)$.
5. Calculate annual losses $z_t^{(j)} = \sum_{s=1}^{n_t^{(j)}} x_s^{(j)}(t)$, $j = 1, \ldots, J$.
6. Repeat steps 1–5 K times to get K random samples of the annual losses $z_t^{(j)}$.

Remark 7.4 Simulation of the random variates from a copula in step 1 is easy for many types of copulas. In the case of Gaussian, t, Clayton and Gumbel copulas it can be done using algorithms from Sect. 7.4; for other types, see McNeil, Frey and Embrechts ([157], chapter 5) and references therein.

Using the above simulation procedure we can examine the strength of dependence between the annual losses if there is a dependence between the risk profiles. Figure 7.4 shows the induced dependence between the annual losses $Z_t^{(1)}$ and $Z_t^{(2)}$ vs the copula dependence parameter for three cases:

- only $\lambda_t^{(1)}$ and $\lambda_t^{(2)}$ are dependent;
- only $\mu_1(t)$ and $\mu_2(t)$ are dependent;
- the dependence between $\lambda_t^{(1)}$ and $\lambda_t^{(2)}$ is the same as between $\mu_1(t)$ and $\mu_2(t)$.

In all cases the dependence is the Gaussian copula (7.12) denoted as $C(u_1, u_2|\rho)$ and parameterised by one parameter ρ which controls the degree of dependence. In the case of the Gaussian copula, ρ is a non-diagonal element of correlation matrix $\boldsymbol{\Sigma}$ in (7.12). The parameter ρ corresponds to θ_ρ in Model Assumptions 7.1.

In each of these examples we vary the parameter of the copula model ρ from weak to strong dependence. The annual losses are not Gaussian distributed and to measure the dependence between the annual losses we use a non-linear rank correlation measure, Spearman's rank correlation, $\rho_S[Z_t^{(1)}, Z_t^{(2)}]$. The Spearman's rank correlation between the annual losses was estimated using 10,000 simulated years for each value of ρ. These numerical experiments show that the range of

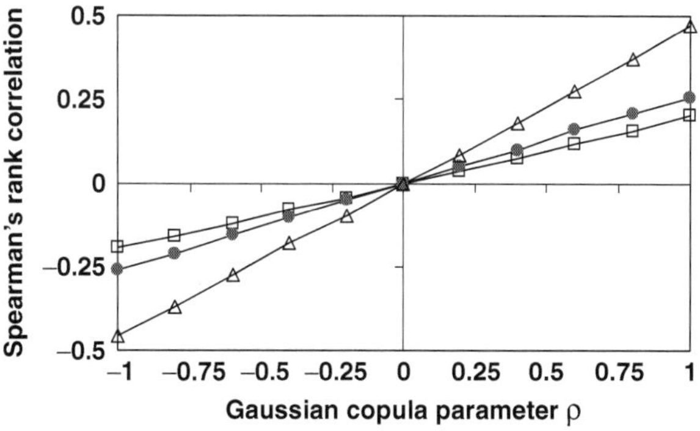

Fig. 7.4 Spearman's rank correlation $\rho_S[Z^{(1)}, Z^{(2)}]$ between annual losses vs the Gaussian copula parameter ρ: (□) – copula for the frequency profiles $\Lambda_t^{(1)}$ and $\Lambda_t^{(2)}$; (•) – copula for the severity profiles $\Psi_t^{(1)}$ and $\Psi_t^{(2)}$ that correspond to μ_1 and μ_2 in the severity distribution respectively; (△) – copula for λ_1 and λ_2 and the same copula for $\Psi_t^{(1)}$ and $\Psi_t^{(2)}$

possible dependence between the annual losses of different risks induced by the dependence between risk profiles is very wide and should be flexible enough to model dependence in practice. Note that the degree of induced correlation can be further extended by working with more flexible copula models at the expense of estimation of a larger number of model parameters.

7.13 Dependence and Combining Different Data Sources

We have noted several times (see Chap. 4) that Basel II operational risk models have to combine information from internal data, external data and expert opinions. We should also note that experts in financial institutions often attempt to specify not only frequency and severity distributions but also correlations between risks.

Combining of expert opinions with internal and external data is a difficult problem and complicated ad-hoc procedures are used in practice. Some prominent risk professionals in industry have argued that statistically consistent combining of these different data sources is one of the most pertinent and challenging aspects of operational risk modelling.

A Bayesian model to combine three data sources (internal data, external data and expert opinion) for the case of a single risk cell was presented in Lambrigger, Shevchenko and Wüthrich [141]. Then Peters, Shevchenko and Wüthrich [187] extended this to a multivariate case. The main idea was to utilise Bayesian inference to estimate the parameters of the model through the combination of expert opinions and observed loss data (internal and external).

To illustrate the approach, consider modelling frequencies only. The estimation procedure is presented for frequencies only. However it is not difficult to extend the actual procedure to include severities. The case of single risk was presented in Sect. 4.5.3. Here we extend this single risk cell frequency model to the general multiple risk cell setting. This will involve formulation of the multivariate posterior distribution.

Model Assumptions 7.2 (Multiple risk cell frequency model) *Consider J risk cells. Assume that every risk cell j has a fixed, deterministic volume $V^{(j)}$.*

1. *The risk characteristic $\boldsymbol{\Theta}_\Lambda = (\Theta_\Lambda^{(1)}, \ldots, \Theta_\Lambda^{(J)})'$ has a J-dimensional prior density $\pi(\boldsymbol{\theta}_\Lambda)$. The copula parameters $\boldsymbol{\theta}_\rho$ are modelled by a random vector $\boldsymbol{\Theta}_\rho$ with the prior density $\pi\left(\boldsymbol{\theta}_\rho\right)$; $\boldsymbol{\Theta}_\Lambda$ and $\boldsymbol{\Theta}_\rho$ are independent.*
2. *Given $\boldsymbol{\Theta}_\Lambda = \boldsymbol{\theta}_\Lambda$ and $\boldsymbol{\Theta}_\rho = \boldsymbol{\theta}_\rho$: the vectors $(\boldsymbol{\Lambda}_1, \mathbf{N}_1)', \ldots, (\boldsymbol{\Lambda}_{T+1}, \mathbf{N}_{T+1})'$ are independent and identically distributed; and the intensities $\boldsymbol{\Lambda}_t = (\Lambda_t^{(1)}, \ldots, \Lambda_t^{(J)})'$ have a J-dimensional conditional density with marginal distributions*

$$\Lambda_t^{(j)} \sim G\left(\cdot|\theta_\Lambda^{(j)}\right) = Gamma\left(\alpha^{(j)}, \theta_\Lambda^{(j)}/\alpha^{(j)}\right)$$

and the copula $c(\cdot|\boldsymbol{\theta}_\rho)$. Thus the joint density of $\boldsymbol{\Lambda}_t$ is given by

$$\pi(\boldsymbol{\lambda}_t|\boldsymbol{\theta}_\Lambda, \boldsymbol{\theta}_\rho) = c\left(G(\lambda_t^{(1)}|\theta_\Lambda^{(1)}), \ldots, G(\lambda_t^{(J)}|\theta_\Lambda^{(J)})|\boldsymbol{\theta}_\rho\right) \prod_{j=1}^{J} \pi(\lambda_t^{(j)}|\theta_\Lambda^{(j)}), \tag{7.45}$$

where $\pi\left(\cdot|\theta_\Lambda^{(j)}\right)$ denotes the marginal density.
3. *Given $\boldsymbol{\Theta}_\Lambda = \boldsymbol{\theta}_\Lambda$ and $\boldsymbol{\Lambda}_t = \boldsymbol{\lambda}_t$, the frequencies are independent with*

$$N_t^{(j)} \sim Poisson(V^{(j)}\lambda_t^{(j)}),\ j = 1, \ldots, J.$$

4. *There are expert opinions $\boldsymbol{\Delta}_k = (\Delta_k^{(1)}, \ldots, \Delta_k^{(J)})'$, $k = 1, \ldots, K$. Given $\boldsymbol{\Theta}_\Lambda = \boldsymbol{\theta}_\Lambda$: $\boldsymbol{\Delta}_k$ and $(\boldsymbol{\Lambda}_t, \mathbf{N}_t)'$ are independent for all k and t; and $\Delta_k^{(j)}$ are all independent with*

$$\Delta_k^{(j)} \sim Gamma\left(\xi^{(j)}, \theta_\Lambda^{(j)}/\xi^{(j)}\right).$$

Prior Structure $\pi(\boldsymbol{\theta}_\Lambda)$ and $\pi\left(\boldsymbol{\theta}_\rho\right)$. In the following examples, a priori, the risk characteristics $\Theta_\Lambda^{(j)}$ are independent gamma distributed: $\Theta_\Lambda^{(j)} \sim Gamma(a^{(j)}, 1/b^{(j)})$ with hyper-parameters $a^{(j)} > 0$ and $b^{(j)} > 0$. This means that a priori the risk characteristics for the different risk classes are independent. That is, if the company has a bad risk profile in risk class j then the risk profile in risk class i is not necessarily bad. Dependence is then modelled through the dependence between the intensities $\Lambda_t^{(1)}, \ldots, \Lambda_t^{(J)}$. If this is not appropriate then, of course, this can easily be changed by assuming dependence within $\boldsymbol{\Theta}_\Lambda$. In the simulation experiments below we consider cases when the copula is parameterised by a scalar θ_ρ. Additionally, we are interested in obtaining inferences on θ_ρ implied by the data only so we use noninformative constant prior on the range $[-1, 1]$ in the case of Gaussian copula.

Posterior density. The marginal posterior density of random vector $(\boldsymbol{\Theta}_\Lambda, \Theta_\rho)'$ given data of counts $\mathbf{N}_{1:T} = \mathbf{n}_{1:T}$ and expert opinions $\boldsymbol{\Delta}_{1:K} = \boldsymbol{\delta}_{1:K}$ is

$$\begin{aligned}
\pi\left(\boldsymbol{\theta}_\Lambda, \theta_\rho | \mathbf{n}_{1:T}, \boldsymbol{\delta}_{1:K}\right) &= \prod_{t=1}^{T} \int \pi\left(\boldsymbol{\theta}_\Lambda, \theta_\rho, \boldsymbol{\lambda}_t | \mathbf{n}_{1:T}, \boldsymbol{\delta}_{1:K}\right) d\boldsymbol{\lambda}_t \\
&\propto \prod_{t=1}^{T} \left(\int \prod_{j=1}^{J} \exp\left\{-V^{(j)} \lambda_t^{(j)}\right\} \frac{(V^{(j)} \lambda_t^{(j)})^{n_t^{(j)}}}{n_t^{(j)}!} \pi(\boldsymbol{\lambda}_t | \boldsymbol{\theta}_\Lambda, \theta_\rho) \, d\boldsymbol{\lambda}_t \right) \\
&\times \prod_{k=1}^{K} \prod_{j=1}^{J} \left(\frac{(\xi^{(j)} / \theta_\Lambda^{(j)})^{\xi^{(j)}}}{\Gamma(\xi^{(j)})} (\delta_k^{(j)})^{\xi^{(j)}-1} \exp\left\{-\delta_k^{(j)} \xi^{(j)} / \theta_\Lambda^{(j)}\right\} \right) \\
&\times \prod_{j=1}^{J} \frac{(b^{(j)})^{a^{(j)}}}{\Gamma(a^{(j)})} (\theta_\Lambda^{(j)})^{a^{(j)}-1} \exp\left\{-b^{(j)} \theta_\Lambda^{(j)}\right\} \pi\left(\theta_\rho\right).
\end{aligned} \tag{7.46}$$

Here, for convenience, we use notation $x_{1:M} = \{x_1, x_2, \ldots, x_M\}$. For example,

$$\mathbf{N}_{1:T} = \left\{ \left(N_1^{(1)}, \ldots, N_1^{(J)}\right)', \left(N_2^{(1)}, \ldots, N_2^{(J)}\right)', \ldots, \left(N_T^{(1)}, \ldots, N_T^{(J)}\right)' \right\}$$

are the annual number of losses for all risk profiles and years; and

$$\boldsymbol{\Delta}_{1:K} = \left\{ \left(\Delta_1^{(1)}, \ldots, \Delta_1^{(J)}\right)', \left(\Delta_2^{(1)}, \ldots, \Delta_2^{(J)}\right)', \ldots, \left(\Delta_K^{(1)}, \ldots, \Delta_K^{(J)}\right)' \right\}$$

are the expert opinions on mean frequency intensities for all experts and risk profiles.

7.13.1 Bayesian Inference Using MCMC

Posterior (7.46) involves integration and sampling from this distribution is difficult. The common trick is to sample from the desired target posterior density $\pi(\boldsymbol{\theta}_\Lambda, \theta_\rho, \boldsymbol{\lambda}_{1:T} | \mathbf{n}_{1:T}, \boldsymbol{\delta}_{1:K})$. Then marginally taken samples of $\boldsymbol{\Theta}_\Lambda$ and Θ_ρ are samples from $\pi\left(\boldsymbol{\theta}_\Lambda, \theta_\rho | \mathbf{n}_{1:T}, \boldsymbol{\delta}_{1:K}\right)$ which can be used to make inferences for required quantities.

Sampling from $\pi\left(\boldsymbol{\theta}_\Lambda, \theta_\rho, \boldsymbol{\lambda}_{1:T} | \mathbf{n}_{1:T}, \boldsymbol{\delta}_{1:K}\right)$ via closed-form inversion or rejection sampling is still not an option. To accomplish this task, one can develop a specialised MCMC method. One possible way is to use Gibbs sampling methodology. This requires the knowledge of full conditional distributions that can be derived for this particular model (see Appendix B in Peters, Shevchenko and Wüthrich [187]) as:

$$\pi(\theta_\Lambda^{(j)}|\boldsymbol{\theta}_\Lambda^{(-j)}, \boldsymbol{\lambda}_{1:T}, \mathbf{n}_{1:T}, \boldsymbol{\delta}_{1:K}, \theta_\rho) \propto \pi(\boldsymbol{\lambda}_{1:T}|\boldsymbol{\theta}_\Lambda^{(-j)}, \theta_\Lambda^{(j)}, \theta_\rho)\pi(\boldsymbol{\delta}_{1:K}|\boldsymbol{\theta}_\Lambda^{(-j)}, \theta_\Lambda^{(j)}) \times\pi(\boldsymbol{\theta}_\Lambda^{(-j)}|\theta_\Lambda^{(j)})\pi(\theta_\Lambda^{(j)}), \tag{7.47}$$

$$\pi(\lambda_t^{(j)}|\boldsymbol{\theta}_\Lambda, \boldsymbol{\lambda}_{1:T}^{(-t,-j)}, \mathbf{n}_{1:T}, \boldsymbol{\delta}_{1:K}, \theta_\rho) \propto \pi(\mathbf{n}_{1:T}|\boldsymbol{\lambda}_{1:T}^{(-t,-j)}, \lambda_t^{(j)}) \times\pi(\boldsymbol{\lambda}_t^{(-j)}, \lambda_t^{(j)}|\boldsymbol{\theta}_\Lambda, \theta_\rho), \tag{7.48}$$

$$\pi\left(\theta_\rho|\boldsymbol{\theta}_\Lambda, \boldsymbol{\lambda}_{1:T}, \mathbf{n}_{1:T}, \boldsymbol{\delta}_{1:K}\right) \propto \pi(\boldsymbol{\lambda}_{1:T}|\boldsymbol{\theta}_\Lambda, \theta_\rho)\pi\left(\theta_\rho\right). \tag{7.49}$$

Here, $\boldsymbol{\lambda}_{1:T}^{(-i,-j)}$, $\boldsymbol{\theta}_\Lambda^{(-j)}$ and $\boldsymbol{\lambda}_t^{(-j)}$ are the exclusion operators:

- $\boldsymbol{\lambda}_{1:T}^{(-2,-1)} = \left\{\left(\lambda_1^{(1)},\ldots,\lambda_1^{(J)}\right)', \left(\lambda_2^{(2)},\ldots,\lambda_2^{(J)}\right)', \ldots, \left(\lambda_T^{(1)},\ldots,\lambda_T^{(J)}\right)'\right\}$ are frequency intensities for all risk profiles and years, excluding risk profile 1 from year 2;
- $\boldsymbol{\theta}_\Lambda^{(-j)} = \left\{\theta_\Lambda^{(1)},\ldots,\theta_\Lambda^{(j-1)},\theta_\Lambda^{(j+1)},\ldots,\theta_\Lambda^{(J)}\right\}$; and similar for $\boldsymbol{\lambda}_t^{(-j)}$.

These full conditionals do not take standard explicit closed forms and typically the normalising constants are not known in closed form. Therefore this will exclude straightforward inversion or basic rejection sampling being used to sample from these distributions. One may adopt a Metropolis-Hastings within Gibbs sampler to obtain samples; see Sect. 2.11.3. To utilise such algorithm it is important to select a suitable proposal distribution. Quite often in high dimensional problems such as ours, this requires tuning of the proposal for a given target distribution. Hence, one incurs a significant additional computational expense in tuning the proposal distribution parameters off-line so that mixing of the resulting Markov chain is sufficient. An alternative not discussed here would include an adaptive Metropolis-Hastings within Gibbs sampling algorithm; see Atchade and Rosenthal [11] and Rosenthal [205]. Here we take a different approach which utilises the full conditional distributions, known as a univariate slice sampler described in Sect. 2.11.5. Note that we only need to know the target full conditional posterior up to normalisation. This is important in this example since solving the normalising constant in this model is not possible analytically.

Algorithm 7.4 (Slice sampling)

1. For $l = 0$, initialise the parameter vector $\left(\boldsymbol{\theta}_{\Lambda,0}, \boldsymbol{\lambda}_{1:T,0}, \theta_{\rho,0}\right)$ randomly or deterministically.
2. Repeat while $l \leq L$
   a. Set $\left(\boldsymbol{\theta}_{\Lambda,l}, \boldsymbol{\lambda}_{1:T,l}, \theta_{\rho,l}\right) = \left(\boldsymbol{\theta}_{\Lambda,l-1}, \boldsymbol{\lambda}_{1:T,l-1}, \theta_{\rho,l-1}\right)$.
   b. Sample j uniformly from set $\{1, 2, \ldots, J\}$.
   Sample new parameter value $\widetilde{\theta}_\Lambda^{(j)}$ from the full conditional posterior distribution $\pi\left(\theta_\Lambda^{(j)}|\boldsymbol{\theta}_{\Lambda,l}^{(-j)}, \boldsymbol{\lambda}_{1:T,l}, \mathbf{n}_{1:T}, \boldsymbol{\delta}_{1:K}, \theta_{\rho,l}\right)$.
   Set $\theta_{\Lambda,l}^{(j)} = \widetilde{\theta}_\Lambda^{(j)}$.

c. Sample j uniformly from set $\{1, 2, \ldots, J\}$ and t uniformly from set $\{1, \ldots, T\}$.
Sample new parameter value $\widetilde{\lambda}_t^{(j)}$ from the full conditional posterior distribution $\pi\left(\lambda_t^{(j)}|\boldsymbol{\theta}_{\Lambda,l}, \boldsymbol{\lambda}_{1:T,l}^{(-t,-j)}, \mathbf{n}_{1:T}, \boldsymbol{\delta}_{1:K}, \theta_{\rho,l}\right)$.
Set $\lambda_{t,l}^{(j)} = \widetilde{\lambda}_t^{(j)}$.

d. Sample new parameter value $\widetilde{\theta}_\rho$ from the full conditional posterior distribution $\pi\left(\theta_\rho|\boldsymbol{\theta}_{\Lambda,l}, \boldsymbol{\lambda}_{1:T,l}, \mathbf{n}_{1:T}, \boldsymbol{\delta}_{1:K}\right)$.
Set $\theta_{\rho,l} = \widetilde{\theta}_\rho$.

3. $l = l + 1$ and return to 2.

The sampling from the full conditional posteriors in stage 2 uses a *univariate slice sampler*. For example, to sample the next iteration of the Markov chain from $\pi\left(\theta_\Lambda^{(j)}|\boldsymbol{\theta}_{\Lambda,l}^{(-j)}, \boldsymbol{\lambda}_{1:T,l}, \mathbf{n}_{1:T}, \boldsymbol{\delta}_{1:K}, \theta_\rho\right)$:

- Sample u from a uniform distribution

$$\mathcal{U}\left(0, \pi\left(\theta_{\Lambda,l}^{(j)}|\boldsymbol{\theta}_{\Lambda,l}^{(-j)}, \boldsymbol{\lambda}_{1:T,l}, \mathbf{n}_{1:T}, \boldsymbol{\delta}_{1:K}, \theta_\rho\right)\right). \tag{7.50}$$

- Sample $\widetilde{\theta}_\Lambda^{(j)}$ uniformly from the intervals (level set)

$$A = \left\{\theta_\Lambda^{(j)} : \pi\left(\theta_\Lambda^{(j)}|\boldsymbol{\theta}_{\Lambda,l}^{(-j)}, \boldsymbol{\lambda}_{1:T,l}, \mathbf{n}_{1:T}, \boldsymbol{\delta}_{1:K}, \theta_\rho\right) > u\right\}. \tag{7.51}$$

The level sets A are determined, for example, by a stepping out and a shrinkage procedure, the details of which can be found in Neal ([170], p. 713, Figure 1); see also Sect. 2.11.5.

7.13.2 Numerical Example

Consider the model with Model Assumptions 7.2 in the case of two risks with dependent intensities and set risk cell volumes $V^{(1)} = V^{(2)} = 1$. Here we estimate $\Theta_\Lambda^{(1)}, \Theta_\Lambda^{(2)}$ and Θ_ρ jointly. We set the true values of $\Theta_\Lambda^{(1)}$ and $\Theta_\Lambda^{(2)}$ to be $\theta_{true}^{(1)} = 5$ and $\theta_{true}^{(2)} = 10$ respectively. Also, we assume a Gaussian copula with $\rho = 0.9$, that is, the true value of Θ_ρ is 0.9. For the expert opinions on the true parameters, assume opinion that underestimates risk profile 1, $\Delta_1^{(1)} = 2$, and opinion that overestimates the risk profile 2, $\Delta_1^{(2)} = 13$. The model parameters were set as follows:

- $\xi^{(1)} = \xi^{(2)} = 2, \alpha^{(1)} = 2, \alpha^{(2)} = 2$ – parameters of the conditional distributions for the intensities and expert opinions,

- $a^{(1)} = a^{(2)} = 2$, $b^{(1)} = 0.4$, $b^{(2)} = 0.2$ – parameters of the prior distribution for $\Theta_\Lambda^{(1)}$ and $\Theta_\Lambda^{(2)}$.

Then, the simulation experiment steps are as follows:

1. Using the true values for the model parameters, simulate a dataset $\mathbf{n}_{1:T}$ of the annual number of events over $T = 20$ years.
2. Obtain correlated MCMC samples from the target posterior distribution after discarding burnin samples, $\{\boldsymbol{\theta}_{\Lambda,l}, \boldsymbol{\lambda}_{1:T,l}, \theta_{\rho,l}\}$, $l = 1{,}001, \ldots, 50{,}000$. Here, we use the slice sampler Algorithm 7.4.
3. Estimate desired posterior quantities such as posterior mean of parameters of interest and posterior standard deviations.

Further analysis can be done by repeating steps 1–3 for independent data realisations and then analysing average of the results; these can be found in Peters, Shevchenko and Wüthrich [187].

Results for this simulation experiment as a function of data size are given in Table 7.1. That is, we study the accuracy of the parameter estimates as the number of observations increases. A typical run with 5 years of data and 1 expert in the bivariate case for 50,000 simulations took approximately 50 s and for the case of 10 risk profiles it took approximately 43 min[2]. The standard errors in the estimates (due to finite number of MCMC iterations) were in the range 1–5% and are not presented in the table.

These results demonstrate that the model and estimation methodology are successfully able to estimate jointly the risk profiles and the correlation parameter. It is also clear that with few observations, for example $T \leq 5$, and a vague prior for the copula parameter, it will be difficult to accurately estimate the copula parameter. This is largely due to the fact that the posterior distribution in this case is diffuse. However, as the number of observations increases the accuracy of the estimate improves and the estimates are reasonable in the case of 15 or 20 years of

Table 7.1 Posterior estimates for $\Theta_\Lambda^{(1)}$, $\Theta_\Lambda^{(2)}$ and copula parameter Θ_ρ. In this case a single data set is generated using Gaussian ($\rho = 0.9$) copula model as specified. Posterior standard deviations are given in brackets next to estimate. Joint estimation was used

Year	1	2	5	10	15	20
$\mathrm{E}[\Theta_\Lambda^{(1)}]$	2.83	4.49	3.31	4.88	4.36	5.07
$\mathrm{stdev}[\Theta_\Lambda^{(1)}]$	1.74	2.02	1.38	1.29	1.10	1.09
$\mathrm{E}[\Theta_\Lambda^{(2)}]$	10.23	10.85	8.72	8.91	8.58	9.94
$\mathrm{stdev}[\Theta_\Lambda^{(2)}]$	3.92	3.52	2.95	2.12	2.04	1.85
$\mathrm{E}[\Theta_\rho]$	0.21	0.47	0.61	0.66	0.70	0.74
$\mathrm{stdev}[\Theta_\rho]$	0.54	0.39	0.30	0.24	0.19	0.15

[2] Computing time is quoted for a standard PC, Intel Core 2 with 2.40 GHz CPU and 2.39 GB of RAM.

data. Additionally, we could further improve the accuracy of this prediction if we incorporated expert opinions into the prior specification of the copula parameter, instead of using a vague prior.

Other results presented in Peters, Shevchenko and Wüthrich [187] demonstrate that, as expected from credibility theory, the joint estimation is better than the marginal, that is, the posterior standard deviations for $\Theta_{\Lambda}^{(1)}$ and $\Theta_{\Lambda}^{(2)}$ are less when joint estimation is used. In addition, the rate of convergence of the posterior mean for Θ_{Λ} to the true value is faster under the joint estimation and there is a strong correlation between $\Theta_{\Lambda}^{(1)}$ and $\Theta_{\Lambda}^{(2)}$. Thus the standard practice in the industry of performing marginal estimation of risk profiles may lead to incorrect results.

Overall, this example demonstrates how the combination of all the relevant sources of data can be achieved and that a sampling methodology has the ability to estimate jointly all the model parameters, including the copula parameter. One can extend this methodology to more sophisticated and flexible copula-based models with more than one parameter. This should be relatively trivial since the methodology developed applies directly. However, the challenge in the case of a more sophisticated copula model relates to finding a relevant choice of prior distribution on the correlation structure.

7.14 Predictive Distribution

Conceptually, quantification of the predictive distribution (accounting both for process and parameter uncertainties) for a bank's annual loss in the case of many risks is similar to the case of single risk considered in Sect. 4.7. If correlation modelling cannot be done then, as required by Basel II, the 0.999 quantile should be quantified for each risk cell as described in Sect. 4.7; the total capital is just a sum of these quantiles. In this section, we assume that the dependence model between risks is developed.

Consider the annual loss in a bank over the next year, Z_{T+1}. Denote the density of the annual loss, conditional on parameters $\boldsymbol{\theta}$, as $f(z_{T+1}|\boldsymbol{\theta})$. Typically, practitioners will take point estimates $\widehat{\boldsymbol{\theta}}$ of all model parameters; conditional on these point estimates construct the predictive density $f(z_{T+1}|\widehat{\boldsymbol{\theta}})$. Then, the latter is used to calculate risk measures such as the 0.999 quantile, $Q_{0.999}(\widehat{\boldsymbol{\theta}})$. Typically, given observations, the MLEs $\widehat{\boldsymbol{\theta}}$ are used as the "best fit" point estimators for $\boldsymbol{\theta}$.

However, the parameters $\boldsymbol{\theta}$ are unknown and it is important to account for this uncertainty when the capital charge is estimated (especially for risks with small datasets) as discussed in Shevchenko [215]. If the Bayesian inference approach is taken, then the parameter $\boldsymbol{\theta}$ is modelled by random variable $\boldsymbol{\Theta}$ and the predictive density (accounting for parameter uncertainty) of Z_{T+1}, given all data $\mathbf{Y}$ used in the estimation procedure, is

$$f(z_{T+1}|\mathbf{y}) = \int f(z_{T+1}|\boldsymbol{\theta})\pi(\boldsymbol{\theta}|\mathbf{y})d\boldsymbol{\theta}. \tag{7.52}$$

Here, $\pi(\boldsymbol{\theta}|\mathbf{y})$ is the posterior density for $\boldsymbol{\Theta}$. Also, it is assumed that, given parameters $\boldsymbol{\Theta}$, Z_{T+1} and $\mathbf{Y}$ are independent. The 0.999 quantile of the predictive distribution

$$Q_q^P = F^{-1}_{Z_{T+1}|\mathbf{Y}}(q) = \inf\{z \in \mathbb{R} : \Pr[Z_{T+1} > z|\mathbf{Y}] \leq 1 - q\}, \tag{7.53}$$

where $q = 0.999$, can be used as a risk measure for capital calculations; also see formula (4.125).

Another approach under a Bayesian framework to account for parameter uncertainty is to consider a quantile $Q_q(\boldsymbol{\theta})$ of the conditional annual loss density $f(\cdot|\boldsymbol{\theta})$:

$$Q_q(\boldsymbol{\Theta}) = F^{-1}_{Z_{T+1}|\boldsymbol{\Theta}}(q) = \inf\{z \in \mathbb{R} : \Pr[Z_{T+1} > z|\boldsymbol{\Theta}] \leq 1 - q\}, \tag{7.54}$$

where we are interested in $q = 0.999$. Then, given that $\boldsymbol{\Theta}$ is distributed from $\pi(\boldsymbol{\theta}|\mathbf{y})$, one can find the distribution of $Q_q = Q_q(\boldsymbol{\Theta})$ and form a predictive interval to contain the true value of Q_q with some probability[3]. Under this approach, one can argue that the conservative estimate of the capital charge accounting for parameter uncertainty should be based on the upper bound of the constructed interval. Note that specification of the confidence level is required and it might be difficult to argue that the commonly used confidence level 0.95 is good enough for estimation of the 0.999 quantile.

In operational risk, it seems that the objective should be to estimate the full predictive density (7.52) for the annual loss Z_{T+1} over next year conditional on all available information and then estimate the capital charge as a quantile $Q^P_{0.999}$ of this distribution (7.53).

Consider all risk cells in the bank. Assume that multivariate model is specified. That is, the frequency $p(\cdot|\boldsymbol{\alpha})$ and severity $f(\cdot|\boldsymbol{\beta})$ densities for each cell are chosen and the dependence structure between risks parameterised by some parameter vector $\boldsymbol{\rho}$ is specified. Also, suppose that the posterior $\pi(\boldsymbol{\theta}|\mathbf{y})$, $\boldsymbol{\theta} = (\boldsymbol{\alpha}, \boldsymbol{\beta}, \boldsymbol{\rho})$ is estimated. Then, the predictive density (7.52) for the annual loss across all risk cells over next year can be calculated using Monte Carlo procedure with the following logical steps.

Algorithm 7.5 (Monte Carlo predictive distribution for many risks)

1. For $k = 1, \ldots, K$
   a. Simulate all model parameters (including the dependence parameters) from their joint posterior $\pi(\boldsymbol{\theta}|\mathbf{y})$. If the posterior is not known in closed form then this simulation can be done using MCMC (see Sect. 2.11). For

---

[3] This is similar to forming a confidence interval in the frequentist approach using the distribution of $Q_{0.999}(\widehat{\boldsymbol{\theta}})$, where $\widehat{\boldsymbol{\theta}}$ is treated as random.

example, one can run MCMC for K iterations beforehand and simply take the k-th iteration parameter values.

b. Given model parameters $\boldsymbol{\theta} = (\boldsymbol{\alpha}, \boldsymbol{\beta}, \boldsymbol{\rho})$, simulate the annual frequencies $N^{(j)}$ and severities $X_s^{(j)}$, $s = 1, \ldots, N^{(j)}$ for all risks $j = 1, \ldots, J$ with a chosen dependence structure. Calculate the bank annual loss $Z_k = Z^{(1)} + \cdots + Z^{(J)}$, where $Z^{(j)} = \sum_{s=1}^{N^{(j)}} X_s^{(j)}$ is the annual loss due to the j-th risk.

2. Next k

Remark 7.5 Obtained annual losses (total across all risks for next year) $Z_1, \ldots, Z_K$ are samples from the predictive density (7.52). Full specification of the dependence model is required. In general, sampling from the joint posterior of all model parameters can be accomplished via MCMC; see Peters, Shevchenko and Wüthrich [187], and Dalla Valle [68]. The 0.999 quantile $Q_{0.999}^P$ and other distribution characteristics can be estimated using the simulated samples in the usual way; see Sect. 3.2.

Note that in the above Monte Carlo procedure the risk profile $\boldsymbol{\theta}$ is simulated from its posterior for each simulation. Thus we model both the process uncertainty, which comes from the fact that frequencies and severities are random variables, and the parameter risk (parameter uncertainty), which comes from the fact that we do not know the true values of $\boldsymbol{\theta}$. Using samples from the joint posterior distribution of the model parameters, we can construct the predictive distribution by removing the parameter uncertainty from the model considered, including the uncertainty arising from the dependence parameters.

Example 7.4 As an example, consider Model Assumptions 7.2. Then the predictive density for the annual loss Z_{T+1} is

$$\pi\left(z_{T+1}|\mathbf{n}_{1:T}, \boldsymbol{\delta}_{1:K}\right) = \int \pi\left(z_{T+1}|\boldsymbol{\theta}_\Lambda, \theta_\rho\right) \pi\left(\boldsymbol{\theta}_\Lambda, \theta_\rho|\mathbf{n}_{1:T}, \boldsymbol{\delta}_{1:K}\right) d\boldsymbol{\theta}_\Lambda d\theta_\rho. \quad (7.55)$$

Here, we used the model assumptions that given $\boldsymbol{\Theta}_\Lambda$ and Θ_ρ we have that Z_{T+1} is independent from the data $(\mathbf{N}_{1:T}, \boldsymbol{\Delta}_{1:K})$. To obtain samples from this predictive distribution, add simulation of $\left(\boldsymbol{\theta}_\Lambda, \theta_\rho\right)$ from the posterior distribution (e.g. using slice sampler methodology) as an extra step before Step 1 in Algorithm 7.3. Specifically, if one wants to simulate L annual losses from the predictive distribution, then this would involve first running the slice sampler for L iterations after *burnin*. Then, for each iteration l one would use the state of the Markov chain $\left(\boldsymbol{\theta}_{\Lambda,l}, \theta_{\rho,l}\right)$ in the simulation Algorithm 7.3.

Problems[4]

7.1 (⋆⋆) Prove that the tail dependence λ for two risks whose dependence is specified by the t-copula with ν degrees of freedom and correlation coefficient ρ is

$$\lambda = 2t_{\nu+1}\left(-\sqrt{(\nu+1)(1-\rho)}/\sqrt{1+\rho}\right),$$

where $t_\nu(\cdot)$ is the standard univariate t-distribution.

7.2 (⋆) Assume that we have two independent risks, $X \sim Pareto(\beta, 1)$ and $Y \sim Pareto(\beta, 1)$, where $Pareto(\beta, a) = 1 - (x/a)^{-\beta}$ and $\beta = 2$. Calculate the $\mathrm{VaR}_q[X+Y]$ using Monte Carlo and find the diversification D_q as defined in (7.7) for several values of the quantile level q in the range $[0.5, 0.999]$.

7.3 (⋆) Assume that there are four independent risks: $X_i \sim \mathcal{LN}(0, 2)$, $i = 1, \ldots, 4$. Calculate the $\mathrm{VaR}_q[X_1 + \cdots + X_4]$ using Monte Carlo and find the diversification D_q as defined in (7.7) for several values of the quantile level q in the range $[0.5, 0.999]$. Repeat calculations for the case when the risk 4 is replaced by $X_4 \sim \mathcal{LN}(0, 4)$.

7.4 (⋆⋆) Simulate 10,000 realisations of two risks

$$Z^{(1)} = \sum_{i=1}^{N^{(1)}} X_i^{(1)} \quad \text{and} \quad Z^{(2)} = \sum_{i=1}^{N^{(2)}} X_i^{(2)},$$

where

- $X_s^{(j)} \sim \mathcal{LN}(0, 2)$, $j = 1, 2$, $s \geq 1$, all independent
- $N^{(1)} \sim Poisson(2)$ and $N^{(2)} \sim Poisson(2)$; and the dependence structure of $(N^{(1)}, N^{(2)})$ is the Gaussian copula with correlation parameter ρ.
- Frequencies $(N^{(1)}, N^{(2)})$ and severities $X_s^{(j)}$ are independent.

Using the obtained sample, estimate: a) linear correlation $\rho[Z^{(1)}, Z^{(2)}]$; b) Spearman's rank correlation $\rho_S[Z^{(1)}, Z^{(2)}]$; c) Kendall's tau $\rho_\tau[Z^{(1)}, Z^{(2)}]$. Perform these calculations for several values of the copula parameter ρ in the range $[-1; 1]$.

7.5 (⋆⋆) Simulate 10,000 realisations of three risks (X_1, X_2, X_3), whose marginal distributions are $X_i \sim \mathcal{LN}(0, 1)$ and dependence structure is Gaussian copula $C^{\boldsymbol{\Sigma}}(\cdot)$. Assume that all non-diagonal coefficients of the copula correlation matrix $\boldsymbol{\Sigma}$ are the same and equal to $\rho = 0.5$. Using the obtained sample, estimate: (a) linear correlations $\rho[X_i, X_j]$; (b) Spearman's rank correlation $\rho_S[X_i, X_j]$; (c) Kendall's tau $\rho_\tau[X_i, X_j]$. Compare the estimated correlations with each other and with $\rho = 0.5$. Explain the observed differences. Are these due to numerical error (due to the finite number of simulations) only?

---

[4] Problem difficulty is indicated by asterisks: (⋆) – low; (⋆⋆) – medium, (⋆⋆⋆) – high.

7.6 (⋆⋆) Simulate 10,000 realisations of three risks (X_1, X_2, X_3), whose marginal distributions are $X_i \sim \mathcal{LN}(0,1)$ and dependence structure is t-copula $C_\nu^{\mathbf{\Sigma}}(\cdot)$. Assume that all non-diagonal coefficients of the copula correlation matrix $\mathbf{\Sigma}$ are the same and equal to $\rho = 0.5$ and degrees-of-freedom parameter $\nu = 2$. Using the obtained sample, estimate: (a) linear correlations $\rho[X_i, X_j]$; (b) Spearman's rank correlation $\rho_S[X_i, X_j]$; (c) Kendall's tau $\rho_\tau[X_i, X_j]$. Compare the estimated correlations with each other and with $\rho = 0.5$. Explain the observed differences. Are these due to numerical error (due to the finite number of simulations) only? Compare the results with corresponding estimates from Problem 7.5.

7.7 (⋆⋆) Suppose that the dependence structure of three risks (X_1, X_2, X_3) is the Gaussian copula $C^{\mathbf{\Sigma}}(\cdot)$ and margins are $X_i \sim \mathcal{LN}(0,1)$. Suppose that Spearman's rank correlation is $\rho_S[X_i, X_j] = 0.5$ for all $i \neq j$. Find the Gaussian copula correlation matrix $\mathbf{\Sigma}$ and simulate 10,000 realisations of these risks. Using simulated samples, estimate $\rho_S[X_i, X_j]$ and compare with the true value $\rho_S[X_i, X_j] = 0.5$.

7.8 (⋆⋆) Suppose that the dependence structure of two risks (X_1, X_2) is the t-copula $C_\nu^{\mathbf{\Sigma}}(\cdot)$ with $\nu = 2$, and margins are $X_i \sim \mathcal{LN}(0,1)$. Suppose that Kendall's rank correlation is $\rho_\tau[X_1, X_2] = 0.5$. Find the t-copula correlation matrix $\mathbf{\Sigma}$ and simulate 10,000 realisations of these risks. Using simulated samples, estimate $\rho_\tau[X_1, X_2]$ and compare with the true value $\rho_\tau[X_1, X_2] = 0.5$.

7.9 (⋆⋆) Simulate 10,000 realisations of two risks (X_1, X_2) whose margins are $X_i \sim \mathcal{LN}(0,1)$ and the dependence structure is Clayton copula with the parameter $\rho = 5$. Using the simulated samples, estimate: (a) linear correlations $\rho[X_i, X_j]$; (b) Spearman's rank correlation $\rho_S[X_i, X_j]$; (c) Kendall's tau $\rho_\tau[X_i, X_j]$; (d) the lower and upper tail dependencies.

7.10 (⋆⋆) Simulate 10,000 realisations of two risks (X_1, X_2) whose margins are $X_i \sim \mathcal{LN}(0,1)$ and the dependence structure is Gumbel copula with the parameter $\rho = 5$. Using the simulated samples, estimate: (a) linear correlations $\rho[X_i, X_j]$; (b) Spearman's rank correlation $\rho_S[X_i, X_j]$; (c) Kendall's tau $\rho_\tau[X_i, X_j]$; (d) the lower and upper tail dependencies.

7.11 (⋆ ⋆ ⋆) Simulate $T = 200$ independent realisations of two risks $(X^{(1)}, X^{(2)})$ whose margins are $X^{(i)} \sim \mathcal{LN}(0,1)$ and the dependence structure is Gaussian copula with the correlation parameter $\rho = 0.5$. Assume now that ρ is unknown while parameters of the marginal distributions are known. Using the simulated samples $(x_t^{(1)}, x_t^{(2)})$, $t = 1, \ldots, T$ as the observed data, estimate parameter ρ and its uncertainty utilising the posterior distribution obtained from the MCMC slice sampler algorithm; see Sect. 2.11.5. Assume that the prior for ρ is $\mathcal{U}(-1,1)$. Repeat estimation using random walk Metropolis-Hastings algorithm; see Sect. 2.11.3.

7.12 (⋆ ⋆ ⋆) Simulate $T = 200$ realisations of two risks $(X^{(1)}, X^{(2)})$ whose margins are $X^{(i)} \sim \mathcal{LN}(\mu_i, \sigma_i)$ with $\mu_i = 0$ and $\sigma_i = 1$; and the dependence structure is Gaussian copula with the correlation parameter $\rho = 0.5$. Assume now that ρ, μ_i and σ_i are unknown. Using the simulated samples $(x_t^{(1)}, x_t^{(2)})$, $t = 1, \ldots, T$ as the data, estimate parameters ρ, μ_i, σ_i and their uncertainties utilising posterior distribution obtained from the random walk Metropolis-Hastings algorithm; see Sect. 2.11.3. Assume constant priors. Estimate predictive distribution of $Z_{T+1} = X^{(1)}_{T+1} + X^{(2)}_{T+1}$ and its 0.999 quantile. Compare the estimated quantile with the true value of the 0.999 quantile.

Appendix A
List of Distributions

Here we list common statistical distributions used throughout the book. The often used indicator symbol $1_{\{.\}}$ and gamma function $\Gamma(\alpha)$ are defined as follows.

Definition A.1 The indicator symbol is defined as

$$1_{\{.\}} = \begin{cases} 1, & \text{if condition in } \{.\} \text{ is true}, \\ 0, & \text{otherwise}. \end{cases} \tag{A.1}$$

Definition A.2 The standard gamma function is defined as

$$\Gamma(\alpha) = \int_0^\infty t^{\alpha-1} e^{-t} dt, \ \alpha > 0. \tag{A.2}$$

A.1 Discrete Distributions

A.1.1 Poisson Distribution, $Poisson(\lambda)$

A Poisson distribution function is denoted as $Poisson(\lambda)$. The random variable N has a Poisson distribution, denoted $N \sim Poisson(\lambda)$, if its probability mass function is

$$p(k) = \Pr[N = k] = \frac{\lambda^k}{k!} e^{-\lambda}, \ \lambda > 0 \tag{A.3}$$

for all $k \in \{0, 1, 2, \ldots\}$. Expectation, variance and variational coefficient of a random variable $N \sim Poisson(\lambda)$ are

$$\mathrm{E}[N] = \lambda, \ \mathrm{Var}[N] = \lambda, \ \mathrm{Vco}[N] = \frac{1}{\sqrt{\lambda}}. \tag{A.4}$$

P. Shevchenko, *Modelling Operational Risk Using Bayesian Inference*,
DOI 10.1007/978-3-642-15923-7,

A.1.2 Binomial Distribution, $Bin(n, p)$

The binomial distribution function is denoted as $Bin(n, p)$. The random variable N has a binomial distribution, denoted $N \sim Bin(n, p)$, if its probability mass function is

$$p(k) = \Pr[N = k] = \binom{n}{k} p^k (1-p)^{n-k}, \ p \in (0,1), \ n \in 1, 2, \ldots \tag{A.5}$$

for all $k \in \{0, 1, 2, \ldots, n\}$. Expectation, variance and variational coefficient of a random variable $N \sim Bin(n, p)$ are

$$\mathrm{E}[N] = np, \ \mathrm{Var}[N] = np(1-p), \ \mathrm{Vco}[N] = \sqrt{\frac{1-p}{np}}. \tag{A.6}$$

Remark A.1 N is the number of successes in n independent trials, where p is the probability of a success in each trial.

A.1.3 Negative Binomial Distribution, $NegBin(r, p)$

A negative binomial distribution function is denoted as $NegBin(r, p)$. The random variable N has a negative binomial distribution, denoted $N \sim NegBin(r, p)$, if its probability mass function is

$$p(k) = \Pr[N = k] = \binom{r+k-1}{k} p^r (1-p)^k, \ p \in (0,1), \ r \in (0, \infty) \tag{A.7}$$

for all $k \in \{0, 1, 2, \ldots\}$. Here, the generalised binomial coefficient is

$$\binom{r+k-1}{k} = \frac{\Gamma(k+r)}{k!\Gamma(r)}, \tag{A.8}$$

where $\Gamma(r)$ is the gamma function.

Expectation, variance and variational coefficient of a random variable $N \sim NegBin(r, p)$ are

$$\mathrm{E}[N] = \frac{r(1-p)}{p}, \ \mathrm{Var}[N] = \frac{r(1-p)}{p^2}, \ \mathrm{Vco}[N] = \frac{1}{\sqrt{r(1-p)}}. \tag{A.9}$$

Remark A.2 If r is a positive integer, N is the number of failures in a sequence of independent trials until r successes, where p is the probability of a success in each trial.

A.2 Continuous Distributions

A.2.1 Uniform Distribution, $\mathcal{U}(a, b)$

A uniform distribution function is denoted as $\mathcal{U}(a, b)$. The random variable X has a uniform distribution, denoted $X \sim \mathcal{U}(a, b)$, if its probability density function is

$$f(x) = \frac{1}{b-a}, \quad a < b \tag{A.10}$$

for $x \in [a, b]$. Expectation, variance and variational coefficient of a random variable $X \sim \mathcal{U}(a, b)$ are

$$\mathrm{E}[X] = \frac{a+b}{2}, \ \mathrm{Var}[X] = \frac{(b-a)^2}{12}, \ \mathrm{Vco}[X] = \frac{b-a}{\sqrt{3}(a+b)}. \tag{A.11}$$

A.2.2 Normal (Gaussian) Distribution, $\mathcal{N}(\mu, \sigma)$

A normal (Gaussian) distribution function is denoted as $\mathcal{N}(\mu, \sigma)$. The random variable X has a normal distribution, denoted $X \sim \mathcal{N}(\mu, \sigma)$, if its probability density function is

$$f(x) = \frac{1}{\sqrt{2\pi\sigma^2}} \exp\left(-\frac{(x-\mu)^2}{2\sigma^2}\right), \quad \sigma^2 > 0, \ \mu \in \mathbb{R} \tag{A.12}$$

for all $x \in \mathbb{R}$. Expectation, variance and variational coefficient of a random variable $X \sim \mathcal{N}(\mu, \sigma)$ are

$$\mathrm{E}[X] = \mu, \ \mathrm{Var}[X] = \sigma^2, \ \mathrm{Vco}[X] = \sigma/\mu. \tag{A.13}$$

A.2.3 Lognormal Distribution, $\mathcal{LN}(\mu, \sigma)$

A lognormal distribution function is denoted as $\mathcal{LN}(\mu, \sigma)$. The random variable X has a lognormal distribution, denoted $X \sim \mathcal{LN}(\mu, \sigma)$, if its probability density function is

$$f(x) = \frac{1}{x\sqrt{2\pi\sigma^2}} \exp\left(-\frac{(\ln(x)-\mu)^2}{2\sigma^2}\right), \quad \sigma^2 > 0, \ \mu \in \mathbb{R} \tag{A.14}$$

for $x > 0$. Expectation, variance and variational coefficient of a random variable $X \sim \mathcal{LN}(\mu, \sigma)$ are

$$\mathrm{E}[X] = e^{\mu+\frac{1}{2}\sigma^2}, \ \mathrm{Var}[X] = e^{2\mu+\sigma^2}(e^{\sigma^2} - 1), \ \mathrm{Vco}[X] = \sqrt{e^{\sigma^2} - 1}. \tag{A.15}$$

A.2.4 *t* Distribution, $\mathcal{T}(\nu, \mu, \sigma^2)$

A *t* distribution function is denoted as $\mathcal{T}(\nu, \mu, \sigma^2)$. The random variable X has a *t* distribution, denoted $X \sim \mathcal{T}(\nu, \mu, \sigma^2)$, if its probability density function is

$$f(x) = \frac{\Gamma((\nu+1)/2)}{\Gamma(\nu/2)} \frac{1}{\sqrt{\nu\pi}} \left(1 + \frac{(x-\mu)^2}{\nu\sigma^2}\right)^{-(\nu+1)/2} \tag{A.16}$$

for $\sigma^2 > 0$, $\mu \in \mathbb{R}$, $\nu = 1, 2, \ldots$ and all $x \in \mathbb{R}$. Expectation, variance and variational coefficient of a random variable $X \sim \mathcal{T}(\nu, \mu, \sigma^2)$ are

$$\begin{aligned} \mathrm{E}[X] &= \mu \text{ if } \nu > 1, \\ \mathrm{Var}[X] &= \sigma^2 \frac{\nu}{\nu-2} \text{ if } \nu > 2, \\ \mathrm{Vco}[X] &= \frac{\sigma}{\mu}\sqrt{\frac{\nu}{\nu-2}} \text{ if } \nu > 2. \end{aligned} \tag{A.17}$$

A.2.5 Gamma Distribution, *Gamma*(α, β)

A gamma distribution function is denoted as $Gamma(\alpha, \beta)$. The random variable X has a gamma distribution, denoted as $X \sim Gamma(\alpha, \beta)$, if its probability density function is

$$f(x) = \frac{x^{\alpha-1}}{\Gamma(\alpha)\beta^\alpha} \exp(-x/\beta), \ \alpha > 0, \ \beta > 0 \tag{A.18}$$

for $x > 0$. Expectation, variance and variational coefficient of a random variable $X \sim Gamma(\alpha, \beta)$ are

$$\mathrm{E}[X] = \alpha\beta, \ \mathrm{Var}[X] = \alpha\beta^2, \ \mathrm{Vco}[X] = 1/\sqrt{\alpha}. \tag{A.19}$$

A.2.6 Weibull Distribution, *Weibull*(α, β)

A Weibull distribution function is denoted as $Weibull(\alpha, \beta)$. The random variable X has a Weibull distribution, denoted as $X \sim Weibull(\alpha, \beta)$, if its probability density function is

$$f(x) = \frac{\alpha}{\beta^\alpha} x^{\alpha-1} \exp(-(x/\beta)^\alpha), \ \alpha > 0, \ \beta > 0 \tag{A.20}$$

for $x > 0$. The corresponding distribution function is

$$F(x) = 1 - \exp\left(-(x/\beta)^{\alpha}\right), \ \alpha > 0, \ \beta > 0. \tag{A.21}$$

Expectation and variance of a random variable $X \sim Weibull(\alpha, \beta)$ are

$$\mathrm{E}[X] = \beta\Gamma(1 + 1/\alpha), \ \mathrm{Var}[X] = \beta^2\left(\Gamma(1 + 2/\alpha) - (\Gamma(1 + 1/\alpha))^2\right).$$

A.2.7 Pareto Distribution (One-Parameter), Pareto(ξ, x_0)

A one-parameter Pareto distribution function is denoted as $Pareto(\xi, x_0)$. The random variable X has a Pareto distribution, denoted as $X \sim Pareto(\xi, x_0)$, if its distribution function is

$$F(x) = 1 - \left(\frac{x}{x_0}\right)^{-\xi}, \quad x \geq x_0, \tag{A.22}$$

where $x_0 > 0$ and $\xi > 0$. The support starts at x_0, which is typically known and not considered as a parameter. Therefore the distribution is referred to as a single parameter Pareto. The corresponding probability density function is

$$f(x) = \frac{\xi}{x_0}\left(\frac{x}{x_0}\right)^{-\xi-1}. \tag{A.23}$$

Expectation, variance and variational coefficient of $X \sim Pareto(\xi, x_0)$ are

$$\begin{aligned}
\mathrm{E}[X] &= x_0\frac{\xi}{\xi - 1} \quad \text{if } \xi > 1, \\
\mathrm{Var}[X^2] &= x_0^2\frac{\xi}{(\xi - 1)^2(\xi - 2)} \quad \text{if } \xi > 2, \\
\mathrm{Vco}[X] &= \frac{1}{\sqrt{\xi(\xi - 2)}} \quad \text{if } \xi > 2.
\end{aligned}$$

A.2.8 Pareto Distribution (Two-Parameter), Pareto$_2$(α, β)

A two-parameter Pareto distribution function is denoted as $Pareto_2(\alpha, \beta)$. The random variable X has a Pareto distribution, denoted as $X \sim Pareto_2(\alpha, \beta)$, if its distribution function is

$$F(x) = 1 - \left(1 + \frac{x}{\beta}\right)^{-\alpha}, \quad x \geq 0, \tag{A.24}$$

where $\alpha > 0$ and $\beta > 0$. The corresponding probability density function is

$$f(x) = \frac{\alpha\beta^{\alpha}}{(x+\beta)^{\alpha+1}}. \tag{A.25}$$

The moments of a random variable $X \sim Pareto_2(\alpha, \beta)$ are

$$\mathrm{E}[X^k] = \frac{\beta^k k!}{\prod_{i=1}^{k}(\alpha - i)}; \quad \alpha > k.$$

A.2.9 Generalised Pareto Distribution, GPD(ξ, β)

A GPD distribution function is denoted as $GPD(\xi, \beta)$. The random variable X has a GPD distribution, denoted as $X \sim GPD(\xi, \beta)$, if its distribution function is

$$H_{\xi,\beta}(x) = \begin{cases} 1 - (1 + \xi x/\beta)^{-1/\xi}, \ \xi \neq 0, \\ 1 - \exp(-x/\beta), \ \xi = 0, \end{cases} \tag{A.26}$$

where $x \geq 0$ when $\xi \geq 0$ and $0 \leq x \leq -\beta/\xi$ when $\xi < 0$. The corresponding probability density function is

$$h(x) = \begin{cases} \frac{1}{\beta}(1 + \xi x/\beta)^{-\frac{1}{\xi}-1}, \ \xi \neq 0, \\ \frac{1}{\beta}\exp(-x/\beta), \ \xi = 0. \end{cases} \tag{A.27}$$

Expectation, variance and variational coefficient of $X \sim GPD(\xi, \beta)$, $\xi \geq 0$, are

$$\begin{aligned} \mathrm{E}[X^n] &= \frac{\beta^n n!}{\prod_{k=1}^{n}(1 - k\xi)}, \ \xi < \frac{1}{n}; \ \mathrm{E}[X] = \frac{\beta}{1-\xi}, \ \xi < 1; \\ \mathrm{Var}[X^2] &= \frac{\beta^2}{(1-\xi)^2(1-2\xi)}, \ \mathrm{Vco}[X] = \frac{1}{\sqrt{1-2\xi}}, \ \xi < \frac{1}{2}. \end{aligned} \tag{A.28}$$

A.2.10 Beta Distribution, Beta(α, β)

A beta distribution function is denoted as $Beta(\alpha, \beta)$. The random variable X has a beta distribution, denoted as $X \sim Beta(\alpha, \beta)$, if its probability density function is

$$f(x) = \frac{\Gamma(\alpha+\beta)}{\Gamma(\alpha)\Gamma(\beta)} x^{\alpha-1}(1-x)^{\beta-1}, \quad 0 \leq x \leq 1, \tag{A.29}$$

for $\alpha > 0$ and $\beta > 0$. Expectation, variance and variational coefficient of a random variable $X \sim Beta(\alpha, \beta)$ are

$$\mathrm{E}[X] = \frac{\alpha}{\alpha+\beta}, \quad \mathrm{Var}[X] = \frac{\alpha\beta}{(\alpha+\beta)^2(1+\alpha+\beta)}, \quad \mathrm{Vco}[X] = \sqrt{\frac{\beta}{\alpha(1+\alpha+\beta)}}.$$

A.2.11 Generalised Inverse Gaussian Distribution, GIG(ω, ϕ, ν)

A GIG distribution function is denoted as $GIG(\omega, \phi, \nu)$. The random variable X has a GIG distribution, denoted as $X \sim GIG(\omega, \phi, \nu)$, if its probability density function is

$$f(x) = \frac{(\omega/\phi)^{(\nu+1)/2}}{2K_{\nu+1}\left(2\sqrt{\omega\phi}\right)} x^{\nu} e^{-x\omega - x^{-1}\phi}, \quad x > 0, \tag{A.30}$$

where $\phi > 0, \omega \geq 0$ if $\nu < -1$; $\phi > 0, \omega > 0$ if $\nu = -1$; $\phi \geq 0, \omega > 0$ if $\nu > -1$; and

$$K_{\nu+1}(z) = \frac{1}{2}\int_0^{\infty} u^{\nu} e^{-z(u+1/u)/2} du.$$

$K_\nu(z)$ is called a modified Bessel function of the third kind; see for instance Abramowitz and Stegun ([3], p. 375).

The moments of a random variable $X \sim GIG(\omega, \phi, \nu)$ are not available in a closed form through elementary functions but can be expressed in terms of Bessel functions:

$$\mathrm{E}[X^\alpha] = \left(\frac{\phi}{\omega}\right)^{\alpha/2} \frac{K_{\nu+1+\alpha}\left(2\sqrt{\omega\phi}\right)}{K_{\nu+1}\left(2\sqrt{\omega\phi}\right)}, \quad \alpha \geq 1, \ \phi > 0, \ \omega > 0.$$

Often, using notation $R_\nu(z) = K_{\nu+1}(z)/K_\nu(z)$, it is written as

$$\mathrm{E}[X^\alpha] = \left(\frac{\phi}{\omega}\right)^{\alpha/2} \prod_{k=1}^{\alpha} R_{\nu+k}\left(2\sqrt{\omega\phi}\right), \quad \alpha = 1, 2, \ldots$$

The mode is easily calculated from $\frac{\partial}{\partial x} x^{\nu} e^{-(\omega x + \phi/x)} = 0$ as

$$\mathrm{mode}[X] = \frac{1}{2\omega}\left(\nu + \sqrt{\nu^2 + 4\omega\phi}\right),$$

that differs only slightly from the expected value for large ν, i.e.

$$\mathrm{mode}[X] \to \mathrm{E}[X] \qquad \text{for} \quad \nu \to \infty.$$

A.2.12 d-variate Normal Distribution, $\mathcal{N}_d(\boldsymbol{\mu}, \boldsymbol{\Sigma})$

A d-variate normal distribution function is denoted as $\mathcal{N}_d(\boldsymbol{\mu}, \boldsymbol{\Sigma})$, where $\boldsymbol{\mu} = (\mu_1, \dots \mu_d)' \in \mathbb{R}^d$ and $\boldsymbol{\Sigma}$ is a positive definite matrix ($d \times d$). The corresponding probability density function is

$$f(\mathbf{x}) = \frac{1}{(2\pi)^{d/2}\sqrt{\det \boldsymbol{\Sigma}}} \exp\left(-\frac{1}{2}(\mathbf{x} - \boldsymbol{\mu})'\boldsymbol{\Sigma}^{-1}(\mathbf{x} - \boldsymbol{\mu})\right), \quad \mathbf{x} \in \mathbb{R}^d, \tag{A.31}$$

where $\boldsymbol{\Sigma}^{-1}$ is the inverse of the matrix $\boldsymbol{\Sigma}$. Expectations and covariances of a random vector $\mathbf{X} = (X_1, \dots, X_d)' \sim \mathcal{N}_d(\boldsymbol{\mu}, \boldsymbol{\Sigma})$ are

$$\mathrm{E}[X_i] = \mu_i, \ \mathrm{Cov}[X_i, X_j] = \Sigma_{i,j}, \ i, j = 1, \dots, d. \tag{A.32}$$

A.2.13 d-variate t-Distribution, $\mathcal{T}_d(\nu, \boldsymbol{\mu}, \boldsymbol{\Sigma})$

A d-variate t-distribution function with ν degrees of freedom is denoted as $\mathcal{T}_d(\nu, \boldsymbol{\mu}, \boldsymbol{\Sigma})$, where $\nu > 0$, $\boldsymbol{\mu} = (\mu_1, \dots \mu_d)' \in \mathbb{R}^d$ is a location vector and $\boldsymbol{\Sigma}$ is a positive definite matrix ($d \times d$). The corresponding probability density function is

$$f(\mathbf{x}) = \frac{\Gamma\left(\frac{\nu+d}{2}\right)}{(\nu\pi)^{d/2}\Gamma\left(\frac{\nu}{2}\right)\sqrt{\det \boldsymbol{\Sigma}}} \left(1 + \frac{(\mathbf{x} - \boldsymbol{\mu})'\boldsymbol{\Sigma}^{-1}(\mathbf{x} - \boldsymbol{\mu})}{\nu}\right)^{-\frac{\nu+d}{2}}, \tag{A.33}$$

where $\mathbf{x} \in \mathbb{R}^d$ and $\boldsymbol{\Sigma}^{-1}$ is the inverse of the matrix $\boldsymbol{\Sigma}$. Expectations and covariances of a random vector $\mathbf{X} = (X_1, \dots, X_d)' \sim \mathcal{T}_d(\nu, \boldsymbol{\mu}, \boldsymbol{\Sigma})$ are

$$\begin{aligned} \mathrm{E}[X_i] &= \mu_i, \ \text{if } \nu > 1, \ i = 1, \dots, d; \\ \mathrm{Cov}[X_i, X_j] &= \nu\Sigma_{i,j}/(\nu - 2), \ \text{if } \nu > 2, \ i, j = 1, \dots, d. \end{aligned} \tag{A.34}$$

Appendix B
Selected Simulation Algorithms

B.1 Simulation from GIG Distribution

To generate realisations of a random variable $X \sim \text{GIG}(\omega, \phi, \nu)$ with $\omega, \phi > 0$, a special algorithm is required because we cannot invert the distribution function in closed form. The following algorithm can be found in Dagpunar [67]:

Algorithm B.1 (Simulation from GIG distribution)

1. $\alpha = \sqrt{\omega/\phi}$; $\beta = 2\sqrt{\omega\phi}$,
   $m = \frac{1}{\beta}\left(\nu + \sqrt{\nu^2 + \beta^2}\right)$,
   $g(y) = \frac{1}{2}\beta y^3 - y^2\left(\frac{1}{2}\beta m + \nu + 2\right) + y\left(\nu m - \frac{\beta}{2}\right) + \frac{1}{2}\beta m.$
2. Set $y_0 = m$,
   While $g(y_0) \leq 0$ do $y_0 = 2y_0$,
   y_+: root of g in the interval (m, y_0),
   y_-: root of g in the interval $(0, m)$.
3. $a = (y_+ - m)\left(\frac{y_+}{m}\right)^{\nu/2}\exp\left(-\frac{\beta}{4}\left(y_+ + \frac{1}{y_+} - m - \frac{1}{m}\right)\right)$,
   $b = (y_- - m)\left(\frac{y_-}{m}\right)^{\nu/2}\exp\left(-\frac{\beta}{4}\left(y_- + \frac{1}{y_-} - m - \frac{1}{m}\right)\right)$,
   $c = -\frac{\beta}{4}\left(m + \frac{1}{m}\right) + \frac{\nu}{2}\ln(m)$.
4. Repeat $U, V \sim \mathcal{U}(0,1)$, $Y = m + a\frac{U}{V} + b\frac{1-V}{U}$,
   until $Y > 0$ and $-\ln U \geq -\frac{\nu}{2}\ln Y + \frac{1}{4}\beta\left(Y + \frac{1}{Y}\right) + c$,
   Then $X = \frac{Y}{\alpha}$ is $\text{GIG}(\omega, \phi, \nu)$.

To generate a sequence of n realisations from a GIG random variable, step 4 is repeated n times.

B.2 Simulation from α-stable Distribution

To generate realisations of a random variable $X \sim \alpha\text{Stable}(\alpha, \beta, \sigma, \mu)$, defined by (6.56), a special algorithm is required because the density of α-stable distribution is not available in closed form. An elegant and efficient solution was proposed in Chambers, Mallows and Stuck [50]; also see Nolan [176].

Algorithm B.2 (Simulation from α-stable distribution)

1. Simulate W from the exponential distribution with mean = 1.
2. Simulate U from $\mathcal{U}\left(-\frac{\pi}{2}, \frac{\pi}{2}\right)$.
3. Calculate

$$Z = \begin{cases} S_{\alpha,\beta} \frac{\sin(\alpha(U+B_{\alpha,\beta}))}{(\cos U)^{1/\alpha}} \left(\frac{\cos(U-\alpha(U+B_{\alpha,\beta}))}{W}\right)^{-1+\frac{1}{\alpha}}, & \alpha \neq 1, \\ \frac{2}{\pi}\left(\left(\frac{\pi}{2} + \beta U\right)\tan U - \beta \ln\left(\frac{\pi W \cos U}{\pi + 2\beta U}\right)\right), & \alpha = 1, \end{cases}$$

   where

$$S_{\alpha,\beta} = (1 + \beta^2 \tan^2(\pi\alpha/2))^{\frac{1}{2\alpha}},$$

$$B_{\alpha,\beta} = \frac{1}{\alpha} \arctan(\beta \tan(\pi\alpha/2)).$$

   The obtained Z is a sample from $\alpha\text{Stable}(\alpha, \beta, 1, 0)$.
4. Then,

$$X = \begin{cases} \mu + \sigma Z, & \alpha \neq 1, \\ \mu + \sigma Z + \frac{2}{\pi}\beta\sigma \ln \sigma, & \alpha = 1, \end{cases}$$

   is a sample from $\alpha\text{Stable}(\alpha, \beta, \sigma, \mu)$.

Note that there are different parameterisations of the α-stable distribution. The algorithm above is for representation (6.56).

Solutions for Selected Problems

Problems of Chapter 2

2.1 The likelihood function for independent data $\mathbf{N} = \{N_1, N_2, \ldots, N_M\}$ from $Poisson(\lambda)$ is

$$\ell_{\mathbf{n}}(\lambda) = \prod_{i=1}^{M} e^{-\lambda} \frac{\lambda^{n_i}}{n_i!},$$

$$\ln \ell_{\mathbf{n}}(\lambda) = -\lambda M + \ln \lambda \sum_{i=1}^{M} n_i - \sum_{i=1}^{M} \ln(n_i!).$$

The MLE $\widehat{\Lambda}$ maximising the log-likelihood function $\ln \ell_{\mathbf{N}}(\lambda)$ is

$$\widehat{\Lambda} = \frac{1}{M} \sum_{i=1}^{M} N_i.$$

Using the properties of the Poisson distribution, $\mathrm{E}[N_i] = \mathrm{Var}[N_i] = \lambda$, it is easy to get

$$\mathrm{E}[\widehat{\Lambda}] = \frac{1}{M} \sum_{i=1}^{M} \mathrm{E}[N_i] = \lambda;$$

$$\mathrm{Var}[\widehat{\Lambda}] = \frac{1}{M^2} \sum_{i=1}^{M} \mathrm{Var}[N_i] = \frac{\lambda}{M}.$$

To estimate the variance of $\widehat{\Lambda}$ using a normal approximation, find the information matrix

$$\mathrm{I}(\lambda) = -\frac{1}{M}\mathrm{E}\left[\frac{\partial^2 \ln \ell_{\mathbf{N}}(\lambda)}{\partial \lambda^2}\right] = \frac{1}{M\lambda^2}\mathrm{E}\left[\sum_{i=1}^{M} N_i\right] = \frac{1}{\lambda}.$$

Thus, using asymptotic normal distribution approximation,

$$\mathrm{Var}[\widehat{\Lambda}] \approx \mathrm{I}^{-1}(\lambda)/M = \lambda/M.$$

In both cases the variance depends on unknown true parameter λ that can be estimated, for a given realisation $\mathbf{n}$, as $\widehat{\lambda}$.

2.4 Consider

$$L(\mathbf{u}) = u_1 L_1 + \cdots + u_J L_J,$$

where $\mathbf{u} \in \mathbb{R}^J$ and set

$$\phi_{\mathbf{u}}(t) = \varrho[tL(\mathbf{u})]], \quad t > 0.$$

Then using homogeneity property $\varrho[tL] = t\varrho[L]$,

$$\frac{d\phi_{\mathbf{u}}(t)}{dt} = \varrho[L(u)].$$

From another side

$$\frac{d\phi_{\mathbf{u}}(t)}{dt} = \sum_{j=1}^{J} \left.\frac{\varrho[L(\mathbf{x})]}{\partial x_j}\right|_{\mathbf{x}=t\mathbf{u}} u_j = \sum_{j=1}^{J} \frac{\varrho[L(\mathbf{u})]}{\partial u_j} u_j,$$

where to get the last equality we used homogeneity property. Thus

$$\varrho[L(\mathbf{1})] = \sum_{j=1}^{J} \left.\frac{\varrho[L_1 + \cdots + L_j + hL_j]}{\partial h}\right|_{h=0}$$

completes the proof.

2.5 The sum of risks is gamma distributed:

$$Z_1 + Z_2 + Z_3 \sim Gamma(\alpha_1 + \alpha_2 + \alpha_3, \beta).$$

Thus $\mathrm{VaR}_{0.999}[Z_i] = F_G^{-1}(0.999|\alpha_i, \beta)$ and

$$\mathrm{VaR}_{0.999}[Z_1 + Z_2 + Z_3] = F_G^{-1}(0.999|\alpha_1 + \alpha_2 + \alpha_3, \beta),$$

where $F_G^{-1}(\cdot|\alpha, \beta)$ is the inverse of the $Gamma(\alpha, \beta)$. Using, for example, MS Excel spreadsheet function GAMMAINV($\cdot$), find

$$\mathrm{VaR}_{0.999}[Z_1] \approx 5.414, \quad \mathrm{VaR}_{0.999}[Z_2] \approx 6.908,$$
$$\mathrm{VaR}_{0.999}[Z_3] \approx 8.133, \quad \mathrm{VaR}_{0.999}[Z_1 + Z_2 + Z_3] \approx 11.229.$$

The sum of VaRs is $\text{VaR}_{0.999}[Z_1] + \text{VaR}_{0.999}[Z_2] + \text{VaR}_{0.999}[Z_3] \approx 20.455$ and thus the diversification is $\approx 45\%$.

Problems of Chapter 3

3.1 By definition of the expected shortfall we have

$$\begin{aligned} \text{E}[Z|Z > L] &= \frac{1}{1 - H(L)} \int_{L}^{\infty} z h(z) dz \\ &= \frac{\text{E}[Z]}{1 - H(L)} - \frac{1}{1 - H(L)} \int_{0}^{L} z h(z) dz. \end{aligned}$$

Substituting $h(z)$ calculated via characteristic function (3.11) and changing variable $x = t \times L$, we obtain

$$\begin{aligned} \int_0^L z h(z) dz &= \frac{2}{\pi} \int_0^L z \int_0^{\infty} \text{Re}[\chi(t)] \cos(tz) dt dz \\ &= \frac{2L}{\pi} \int_0^{\infty} \text{Re}[\chi\,(x/L)] \left[\frac{\sin(x)}{x} - \frac{1 - \cos(x)}{x^2} \right] dx. \end{aligned}$$

Recognizing that the term involving $\sin(x)/x$ corresponds to $H(L)$, we obtain

$$\text{E}[Z|Z > L] = \frac{1}{1 - H(L)} \left[\text{E}[Z] - H(L)L + \frac{2L}{\pi} \int_0^{\infty} \text{Re}[\chi(x/L)] \frac{1 - \cos x}{x^2} dx \right].$$

Problems of Chapter 4

4.1 The linear estimator $\widehat{\theta}_{tot} = w_1 \widehat{\theta}_1 + \cdots + w_K \widehat{\theta}_K$ is unbiased, i.e. $\text{E}[\widehat{\theta}_{tot}] = \theta$, if $w_1 + \cdots + w_K = 1$ because $\text{E}[\widehat{\theta}_k] = \theta$. Minimisation of the variance

$$\text{Var}[\widehat{\theta}_{tot}] = w_1^2 \sigma_1^2 + \cdots + w_K^2 \sigma_K^2$$

under the constraint $w_1 + \cdots + w_K$ is equivalent to unconstrained minimisation of the

$$\Psi = \text{Var}[\widehat{\theta}_{tot}] - \lambda(w_1 + \cdots + w_K),$$

which is a well-known method of Lagrange multipliers. Optimisation of the above requires solution of the following equations:

$$\frac{\partial \Psi}{\partial w_i} = 2w_i\sigma_i^2 - \lambda = 0, \quad i = 1, \ldots, K;$$
$$\frac{\partial \Psi}{\partial \lambda} = -(w_1 + \cdots + w_K) = 0.$$

That gives

$$\frac{1}{2}\lambda = \left(\sum_{k=1}^{K}\left(1/\sigma_k^2\right)\right)^{-1}, \quad w_i = \frac{1/\sigma_i^2}{\sum_{k=1}^{K}\left(1/\sigma_k^2\right)}.$$

4.2 Given $\Theta = \theta$, the joint density of the data at $\mathbf{N} = \mathbf{n}$ is

$$f(\mathbf{n}|\theta) \propto \prod_{i=1}^{T} \theta^{n_i}(1-\theta)^{V_i - n_i}.$$

From Bayes's theorem, the posterior density of θ is $\pi(\theta|\mathbf{n}) \propto f(\mathbf{n}|\theta)\pi(\theta)$, where $\pi(\theta)$ is the prior density. Thus

$$\pi(\theta|\mathbf{n}) \propto \theta^{\alpha-1}(1-\theta)^{\beta-1}\prod_{i=1}^{T}\theta^{n_i}(1-\theta)^{V_i-n_i}$$
$$= \theta^{\alpha_T - 1}(1-\theta)^{\beta_T - 1},$$

where

$$\alpha_T = \alpha + \sum_{i=1}^{T} n_i, \quad \beta_T = \beta + \sum_{i=1}^{T} V_i - \sum_{i=1}^{T} n_i.$$

Thus the posterior distribution of Θ is $Beta(\alpha_T, \beta_T)$.

Problems of Chapter 5

5.1 Denote the data above L as $\widetilde{\mathbf{X}} = (\widetilde{X}_1, \widetilde{X}_2, \ldots, \widetilde{X}_k)'$. These random variables are independent with a common density $f(x|\xi)/(1 - F(L|\xi))$, $x \geq L$, where $f(x|\xi)$ is the density of the Pareto distribution $F(x|\xi) = 1 - (x/a)^{-\xi}$, $x \geq a > 0$. Thus the likelihood function for given data above L is

$$\ell_{\widetilde{\mathbf{x}}}(\xi) = \prod_{i=1}^{k} \frac{f(\widetilde{x}_i|\xi)}{1 - F(L|\xi)}.$$

Substituting the Pareto density

$$f(x|\xi) = \frac{\xi}{a}\left(\frac{x}{a}\right)^{-\xi-1}$$

gives

$$\ln \ell_{\widetilde{\mathbf{x}}}(\xi) = K\xi \ln(L/a) + K \ln(\xi/a) - (\xi + 1) \sum_{i=1}^{K} \ln(\widetilde{x}_i/a).$$

Then, solving $\partial \ln \ell_{\widetilde{\mathbf{x}}}(\xi)/\partial\xi = 0$, we obtain

$$\widehat{\xi}^{\mathrm{MLE}} = \left(-\ln(L/a) + \frac{1}{K}\sum_{i=1}^{K} \ln(\widetilde{x}_i/a)\right)^{-1}.$$

Problems of Chapter 6

6.1 The probability generating function of the negative binomial, $NegBin(r, p)$, is $\psi(t) = (1 - (t-1)(1-p)/p)^{-r}$. Then, using formula (6.29), we obtain that the distribution of the maximum loss over one year is

$$F_M(x) = \psi(F(x)) = \left(1 + \frac{1-p}{p}(1 - F(x))\right)^{-r},$$

where $F(x) = 1 - \exp(-x/\beta)$ is the severity distribution. The distribution of the maximum loss over m years is simply

$$(F_M(x))^m = \left(1 + \frac{1-p}{p}(1 - F(x))\right)^{-r\times m}.$$

Problems of Chapter 7

7.1 Consider random variables U_1 and U_2 from the t-copula $C^{(t)}_{\nu,\rho}(u_1, u_2)$. By definition, the lower tail dependence is

$$\lambda_L = \lim_{q\to 0+} \frac{C^{(t)}_{\nu,\rho}(q, q)}{q}.$$

Due to the radial symmetry of the t-copula, the upper tail dependence λ_U is the same as λ_L. Applying L'Hôpital's rule, that is, taking derivatives of the nominator and denominator,

$$\lambda_L = \lim_{q \to 0+} \frac{dC^{(t)}_{\nu,\rho}(q,q)}{dq} = \lim_{q \to 0+} \{\Pr[U_2 \leq q | U_1 = q] + \Pr[U_1 \leq q | U_2 = q]\}.$$

Let $X_1 = F_\nu^{(-1)}(U_1)$ and $X_2 = F_\nu^{(-1)}(U_2)$, where $F_\nu(\cdot)$ is a standard univariate t-distribution with ν degrees of freedom, $\mathcal{T}(\nu, 0, 1)$. Thus $(X_1, X_2)'$ is from a bivariate t-distribution $\mathcal{T}_2(\nu, 0, \mathbf{\Sigma})$, where $\mathbf{\Sigma}$ is a correlation matrix with off-diagonal element ρ. Then, one can calculate the conditional density of X_2 given $X_1 = x_1$:

$$f(x_2|x_1) = \frac{f(x_1, x_2)}{f(x_1)} \propto \left(1 + \frac{\nu+1}{(1-\rho^2)(\nu + x_1^2)} \frac{(x_2 - \rho x_1)^2}{\nu+1}\right)^{-(\nu+2)/2}.$$

This can be recognised as a univariate t distribution $\mathcal{T}(\nu+1, \mu, \sigma^2)$ with the mean $\mu = \rho x_1, \sigma^2 = \frac{(1-\rho^2)(\nu+x_1^2)}{\nu+1}$ and $\nu + 1$ degrees of freedom. Thus

$$\Pr[X_2 \leq x | X_1 = x] = F_{\nu+1}\left(\frac{(x - x\rho)\sqrt{\nu+1}}{\sqrt{(1-\rho^2)(\nu+x^2)}}\right).$$

Finally, using that $\Pr[X_1 \leq x | X_2 = x] = \Pr[X_2 \leq x | X_1 = x]$ and taking limit $x \to -\infty$ we get

$$\lambda = 2F_{\nu+1}\left(-\sqrt{\frac{(\nu+1)(1-\rho)}{1+\rho}}\right).$$

References

1. Abate, J., Whitt, W.: Numerical inversion of Laplace transforms of probability distributions. ORSA Journal of Computing **7**, 36–43 (1992)
2. Abate, J., Whitt, W.: Numerical inversion of probability generating functions. Operations Research Letters **12**, 245–251 (1995)
3. Abramowitz, M., Stegun, I.A.: Handbook of Mathematical Functions. Dover Publications, New York, NY (1965)
4. Acerbi, C., Tasche, D.: On the coherence of expected shortfall. Journal of Banking and Finance **26**, 1487–1503 (2002)
5. Akaike, H.: Information measure and model selection. Bulletin of the International Statistical Institute **50**, 277–290 (1983)
6. Alderweireld, T., Garcia, J., Léonard, L.: A practical operational risk scenario analysis quantification. Risk Magazine **19**(2), 93–95 (2006)
7. Ale, B.J.M., Bellamy, L.J., van der Boom, R., Cooper, J., Cooke, R.M., Goossens, L.H.J., Hale, A.R., Kurowicka, D., Morales, O., Roelen, A.L.C., Spouge, J.: Further development of a causal model for air transport safety (CATS); building the mathematical heart. Reliability Engineering and System Safety **94**(9), 1433–1441 (2009)
8. Allen, L., Bali, T.: Cyclicality in catastrophic and operational risk measurements. Baruch College (2004). Technical report
9. Allen, L., Boudoukh, J., Saunders, A.: Understanding Market, Credit and Operational Risk: The Value-at-Risk Approach. Blackwell Publishing, Oxford (2005)
10. Artzner, P., Delbaen, F., Eber, J.M., Heath, D.: Coherent measures of risk. Mathematical Finance **9**, 203–228 (1999)
11. Atchade, Y., Rosenthal, J.: On adaptive Markov chain Monte Carlo algorithms. Bernoulli **11**(5), 815–828 (2005)
12. Aue, F., Klakbrener, M.: LDA at work: Deutsche Bank's approach to quantify operational risk. The Journal of Operational Risk **1**(4), 49–95 (2006)
13. Basel Committee on Banking Supervision: Operational Risk Management. Bank for International Settlements (September 1998). URL www.bis.org
14. Basel Committee on Banking Supervision: Quantitative Impact Study for Operational Risk: Overview of Individual Loss Data and Lessons Learned. Bank for International Settlements (January 2002). URL www.bis.org/bcbs/qis/qishist.htm
15. Basel Committee on Banking Supervision: The 2002 Loss Data Collection Exercise for Operational Risk: Summary of the Data Collected. Bank for International Settlements (March 2003). URL www.bis.org/bcbs/qis/ldce2002.htm
16. Basel Committee on Banking Supervision: International Convergence of Capital Measurement and Capital Standards: a revised framework. Bank for International Settlements, Basel (June 2004). URL www.bis.org
17. Basel Committee on Banking Supervision: International Convergence of Capital Measurement and Capital Standards: a revised framework. Bank for International Settlements, Basel (June 2006). URL www.bis.org

18. Basel Committee on Banking Supervision: Results from the 2008 Loss Data Collection Exercise for Operational Risk. Bank for International Settlements (July 2009). URL www.bis.org
19. Basel Committee on Banking Supervision: Working Paper on the Regulatory Treatment of Operational Risk. Bank for International Settlements (September 2001). URL www.bis.org
20. Baud, N., Frachot, A., Roncalli, T.: How to avoid over-estimating capital charge for operational risk? OpRisk&Compliance (February 2003)
21. Bayes, T.: An essay towards solving a problem in the doctrine of chances. Philosophical Transactions of the Royal Society of London **53**, 370–418 (1763)
22. Bazzarello, D., Crielaard, B., Piacenza, F., Soprano, A.: Modeling insurance mitigation on operational risk capital. The Journal of Operational Risk **1**(1), 57–65 (2006)
23. Beaumont, M.A., Zhang, W., Balding, D.J.: Approximate Bayesian computation in population genetics. Genetics **162**, 2025–2035 (2002)
24. Bedard, M., Rosenthal, J.S.: Optimal scaling of Metropolis algorithms: heading towards general target distributions. The Canadian Journal of Statistics **36**(4), 483–503 (2008)
25. Bee, M.: Copula-based multivariate models with applications to risk management and insurance. Dipartimento di Economia, Universita degli Studi di Trento (2005). URL www.gloriamundi.org. Working paper
26. Bee, M.: On Maximum Likelihood Estimation of Operational Loss Distributions. Dipartimento di Economia, Universita degli Studi di Trento (2005). Discussion paper No.3
27. Berger, J.O.: Statistical Decision Theory and Bayesian Analysis, 2nd edn. Springer, New York, NY (1985)
28. Bladt, M.: A review of phase-type distributions and their use in risk theory. ASTIN Bulletin **35**(1), 145–167 (2005)
29. Böcker, K., Klüppelberg, C.: Operational VAR: a closed-form approximation. Risk Magazine **12**, 90–93 (2005)
30. Böcker, K., Klüppelberg, C.: Modelling and measuring multivariate operational risk with Lévy copulas. The Journal of Operational Risk **3**(2), 3–27 (2008)
31. Böcker, K., Klüppelberg, C.: Multivariate models for Operational Risk. Quantitative Finance **10**(8), 855–869 (2010)
32. Böcker, K., Sprittulla, J.: Operational VAR: meaningful means. Risk Magazine **12**, 96–98 (2006)
33. Bohman, H.: Numerical inversion of characteristic functions. Scandinavian Actuarial Journal pp. 121–124 (1975)
34. Bookstabber, R.M., McDonald, J.B.: A general distribution for describing security price returns. The Journal of Business **60**(3), 401–424 (1987)
35. Brass, H., Förster, K.J.: On the estimation of linear functionals. Analysis **7**, 237–258 (1987)
36. Brewer, M.J., Aitken, C.G.G., Talbot, M.: A comparison of hybrid strategies for Gibbs sampling in mixed graphical models. Computational Statistics and Data Analysis **21**(3), 343–365 (1996)
37. Brigham, E.O.: The Fast Fourier Transform. Prentice-Hall, Englewood Cliffs, NJ (1974)
38. van den Brink, J.: Operational Risk: The New Challenge for Banks. Pulgrave, London (2002)
39. British Bankers Association: Operational Risk Management Survey (December 1999). URL www.bba.org.uk
40. Buch-Kromann, T.: Comparison of tail performance of the Champernowne transformed kernel density estimator, the generalized Pareto distribution and the g-and-h distribution. The Journal of Operational Risk **4**(2), 43–67 (2009)
41. Buch-Larsen, T., Nielsen, J.P., Guillen, M., Bolance, C.: Kernel density estimation for heavy-tailed distributions using the Champernowne transformation. Statistics **39**(6), 503–518 (2005)
42. Bühlmann, H.: Mathematical Methods in Risk Theory. Springer, New York, NY (1970)
43. Bühlmann, H.: Numerical evaluation of the compound Poisson distribution: recursion or Fast Fourier Transform? Scandinavian Actuarial Journal **2**, 116–126 (1984)
44. Bühlmann, H., Gisler, A.: A Course in Credibility Theory and its Applications. Springer, Berlin (2005)

45. Bühlmann, H., Shevchenko, P.V., Wüthrich, M.V.: A "toy" model for operational risk quantification using credibility theory. The Journal of Operational Risk **2**(1), 3–19 (2007)
46. Bühlmann, H., Straub, E.: Glaubwürdigkeit für schadensätze. Bulletin of the Swiss Association of Actuaries **70**, 111–133 (1970)
47. Burnecki, K., Kukla, G., Taylor, D.: Pricing of catastrophic bonds. In: P. Cízek, W. Härdle, R. Weron (eds.) Statistical Tools for Finance and Insurance. Springer, New York, NY (2005)
48. Carter, M., Van Brunt, B.: The Lebesgue-Stieltjes Integral. A Practical Introduction. Springer, New York, NY (2000)
49. Casella, G., Berger, R.L.: Statistical Inference. Duxbury, Pasific Grove (2002)
50. Chambers, J.M., Mallows, C.L., Stuck, B.W.: A method for simulating stable random variables. Journal of the American Statistical Association **71**, 340–344 (1976)
51. Champernowne, D.G.: The graduation of income distributions. Econometrica **20**, 591–615 (1952)
52. Chavez-Demoulin, V., Embrechts, P., Nešlehová, J.: Quantitative models for operational risk: extremes, dependence and aggregation. Journal of Banking and Finance **30**(9), 2635–2658 (2006)
53. Chernobai, A., Menn, C., Trück, S., Rachev, S.T.: A note on the estimation of the frequency and severity distribution of operational losses. The Mathematical Scientist **30**(2) (2005)
54. Chernobai, A., Rachev, S.T.: Stable modelling of operational risk. In: M.G. Cruz (ed.) Operational Risk Modelling and Analysis. Theory and Practice. Risk Books, London (2004)
55. Chernobai, A.S., Rachev, S.T., Fabozzi, F.J.: Operational Risk: A Guide to Basel II Capital Requirements, Models, and Analysis. Wiley, Hoboken, NJ (2007)
56. Chib, S.: Marginal likelihood from the Gibbs output. Journal of the American Statistical Association **90**(432), 1313–1321 (1995)
57. Clements, A.E., Hurn, A.S., Lindsay, K.A.: Möbius-like mappings and their use in kernel density estimation. Journal of the American Statistical Association **98**, 993–1000 (2003)
58. Clenshaw, C.W., Curtis, A.R.: A method for numerical integration on an automatic computer. Num. Math **2**, 197–205 (1960)
59. Committee of European Banking Supervisors: Guidelines on Operational Risk Mitigation Techniques (December 2009). URL www.c-ebs.org
60. Congdon, P.: Bayesian Statistical Modelling, 2nd edn. Wiley, Chichester, England (2006)
61. Cont, R., Tankov, P.: Financial Modelling With Jump Processes. Chapman & Hall/CRC, Boca Raton, FL (2004)
62. Cope, E.W., Antonini, G., Mignola, G., Ugoccioni, R.: Challenges and pitfalls in measuring operational risk from loss data. The Journal of Operational Risk **4**(4), 3–27 (2009)
63. Cowles, M.K., Carlin, B.P.: Markov chain Monte Carlo convergence diagnostics: a comparative review. Technical report 94-008, Division of Biostatistics, School of Public Health, University of Minessota (1994)
64. Craddock, M., Heath, D., Platen, E.: Numerical inversion of Laplace transforms: a survey of techniques with applications to derivative pricing. Computational Finance **4**(1), 57–81 (2000)
65. Cruz, M.G.: Modeling, Measuring and Hedging Operational Risk. Wiley, Chichester (2002)
66. Cruz, M.G. (ed.): Operational Risk Modelling and Analysis: Theory and Practice. Risk Books, London (2004)
67. Dagpunar, J.S.: An easily implemented generalised inverse Gaussian generator. Communications in Statistics, Simulation and Computation **18**, 703–710 (1989)
68. Dalla Valle, L.: Bayesian copulae distributions, with application to operational risk management. Methodology and Computing in Applied Probability **11**(1), 95–115 (2009)
69. Daul, S., De Giorgi, E., Lindskog, F., McNeil, A.: The grouped t-copula with an application to credit risk. Risk **16**, 73–76 (2003)
70. Davis, E.: Theory vs Reality. OpRisk&Compliance (1 September 2006). URL www.opriskandcompliance.com/public/showPage.html?page=345305
71. Degen, M., Embrechts, P., Lambrigger, D.D.: The quantitative modeling of operational risk: between g-and-h and EVT. ASTIN Bulletin **37**(2), 265–291 (2007)

72. Demarta, S., McNeil, A.: The t copula and related copulas. International Statistical Review **73**, 111–129 (2005)
73. Dempster, A.P.: A generalization of Bayesian inference. Journal of the Royal Statistical Society, Series B **30**, 205–247 (1968)
74. Den Iseger, P.W.: Numerical Laplace inversion using Gaussian quadrature. Probability in the Engineering and Informational Sciences **20**, 1–44 (2006)
75. Denaut, D.: Coherent allocation of risk capital. Journal of Risk **4**(1), 1–34 (2001)
76. Devroye, L.: Non-Uniform Random Variate Generation. Springer, New York, NY (1986)
77. Dutta, K., Perry, J.: A tale of tails: an empirical analysis of loss distribution models for estimating operational risk capital. Federal Reserve Bank of Boston (2006). URL http://www.bos.frb.org/economic/wp/index.htm. Working paper No. 06–13
78. Efron, B.F., Hinkley, D.V.: Assessing the accuracy of the maximum likelihood estimator: observed versus expected Fisher information. Biometrika **65**, 457–487 (1978)
79. Efron, B.F., Tibshirani, R.J.: An Introduction to the Bootstrap. Chapman and Hall, London (1993)
80. Embrechts, P.: A property of the generalized inverse Gaussian distribution with some applications. Journal of Applied Probability **20**, 537–544 (1983)
81. Embrechts, P., Frei, M.: Panjer recursion versus FFT for compound distributions. Mathematical Methods of Operations Research **69**(3), 497–508 (2009)
82. Embrechts, P., Goldie, C., Veraverbeke, N.: Subexponentiality and infinite divisibility. Zeitschrift fur Wahrscheinlichkeitstheorie und Verwandte Gebiete **49**, 335–347 (1979)
83. Embrechts, P., Klüppelberg, C., Mikosch, T.: Modelling Extremal Events for Insurance and Finance. Springer, Berlin (1997). Corrected fourth printing 2003
84. Embrechts, P., Lambrigger, D.D., Wüthrich, M.V.: Multivariate extremes and the aggregation of dependent risks: examples and counter-examples. Extremes **12**(2), 107–127 (2009)
85. Embrechts, P., McNeil, A., Straumann, D.: Correlation and dependence in risk management: properties and pitfalls. In: M. Dempster, H. Moffatt (eds.) Risk Management: Value at Risk and Beyond, pp. 176–223. Cambridge University Press, Cambridge (2002)
86. Embrechts, P., Nešlehová, J., Wüthrich, M.V.: Additivity properties for Value-at-Risk under Archimedean dependence and heavy-tailedness. Insurance: Mathematics and Economics **44**, 164–169 (2009)
87. Embrechts, P., Puccetti, G.: Aggregating risk capital, with an application to operational risk. The Geneva Risk and Insurance Review **31**(2), 71–90 (2006)
88. Embrechts, P., Puccetti, G.: Aggregation operational risk across matrix structured loss data. The Journal of Operational Risk **3**(2), 29–44 (2008)
89. Ergashev, B.: Should risk managers rely on the maximum likelihood estimation method while quantifying operational risk? The Journal of Operational Risk **3**(2), 63–86 (2008)
90. Ergashev, B.: Estimating the lognormal-gamma model of operational risk using the Markov chain Monte Carlo method. The Journal of Operational Risk **4**(1), 35–57 (2009)
91. Fang, H., Fang, K., Kotz, S.: The meta-elliptical distributions with given marginals. Journal of Multivariate Analysis **82**, 1–16 (2002)
92. Federal Reserve System, Office of the Comptroller of the Currency, Office of Thrift Supervision and Federal Deposit Insurance Corporation: Results of the 2004 Loss Data Collection Exercise for Operational Risk (May 2005). URL www.bos.frb.org/bankinfo/qau/papers/pd051205.pdf
93. Ferguson, T.S.: A Course in Large Sample Theory. Chapman and Hall, London (1996)
94. Ferson, S., Kreinovich, V., Ginzburg, L., Myers, D.S., Sentz, K.: Constructing Probability Boxes and Dempster-Shafer Structures. Sandia National Laboratories, Albuquerque, New Mexico 87185 and Livermore, California 94550 (January 2003). SAND report: SAND2002-4015
95. Frachot, A., Moudoulaud, O., Roncalli, T.: Loss distribution approach in practice. In: M. Ong (ed.) The Basel Handbook: A Guide for Financial Practitioners. Risk Books, London (2004)
96. Frachot, A., Roncalli, T.: Mixing internal and external data for managing operational risk. Working paper (2002). Groupe de Recherche Opérationnelle, Crédit Lyonnais, France

97. Frachot, A., Roncalli, T., Salomon, E.: The correlation problem in operational risk. Working paper (2004). Groupe de Recherche Opérationnelle, France
98. Frees, E., Valdez, E.: Understanding relationships using copulas. North American Journal **2**, 1–25 (1998)
99. Gelfand, A., Dey, D.: Bayesian model choice: Asymptotic and exact calculations. Journal of the Royal Statistical Society, series B **56**, 501–514 (1994)
100. Gelman, A., Gilks, W.R., Roberts, G.O.: Weak convergence and optimal scaling of random walks Metropolis algorithm. Annals of Applied Probability **7**, 110–120 (1997)
101. Geman, S., Geman, D.: Stochastic relaxation, Gibbs distributions, and the Bayesian restoration of images. IEEE Transactions on Pattern Analysis and Machine Intelligence **6**, 721–741 (1984)
102. Gerhold, S., Schmock, U., Warnung, R.: A generalization of Panjer's recursion and numerically stable risk aggregation. Finance and Stochastics **14**(1), 81–128 (2010)
103. Geyer, C.J., Thompson, E.A.: Annealing Markov chain Monte Carlo with applications to ancestral inference. Journal of the American Statistical Association **90**(431), 909–920 (1995)
104. Giacometti, R., Rachev, S.T., Chernobai, A., Bertocchi, M.: Aggregation issues in operational risk. The Journal of Operational Risk **3**(3), 3–23 (2008)
105. Gilks, W.R., Best, N.G., Tan, K.C.: Adaptive rejection Metropolis sampling within Gibbs sampling. Applied Statistics **44**, 455–472 (1995)
106. Gilks, W.R., Richardson, S., Spiegelhalter, D.J.: Markov Chain Monte Carlo in practice. Chapman & Hall, London (1996)
107. Gilks, W.R., Wild, P.: Adaptive rejection sampling for Gibbs sampling. Applied Statistics **41**(2), 337–348 (1992)
108. Glasserman, P.: Monte Carlo Methods in Financial Engineering. Springer, New York, NY (2004)
109. Glasserman, P.: Measuring marginal risk contributions in credit portfolios. Journal Computational Finance **9**(2), 1–41 (2005)
110. Golub, G.H., Welsch, J.H.: Calculation of Gaussian quadrature rules. Mathematics of Computation **23**, 221–230 (1969)
111. Gramacy, R.B., Samworth, R., King, R.: Importance tempering. Technical report, Cambridge University Statistical Laboratory Technical Report Series (2007). Preprint on arXiv:0707.4242
112. Green, P.: Reversible jump MCMC computation and Bayesian model determination. Biometrika **82**, 711–732 (1995)
113. Grønneberg, S., Hjort, N.L.: The copula information criterion. Statistical research report no. 7, Department of Mathematics, University of Oslo (July 2008). ISSN 0806–3842
114. Grubel, R., Hermesmeier, R.: Computation of compound distributions I: aliasing errors and exponential tilting. ASTIN Bulletin **29**(2), 197–214 (1999)
115. Gustafsson, J., Thuring, F.: A suitable parametric model for operational risk applications (February 2008). URL http://ssrn.com/abstract=926309
116. Hastings, W.: Monte Carlo sampling methods using Markov chains and their applications. Biometrika **57**, 97–109 (1970)
117. Haubenstock, M.: The operational risk framework. In: C. Alexander (ed.) Operational Risk: Regulation, Analysis and Management, pp. 241–261. Prentice Hall, New York, NY (2003)
118. Heckman, P.E., Meyers, G.N.: The calculation of aggregate loss distributions from claim severity and claim count distributions. Proceedings of the Casualty Actuarial Society **LXX**, 22–61 (1983)
119. Hess, K.T., Liewald, A., Schmidt, K.D.: An extension of Panjer's recursion. ASTIN Bulletin **32**(2), 283–297 (2002)
120. Hesselager, O.: Recursions for certain bivariate counting distributions and their compound distributions. ASTIN Bulletin **26**(1), 35–52 (1996)
121. Hipp, C.: Speedy convolution algorithms and Panjer recursions for phase-type distributions. Insurance: Mathematics and Economics **38**(1), 176–188 (2006)

122. Hiwatashi, J., Ashida, H.: Advancing Operational Risk Management using Japanese Banking Experiences. Federal Reserve Bank of Chicago (2002)
123. Hoaglin, D.C.: Using quantiles to study shape. In: D.C. Hoaglin, F. Mosteller, J.W. Tukey (eds.) Exploring Data Tables, Trends, and Shapes, pp. 417–460. Wiley, New York, NY (1985a)
124. Hoaglin, D.C.: Summarizing shape numerically: The g-and-h distributions. In: D.C. Hoaglin, F. Mosteller, J.W. Tukey (eds.) Exploring Data Tables, Trends, and Shapes, pp. 461–513. John Wiley & Sons, New York, NY (1985b)
125. Ibragimov, R., Walden, J.: The limits of diversification when losses may be large. Journal of Banking and Finance **31**, 2251–2569 (2007)
126. International Actuarial Association: A Global Framework for Insurer Solvency Assessment (2004). URL www.actuaries.org. A Report by the Insurer Solvency Assessment Working Party of the International Actuarial Association
127. Jeffreys, H.: Theory of Probability, 3rd edn. Oxford University Press, London (1961)
128. Johnson, N.L., Kotz, S., Balakrishnan, N.: Discrete Multivariate Distributions. John Wiley & Sons, New York (1997)
129. Jørgensen, B.: Statistical Properties of the Generalized Inverse Gaussian Distribution. Springer, New York (1982)
130. Kaas, R., Goovaerts, M.J., Dhaene, J., Denuit, M.: Modern Actuarial Risk Theory. Kluwer Academic Publishers, Dordrecht (2001)
131. Kass, R., Raftery A.: Bayes factor. Journal of the American Statistical Association **90**, 773–792 (1995)
132. Kahaner, D., Moler, C., Nash, S.: Numerical Methods and Software. Prentice-Hall, Englewood Cliffs, NJ (1989)
133. Kass, R.E., Carlin, B.P., Gelman, A., Neal, R.M.: Markov chain Monte Carlo in practice: a roundtable discussion. The American Statistician **52**, 93–100 (1998)
134. King, J.L.: Operational Risk: Measurements and Modelling. Wiley, Chichester, England (2001)
135. Klugman, S., Parsa, A.: Fitting bivariate distributions with copulas. Insurance: Mathematics and Economics **24**, 139–148 (1999)
136. Klugman, S.A., Panjer, H.H., Willmot, G.E.: Loss Models: From Data to Decisions. Wiley, New York, NY (1998)
137. Kogon, S.M., William, B.W.: Characteristic function based estimation of stable distribution parameters. In: R.J. Adler, R.E. Feldman, M.S. Taqqu (eds.) A Practical Guide to Heavy Tails: Statistical Techniques and Applications. Birkhäuser, Boston, MA (1998)
138. Kolmogorov, A.N.: Grundbegriffe der Wahrscheinlichkeistrechung. Ergebnisse der Mathematik, Springer, Berlin (1933)
139. Kronrod, A.S.: Nodes and weights of quadrature formulas. Sixteen-place tables. New York: Consultants Bureau Authorized translation from Russian Doklady Akad. Nauk SSSR **154**, 283–286 (1965)
140. Lam, J.: Enterprise Risk Management: From Incentives to Controls. Wiley, Hoboken, NJ (2003)
141. Lambrigger, D.D., Shevchenko, P.V., Wüthrich, M.V.: The quantification of operational risk using internal data, relevant external data and expert opinions. The Journal of Operational Risk **2**(3), 3–27 (2007)
142. Lavin, M., Scherrish, M.: Bayes factors: what they are and what they are not. The American Statistician **53**, 119–122 (1999)
143. Lehmann, E.L.: Theory of Point Estimation. Wiley, New York, NY (1983)
144. Lehmann, E.L., Casella, G.: Theory of Point Estimation, 2nd edn. Springer, New York, NY (1998)
145. Lindskog, F., McNeil, A.: Common Poisson shock models: application to insurance and credit risk modelling. ASTIN Bulletin **33**, 209–238 (2003)
146. Litterman, R.: Hot spots^{TM} and hedges. The Journal of Portfolio Management **22**, 52–75 (1996)

147. Luo, X., Shevchenko, P.V.: Computing tails of compound distributions using direct numerical integration. The Journal of Computational Finance **13**(2), 73–111 (2009)
148. Luo, X., Shevchenko, P.V.: The t copula with multiple parameters of degrees of freedom: bivariate characteristics and application to risk management. Quantitative Finance **10**(9), 1039–1054 (2010)
149. Luo, X., Shevchenko, P.V.: A short tale of long tail integration. Numerical Algorithms (2010). DOI: 10.1007/s11075-010-9406-9
150. Luo, X., Shevchenko, P.V.: Bayesian model choice of grouped t-copula (2011). To appear in Methodology and Computing in Applied Probability
151. Luo, X., Shevchenko, P.V., Donnelly, J.: Addressing impact of truncation and parameter uncertainty on operational risk estimates. The Journal of Operational Risk **2**(4), 3–26 (2007)
152. MacEachern, S.N., Berliner, L.M.: Subsampling the Gibbs sampler. The American Statistician **48**, 188–190 (1994)
153. Marinari, E., Parisi, G.: Simulated tempering: a new Monte Carlo scheme. Europhysics Letters **19**(6), 451–458 (1992)
154. Marshall, C.L.: Measuring and Managing Operational Risks in Financial Institutions. Wiley, Singapore (2001)
155. Martinez, J., Iglewicz, B.: Some properties of the tukey g and h family of distributions. Communications in Statistics – Theory and Methods **13**(3), 353–369 (1984)
156. McDonald, J.B., Xu, Y.J.: A generalization of the beta distribution with applications. The Journal of Econometrics **66**, 133–152 (1995)
157. McNeil, A.J., Frey, R., Embrechts, P.: Quantitative Risk Management: Concepts, Techniques and Tools. Princeton University Press, Princeton, NJ (2005)
158. Melchiori, M.: Tools for sampling multivariate Archimedian copulas. Preprint, www.yieldcurve.com
159. Meng, X., Wong, W.: Simulating ratios of normalizing constants via a simple identity. Statistical Sinica **6**, 831–860 (1996)
160. Menn, C., Rachev, S.T.: Calibrated FFT-based density approximation for alpha stable distributions. Computational Statistics and data Analysis **50**, 1891–1904 (2006)
161. Metropolis, N., Rosenbluth, A.W., Rosenbluth, M.N., Teller, A.H., Teller, E.: Equations of state calculations by fast computing machines. Journal of Chemical Physics **21**, 1087–1091 (1953)
162. Miazhynskaia, T., Dorffner, G.: A comparison of Bayesian model selection based on MCMC with an application to GARCH-type models. Statistical Papers **47**, 525–549 (2006)
163. Mignola, G., Ugoccioni, R.: Effect of a data collection threshold in the loss distribution approach. The Journal of Operational Risk **1**(4), 35–47 (2006)
164. Mignola, G., Ugoccioni, R.: Tests for extreme value theory. Operational Risk&Compliance pp. 32–35 (October 2005)
165. Mira, A., Tierney, L.: Efficiency and convergence properties of slice samplers. Scandinavian Journal of Statistics **29**(1), 1–12 (2002)
166. Moscadelli, M.: The modelling of operational risk: experiences with the analysis of the data collected by the Basel Committee. Bank of Italy (2004). Working paper No. 517
167. Muermann, A., Oktem, U.: The near-miss management of operational risk. The Journal of Risk Finance **4**(1), 25–36 (2002)
168. Neal, R.M.: Probabilistic inference using Markov chain samplers. Technical report, Department of Computer Science, University of Toronto (1993)
169. Neal, R.M.: Sampling from multimodal distributions using tempered transitions. Statistics and Computing **6**, 353–366 (1996)
170. Neal, R.M.: Slice sampling (with discussions). Annals of Statistics **31**, 705–767 (2003)
171. Neil, M., Fenton, N.E., Tailor, M.: Using Bayesian networks to model expected and unexpected operational losses. Risk Analysis **25**(4), 963–972 (2005)
172. Neil, M., Häger, D., Andersen, L.B.: Modeling operational risk in financial institutions using hybrid dynamic Bayesian networks. Journal of Operational Risk **4**(1), 3–33 (2009)

173. Neil, M., Malcolm, B., Shaw, R.: Modelling an air traffic control environment using Bayesian belief networks. Ottawa, Ontario, Canada (August 2003). 21st International System Safety Conference
174. Nešlehová, J., Embrechts, P., Chavez-Demoulin, V.: Infinite mean models and the LDA for operational risk. Journal of Operational Risk **1**(1), 3–25 (2006)
175. Newton, M., Raftery, A.: Approximate Bayesian inference by the weighted likelihood bootstrap. Journal of the Royal Statistical Society, series B **56**, 1–48 (1994)
176. Nolan, J.: Stable Distributions – Models for Heavy Tailed Data. Birkhauser, Boston, MA (2007)
177. Nolan, J.P.: Numerical calculation of stable densities and distribution functions. Communications in Statistics – Stochastic Models **13**, 759–774 (1997)
178. O'Hagan, A., Buck, C.E., Daneshkhah, A., Eiser, J.R., Garthwaite, P.H., Jenkinson, D., Oakley, J.E., Rakov, T.: Uncertain Judgements: Eliciting Expert's Probabilities. Wiley, Chichester (2006)
179. Panjer, H., Willmot, G.: Insurance Risk Models. Society of Actuaries, Chicago, IL (1992)
180. Panjer, H.H.: Recursive evaluation of a family of compound distribution. ASTIN Bulletin **12**(1), 22–26 (1981)
181. Panjer, H.H.: Operational Risks: Modeling Analytics. Wiley, New York, NY (2006)
182. Panjer, H.H., Wang, S.: On the stability of recursive formulas. ASTIN Bulletin **23**(2), 227–258 (1993)
183. Panjer, H.H., Willmot, G.E.: Computational aspects of recursive evaluation of compound distributions. Insurance: Mathematics and Economics **5**, 113–116 (1986)
184. Peters, G.W., Byrnes, A.D., Shevchenko, P.V.: Impact of insurance for operational risk: Is it worthwhile to insure or be insured for severe losses? Insurance: Mathematics and Economics (2010), doi:10.1016/j.insmatheco.2010.12.001
185. Peters, G.W., Johansen, A.M., Doucet, A.: Simulation of the annual loss distribution in operational risk via Panjer recursions and Volterra integral equations for value-at-risk and expected shortfall estimation. The Journal of Operational Risk **2**(3), 29–58 (2007)
186. Peters, G.W., Shevchenko, P.V., Wüthrich, M.V.: Model uncertainty in claims reserving within Tweedie's compound Poisson models. ASTIN Bulletin **39**(1), 1–33 (2009)
187. Peters, G.W., Shevchenko, P.V., Wüthrich, M.V.: Dynamic operational risk: modeling dependence and combining different data sources of information. The Journal of Operational Risk **4**(2), 69–104 (2009)
188. Peters, G.W., Sisson, S.A.: Bayesian inference, Monte Carlo sampling and operational risk. The Journal of Operational Risk **1**(3), 27–50 (2006)
189. Peters, G.W., Sisson, S.A., Fan, Y.: Simulation for Bayesian models constructed with alpha stable distributions utilising approximate Bayesian computation (2009)
190. Peters, G.W., Wüthrich, M.V., Shevchenko, P.V.: Chain ladder method: Bayesian bootstrap versus classical bootstrap. Insurance: Mathematics and Economics **47**(1), 36–51 (2010)
191. Peters, J-P., Hübner, G.: Modeling operational risk based on multiple experts' opinions. In: G.N. Gregoriou (ed.) Operational Risk Toward Basel III: Best Practices and Issues in Modeling, Management, and Regulation. Wiley, New York (2009)
192. Piessens, R., Doncker-Kapenga, E.D., Überhuber, C.W., Kahaner, D.K.: QUADPACK – a Subroutine Package for Automatic Integration. Springer, New York, NY (1983)
193. Planning and Coordination Bureau, Financial Service Agency, Financial Systems and Bank Examination Department, Bank of Japan: Results of the 2007 Operational Risk Data Collection Exercise (August 2007). URL www.boj.or.jp/en/type/ ronbun/ron/research07/ron0709a.htm
194. Powojowski, M.R., Reynolds, D., Tuenter, H.J.H.: Dependent events and operational risk. ALGO Research Quarterly **5**(2), 65–73 (2002)
195. Press, W.H., Teukolsky, S.A., Vetterling, W.T., Flannery, B.P.: Numerical Recipes in C. Cambridge University Press, New York, NY (2002)
196. Pugachev, V.S.: Theory of Random Functions and its applications to control problems, 1st edn. Pergamon Press, London (1965)

197. Rachev, S.T., Mittnik, S.: Stable Paretian Models in Finance. Wiley, New York, NY (2000)
198. Rayner, G.D., MacGillivray, H.L.: Numerical maximum likelihood estimation for the g-and-h and generalized g-and-h distributions. Statistics and Computing **12**, 57–75 (2002)
199. Ripley, B.D.: Stochastic Simulation. Wiley, New York, NY (1987)
200. Robert, C.P.: The Bayesian Choice. Springer, New York, NY (2001)
201. Robert, C.P., Casella, G.: Monte Carlo Statistical Methods, 2nd edn. Springer Texts in Statistics, New York, NY (2004)
202. Roberts, G.O., Rosenthal, J.S.: Optimal scaling for various Metropolis-Hastings algorithms. Statistical Science **16**, 351–367 (2001)
203. Robertson, J.: The computation of aggregate loss distributions. Proceedings of the Casuality Actuarial Society **79**, 57–133 (1992)
204. Rosenthal, J.S.: AMCMC: An R interface for adaptive MCMC. Computational Statistics and Data Analysis **51**(12), 5467–5470 (2007)
205. Rosenthal, J.S.: Optimal proposal distributions and adaptive mcmc. In: A. Gelman, J. G., X.L. Meng, S. Brooks (eds.) Handbook of Markov chain Monte Carlo: Methods and Applications. Chapman & Hall /CRC Press, Florida, FL (2009)
206. Rytgaard, M.: Estimation in Pareto distribution. ASTIN Bulletin **20**, 201–216 (1990)
207. Sandström, A.: Solvency: Models, Assessment and Regulation. Chapman & Hall/CRC, Boca Raton, FL (2006)
208. Savage, L.J.: The subjective basis of statistical practice. Department of Statistics, University of Michigan, Ann Arbor (1961). Technical report
209. Schwarz, G.: Estimation the dimension of a model. Annals of Statistics **6**, 461–464 (1978)
210. Schmock, U.: Estimating the value of the Wincat coupons of the Winterthur insurance convertible bond: a study of the model risk. ASTIN Bulletin **29**(1), 101–163 (1999)
211. Seal, H.L.: Numerical inversion of characteristic functions. Scandinavian Actuarial Journal pp. 48–53 (1977)
212. Shafer, G.: A Mathematical Theory of Evidence. Princeton University Press, Princeton, NJ (1976)
213. Sharpe, W.F.: Capital asset prices: a theory of market equilibrium under conditions of risk. Journal of Finance **19**, 425–442 (1964)
214. Shephard, N.G.: From characteristic function to distribution function: a simple framework for the theory. Econometric Theory **7**, 519–529 (1991)
215. Shevchenko, P.V.: Estimation of operational risk capital charge under parameter uncertainty. The Journal of Operational Risk **3**(1), 51–63 (2008)
216. Shevchenko, P.V.: Implementing loss distribution approach for operational risk. Applied Stochastic Models in Business and Industry **26**(3), 277–307 (2010)
217. Shevchenko, P.V., Temnov, G.: Modeling operational risk data reported above a time-varying threshold. The Journal of Operational Risk **4**(2), 19–42 (2009)
218. Shevchenko, P.V., Wüthrich, M.V.: The structural modeling of operational risk via Bayesian inference: combining loss data with expert opinions. The Journal of Operational Risk **1**(3), 3–26 (2006)
219. Sidi, A.: Extrapolation methods for oscillatory infinite integrals. Journal of the Institute of Mathematics and Its Applications **26**, 1–20 (1980)
220. Sidi, A.: A user friendly extrapolation method for oscillatory infinite integrals. Mathematics of Computation **51**, 249–266 (1988)
221. Smith, R.L.: Estimating tails of probability distributions. Annals of Statistics **15**, 1174–1207 (1987)
222. Steinhoff, C., Baule, R.: How to validate op risk distributions. OpRisk&Compliance pp. 36–39 (August 2006)
223. Stoer, J., Bulirsch, R.: Introduction to Numerical Analysis, 3rd edn. Springer, New York, NY (2002)
224. Stuart, A., Ord, J.K.: Kendall's Advanced Theory of Statistics: Volume 1, Distribution Theory, Sixth Edition. Edward Arnold, London/Melbourne/Auckland (1994)

225. Stuart, A., Ord, J.K., Arnold, S.: Advanced Theory of Statistics, volume 2A: Classical Inference and the Linear Models, 6th edn. Oxford University Press, London (1999)
226. Sundt, B.: On some extensions of Panjer's class of counting distributions. ASTIN Bulletin **22**(1), 61–80 (1992)
227. Sundt, B.: On multivariate Panjer recursions. ASTIN Bulletin **29**(1), 29–45 (1999)
228. Sundt, B., Jewell, W.S.: Further results on recursive evaluation of compound distributions. ASTIN Bulletin **12**(1), 27–39 (1981)
229. Sundt, B., Vernic, R.: Recursions for Convolutions and Compound Distributions with Insurance Applications. Springer, Berlin (2009)
230. Swiss Financial Market Supervisory Authority (FINMA), Bern, Switzerland: Swiss Solvency Test, Technical Document (2006)
231. Szegö, G.: Orthogonal Polynomials, 4th edn. American Mathematical Society, Providence, RI (1975)
232. Tasche, D.: Risk contributions and performance measurement (1999). URL http://www-m4.ma.tum.de/pers/tasche. Preprint, Department of Mathematics, TU München
233. Tasche, D.: Euler Allocation: Theory and Practice (2008). Preprint arXiv:0708.2542v2 available on http://arxiv.org
234. Tavaré, S., Marjoram, P., Molitor, J., Plagnol, V.: Markov chain Monte Carlo without likelihoods. Proceedings of the National Academy of Science, USA **100**, 15,324–15,328 (2003)
235. Tukey, J.W.: Exploratory Data Analysis. Addison-Wesley, Boston, MA (1977)
236. Vernic, R.: Recursive evaluation of some bivariate compound distributions. ASTIN Bulletin **29**(2), 315–325 (1999)
237. Waller, L.A., Turnbull, B.G., Hardin, J.M.: Obtaining distribution functions by numerical inversion of characteristic functions with applications. The American Statistician **49**(4), 346–350 (1995)
238. Wasserman, L.: Bayesian model selection and model averaging. Technical report, Statistics Department, Carnegie Mellon University (1997)
239. Wüthrich, M.V.: Premium liability risks: modelling small claims. Bulletin of the Swiss Association of Actuaries **1**, 27–38 (2006)
240. Wüthrich, M.V., Merz, M.: Stochastic Claims Reserving Methods in Insurance. Wiley, Chichester, England (2008)
241. Wynn, P.: On a device for computing the $e_m(s_n)$ tranformation. Mathematical Tables and Other Aids to Computation **10**, 91–96 (1956)
242. Yamai, Y., Yoshiba, T.: Comparative analyses of expected shortfall and Value-at-Risk: Their estimation error, decomposition, and optimization. Monetary and Economic Studies pp. 87–121 (January 2002)

Index